Cyclin Dependent Kinase 5 (Cdk5)

Nancy Y. Ip · Li-Huei Tsai
Editors

Cyclin Dependent Kinase 5 (Cdk5)

 Springer

Editors

Nancy Y. Ip
Department of Biochemistry
Hong Kong University of Science
 and Technology
Clear Water Bay
Kowloon
Hong Kong
China
biop@ust.hk

Li-Huei Tsai
Howard Hughes Medical Institute
Department of Brain and Cognitive
 Science
Massachusetts Institute of Technology
77 Massachusetts Avenue
Cambridge, MA 02139
USA
lhtsai@mit.edu

ISBN: 978-0-387-78886-9   e-ISBN: 978-0-387-78887-6
DOI: 10.1007/978-0-387-78887-6

Library of Congress Control Number: 2008924613

Cover illustration: A hippocampal neuron prepared from p35-deficient mouse.

Printed on acid-free paper

springer.com

Preface

When cyclin-dependent kinase 5 (Cdk5) was identified 15 years ago, little was known about its physiological functions. While other cyclin-dependent kinases cooperate to control the cell cycle, this new addition to the family exceeds all expectations. As you will learn through the chapters in this book, although Cdk5 is structurally homologous to other Cdks, Cdk5 is unlike its family members in every way—a different set of activators, different localization of its activity, and importantly, its unexpected involvement in such a diverse plethora of biological actions. Through the extensive efforts in identifying interacting proteins and substrates of Cdk5/p35, it is now recognized that Cdk5 is implicated in almost every aspect of neuronal development and neural functions. Right from when the newly born neurons leave the ventricular zone for their long migratory journey to their final destination, Cdk5 activity is pivotal for proper neuronal migration. Indeed, newly generated neurons fail to display the "inside-out" organization in Cdk5-deficient brain. Upon reaching the target, neurons send out dendrites and axons to form functional connections with neighboring neurons. During this stage, Cdk5 is implicated in the regulation of neurite extension, synapse formation, and synaptic transmission. Cdk5 also plays a role in the control of apoptosis during development, which is important for the pruning and fine tuning of neural connections. As the neuronal network matures, Cdk5 activity remains essential and contributes to the regulation of synaptic plasticity. Indeed, Cdk5 is increasingly implicated in higher cognitive functions such as learning, memory formation, and drug addiction. More importantly, deregulation of Cdk5 activity is also associated with neuronal death in neurodegenerative diseases.

Through the years of research, it is now becoming clear that this kinase plays monumental role in essentially every facet of neuronal development and functions. It is also increasingly recognized that the action of Cdk5 is not limited to the nervous system. Cdk5 was also observed to play an important role at the neuromuscular junction, and in the control of insulin secretion in the pancreas. These findings provide novel grounds for investigating the function of Cdk5 in other systems. Indeed, given the involvement of Cdk5 in such diverse processes and its extensive crosstalk with other signaling pathways, it would not be surprising that Cdk5 is involved in yet unknown aspects of cellular functions.

We are very grateful to each and every contributor of this book for putting in extensive hours to bring these chapters into reality. We would also like to thank Miss Anna So and Dr. Zelda Cheung of the Hong Kong University of Science and Technology, whose efforts have made this book possible. Our collective effort will bring to the awareness of the scientific community the augmenting importance of this kinase as an essential regulator of neuronal development and functions. We believe that this book will also lay the groundwork of many more years of Cdk5 research to come.

Hong Kong, China   Nancy Y. Ip
Cambridge, Massachusetts   Li-Huei Tsai

Contents

Contributors

Marco Angelo
Institute of Neurology, University College London, London, UK,
e-mail: m.angelo@ucl.ac.uk

Mathias Bähr
University of Göttingen, Department of Neurology, Robert-Koch-Str. 40,
37075 Göttingen, Germany
e-mail: mbaehr@gwdg.de

Christina Bark
The Rolf Luft Research Center for Diabetes and Endocrinology, Karolinska
Institutet, Stockholm, Sweden, Phone: 46-8-517 794 52, Fax: 46-8-517 794 50,
e-mail: christina.bark@ki.se

Per-Olof Berggren
The Rolf Luft Research Center for Diabetes and Endocrinology, Karolinska
Institutet, Stockholm, Sweden,
e-mail: per-olof.berggren@ki.se

Zelda H. Cheung
Department of Biochemistry. Hong Kong University of Science and
Technology, Clear Water Bay, Kowloon, Hong Kong, China,
e-mail: zelda@ust.hk

Sul-Hee Chung
Graduate Program in Neuroscience, Institute for Brain Science and
Technology, Inje University, 633-146 Gaegeum 2-dong, Busanjin-gu,
Busan, South Korea 614-735, Phone: 82-51-892-4185, Fax: 82-51-892-0059,
e-mail: sulchung@inje.ac.kr

Karen Duff
Department of Pathology, Taub Institute for Alzheimer's Disease research,
Columbia University/NYS Psychiatric Institute, BB-513, Black Building,
650 W 168th St., New York 10032, USA, Phone: 212-305 8790,
e-mail: ked2115@columbia.edu

Andre Fischer
European Neuroscience Institute (ENI), Department for Experimental
Neuropathology, Medical School University Goettingen, Max Planck Society,
Germany,
e-mail: andre.fischer@mpi-mail.mpg.de

Peter K. Giese
Institute of Psychiatry, King's College London, London, UK,
e-mail: peter.giese@iop.kcl.ac.uk

Yoshio Goshima
Department of Molecular Pharmacology & Neurobiology, Yokohama City
University Graduate School of Medicine, 236-0004, Japan, Phone: 81-45-787
2593, Fax: 81-45-785 3645,
e-mail: goshima@med.yokohama-cu.ac.jp

Lisheng He
Department of Biochemistry, Hong Kong University of Science
and Technology, Kowloon, Hong Kong, China,
e-mail: hlxaa@ust.hk

Karl Herrup
Department Cell Biology and Neuroscience, Nelson Biological Laboratories,
Rutgers University, 604 Allison Road, Piscataway, NJ 08854, USA, Phone:
732-445-3306, Fax: 732-445-2165,
e-mail: herrup@biology.rutgers.edu

Carol D. Hicks
CNS Discovery, Pfizer Global Research and Development, Groton,
CT 06340, USA

Shin-ichi Hisanaga
Molecular Neuroscience, Department of Biological Sciences, Graduate School
of Science, Tokyo Metropolitan University, Hachioji, Tokyo 192-0397,
Phone: 81-42-677 2577, Fax: 81-42-677 2559, Japan,
e-mail: hisanaga-shinichi@c.metro-u.ac.jp

Nancy Y. Ip
Department of Biochemistry, Biotechnology Research Institute and Molecular
Neuroscience Center, Hong Kong University of Science and Technology, Clear
Water Bay, Hong Kong, China, Phone: 852-2358 7304, Fax: 852-2358 1552,
e-mail: boip@ust.hk

Koichi Ishiguro
Mitsubishi Kagaku Institute of Life Sciences, Machida, Tokyo 194-8511, Japan

Marko Jevsek
Institute of Physiology, Faculty of Medicine, University of Maribor, Maribor,
Slovenia

Jyotshna Kanungo
Laboratory of Neurochemistry, National Institute of Neurological Disorders
and Stroke, National Institutes of Health, Bethesda, Maryland 20814, USA
e-mail: kanungoJ@mail.nih.gov

Sashi Kesavapany
National University of Singapore, Singapore 117597

Niranjana D. Amin
Laboratory of Neurochemistry, National Institute of Neurological Disorders
and Stroke, National Institutes of Health, Bethesda, Maryland 20814, USA,
e-mail: bchsk@nus.edu.sg

Ashok B. Kulkarni
Functional Genomics Section, Laboratory of Cell and Developmental
Biology, National Institute of Dental and Craniofacial Research, National
Institutes of Health, Bethesda, MD, USA, Phone: 39-02-57489829/57489828,
Fax: 39-02-57489851,
e-mail: andrea.musacchio@ifom-ieo-campus.it

Lit-Fui Lau
CNS Discovery, Pfizer Global Research and Development, MS 8220-4013,
Eastern Point Road, Groton, CT 06340, USA, Phone: 860-715-1921,
Fax: 860-715-2349,
e-mail: lit-fui.lau@pfizer.com

Slavena A. Mandic
The Rolf Luft Research Center for Diabetes and Endocrinology, Karolinska
Institutet, Stockholm, Sweden

Zixu Mao
Departments of Pharmacology and Neurology, Center for Neurodegenerative
Disease, Emory University School of Medicine, 615 Michael St., Atlanta,
GA 30322 USA,
e-mail: zmao@pharm.emory.edu

Katrin Meuer
University of Göttingen, Department of Neurology, Robert-Koch-Str. 40,
37075 Göttingen, Germany

Andrea Musacchio
Department of Experimental Oncology, European Institute of Oncology,
Via Adamello 16, I-20139 Milan, Italy,
e-mail: andrea.musacchio@ifom-ieo-campus.it

Fumio Nakamura
Department of Molecular Pharmacology & Neurobiology, Yokohama City
University Graduate School of Medicine, 236-0004, Japan

Gary Kar Ho Ng
Department of Biochemistry, Hong Kong University of Science
and Technology, Hong Kong, China

Toshio Ohshima
Laboratory for Molecular Brain Science, Department of Life Science and
Medical Bio-Science, Science and Engineering, Waseda University, Tokyo
169-8555, Japan, Phone: -81-3-5286-3358, Fax: -81-3-5286-3382,
e-mail: ohshima@waseda.jp

Harish C. Pant
Laboratory of Neurochemistry, National Institute of Neurological Disorders
and Stroke, National Institutes of Health, Bethesda, Maryland 20814, USA,
Phone: 301-402 2124, Fax: 301-496 1339,
e-mail: panth@ninds.nih.gov

Tej K. Pareek
Rainbow Babies and Children Hospital, Department of Pediatrics, Case
Western Reserve University, Cleveland, OH,
e-mail: tkp5@case.edu

Florian Plattner
Institute of Neurology, University College London, London, UK,
Phone: 44-0-207-837 8370, Fax: 44-0-207-278 4993,
e-mail: f.plattner@ucl.ac.uk

Robert Z. Qi
Department of Biochemistry, Hong Kong University of Science and
Technology, Hong Kong, China, Phone: 852-2358 7273, Fax: 852-2358 1552,
e-mail: qirz@ust.hk

Parvathi Rudrabhatla
Laboratory of Neurochemistry, National Institute of Neurological
Disorders and Stroke, National Institutes of Health, Bethesda, Maryland
20814, USA

Marjan Rupnik
Institute of Physiology, Faculty of Medicine, University of Maribor,
Maribor, Slovenia

Yukio Sasaki
Department of Molecular Pharmacology & Neurobiology, Yokohama City
University Graduate School of Medicine, 236-0004, Japan

Kazuhito Tomizawa
Department of Physiology, Okayama University Graduate School of
Medicine, Dentistry and Pharmaceutical Sciences, 2-5-1 Shikata-cho,
Okayama 700-8558, Japan, Phone: 81-86-2357107, Fax: 81-86-235-7111,
e-mail: tomikt@md.okayama-u.ac.jp

Li-Huei Tsai
Picower Institute for Learning and Memory, Department of Brain and
Cognitive Sciences, Massachusetts Institute of Technology, Howard Hughes
Medical Institute, RIKEN-MIT center for Neuroscience research, Stanley
Center for Psychiatric Research, Cambridge, MA, USA, Phone: 617.324.1660,
Fax: 617.324.1657, e-mail: lhtsai@mit.edu

Yutaka Uchida
Department of Molecular Pharmacology & Neurobiology, Yokohama City
University Graduate School of Medicine, 236-0004, Japan

Li Wang
Department Genetics, Case Western Reserve University, School of Medicine,
10900 Euclid Ave., Cleveland, OH 44016

Fan-Yan Wei
Department of Physiology, Okayama University Graduate School of Medicine,
Dentistry and Pharmaceutical Sciences, 2-5-1 Shikata-cho, Okayama 700-8558,
Japan

Jochen H. Weishaupt
University of Göttingen, Department of Neurology, Robert-Koch-Str. 40,
37075 Göttingen, Germany, Phone: 49-551-39-14343,
Mobile: 49-177-414-6790, e-mail: jweisha@gwdg.de

Yi Wen
Department of Pathology, Taub Institute for Alzheimer's Disease research,
Columbia University/NYS Psychiatric Institute, BB-513, Black Building,
650 W 168th St., New York 10032, USA

Naoya Yamashita
Department of Molecular Pharmacology & Neurobiology, Yokohama City
University Graduate School of Medicine, 236-0004, Japan

Qian Yang
Departments of Pharmacology and Neurology, Center for Neurodegenerative
Disease, Emory University School of Medicine, 615 Michael St., Atlanta,
GA 30322 USA

Haung Yu
Department of Pathology, Taub Institute for Alzheimer's Disease research,
Columbia University/NYS Psychiatric Institute, BB-513, Black Building,
650 W 168th St., New York 10032, USA

Jie Zhang
Department Cell Biology and Neuroscience, Nelson Biological Laboratories, Rutgers University, 604 Allison Road, Piscataway, NJ 08854, USA

Ya-Li Zheng
Laboratory of Neurochemistry, National Institute of Neurological Disorders and Stroke, National Institutes of Health, Bethesda, Maryland 20814, USA

About the Editors

Prof. Nancy Ip was born in Hong Kong, and spent the early years of her scientific career in the United States. After obtaining her PhD in pharmacology at Harvard Medical School, she continued her scientific training as a post-doctoral fellow at Harvard and Sloan-Kettering Institute. Thereafter, she left academia to work in the biopharmaceutical industry as the Laboratory Head at Lifecodes Corporation and then at Regeneron Pharmaceuticals Inc, both based in New York. In 1993, Nancy Ip returned to an academic career at the newly established Hong Kong University of Science and Technology. She currently serves as Chair Professor and Head of the Department of Biochemistry, as well as Director of the Biotechnology Research Institute and the Molecular Neuroscience Center.

Nancy Ip is well known for her seminal discoveries on the biology of neurotrophic factors, proteins that promote the survival, development and maintenance of neurons in the nervous system. Her research has identified neurotrophic factors as potential pharmaceutical agents for the treatment of neurodegenerative diseases such as Alzheimer's disease. She is also internationally recognized for her work on elucidating the signaling mechanisms of neuronal plasticity, such as establishing the pivotal involvement of cyclin-dependent kinase 5 in the regulation of synapse development and maintenance. Given the association of synaptic dysfunction with the pathophysiology of neurodegenerative diseases, her findings provide important clues in the development of therapeutics for these disorders.

As a highly accomplished researcher, Ip has published over 170 scientific papers with more than 12,000 SCI citations, and holds 18 patents. She is the Editor-in-Chief of *NeuroSignals* and is on the editorial board of journals such as *J. Biol. Chem.*, *J. Neurosci.*, and *Dev. Neurobiol*. She is also an Academician of the Chinese Academy of Sciences, a fellow of the Academy of Sciences for the Developing World, and a founding member of the Asia-Pacific International Molecular Biology Network. Additionally, she has received numerous awards including the National Natural Science Award and the L'OREAL-UNESCO for Women in Science Award.

Dr. Li-Huei Tsai was born in Taipei, Taiwan. In 1986, she started her Ph.D. at the University of Texas Southwestern. Under the direction of Bradford

Ozanne, she graduated in 1990 and joined Ed Harlow's laboratory at Cold Spring Harbor Laboratory and Massachusetts General Hospital for post-doctoral training. During her time in the Harlow lab, she isolated two proteins prominently expressed in the nervous system: cyclin-dependent kinase 5 (Cdk5) and its regulatory activator p35. She was appointed Assistant Professor of Pathology at Harvard Medical School in 1994, elected Investigator of Howard Hughes Medical Institute in 1997, and promoted to Professor of Pathology in 2002. In 2006, she relocated her lab to MIT and became the Picower Professor of Neuroscience in the Picower Institute for Learning and Memory. She began directing the Neurobiology Program at the Stanley Center for Psychiatric Research in 2007.

A major research interest of the Tsai lab is to understand neurodegenerative diseases associated with cognitive decline such as Alzheimer's disease. Her findings have led to the hypothesis that deregulation of Cdk5, through conversion of p35 to p25, plays an important role in the pathogenesis of Alzheimer's disease. Recently, she found that chromatin remodeling via increased histone acetylation is beneficial for learning impairment and memory loss caused by severe neurodegenreation in the inducible p25 mouse model.

Li-Huei Tsai is on the editorial boards of the journal *Neuron, Journal of Neuroscience and NeuroSignals*, and has been awarded the Young Investigator Award from Metropolitan Life Foundation and the Outstanding Contributor Award from the Alzheimer Research Forum. She sits on the scientific advisory boards and committees for NINDS, Gruber Foundation, Alzheimer Research Forum, Hotchkiss Brain Institute at the University of Calgary, among other organizations.

Cdk5/p35 Regulates Neuronal Migration

Toshio Ohshima

Abstract Neurons migrate from proliferative zone to their final position during brain development. Cyclin-dependent kinase 5 (Cdk5) plays an important role in neuronal migration to establish a proper structure of the brain.

Analyses of Cdk5/p35-deficient mice have provided the knowledge about the role of Cdk5/p35 in neuronal migration. Over the past years, migration-related substrates of Cdk5 have been identified. Imaging analyses of neuronal migration of Cdk5/p35-deficient neurons have begun to elucidate how proper phosphorylations of these proteins by Cdk5/p35 are required for the regulation of cytoskeletal dynamics and cellular adhesion during neuronal migration.

Expression of Cdk5, p35, and p39 During Brain Development

Cyclin-dependent kinase 5 (Cdk5), a proline-directed serine (Ser)/threonine (Thr) kinase, had been identified as a member of the CDK family because of its close sequence homology to human CDC2 (Meyerson et al., 1992; Hellmich et al., 1992; Lew et al., 1992). Since the activity of Cdk5 is regulated by binding it with one of its neuron-specific regulatory subunits, either p35 (Lew et al., 1994; Tsai et al., 1994) or its isoform p39 (Tang et al., 1995), its activity is correlated with the level of expression of p35 and p39. Cdk5 expression is basically ubiquitous, and it is abundant in neuronal cells (Tsai et al., 1993). Expression of p35 and p39 overlaps throughout the central nervous system (CNS) during brain development, except for their expression in the cerebral cortex in the early stage, where only p35 is expressed till around E16 (Ohshima et al., 2001). High Cdk5 activity during neuronal differentiation and brain development reflects high-level expression of p35 and p39 (Tsai et al., 1993).

T. Ohshima
Laboratory for Molecular Brain Science, Department of Life Science and Medical Bio-Science, Science and Engineering, Waseda University, Tokyo 169-8555, Japan
e-mail: ohshima@waseda.jp

N.Y. Ip, L.-H. Tsai (eds.), *Cyclin Dependent Kinase 5 (Cdk5)*,
DOI: 10.1007/978-0-387-78887-6_1, © Springer Science+Business Media, LLC 2008

Cdk5/p35 Deficiency Causes Neuronal Migration Defects in CNS

Studies of the phenotypes of knockout (KO) mice have shown that Cdk5 and p35 are critical for migration of neurons to their final positions in the developing brain (Ohshima and Mikoshiba, 2002; Dhavan and Tsai, 2002). Migration defects in the cortical neurons of Cdk5 KO mice result in disruption of the laminar structures in the cerebral cortex, olfactory bulb, hippocampus, and cerebellum (Ohshima et al., 1996). p35 KO mice display a milder phenotype than Cdk5 KO mice because of the redundancy of p39 (Chae et al., 1997; Ohshima et al., 2001). p39 KO mice display no phenotype; however, p35 and p39 double KO mice display a phenotype identical to that of Cdk5 KO mice (Ko et al., 2001), confirming the redundancy of these subunits. Neuronal migration defects in Cdk5 KO mice are observed in many types of neuronal migration, but not in all types. These observations about Cdk5/p35 mutant mice indicate the occurrence of Cdk5-dependent and -independent neuronal migration. For example, radial migration of cortical neurons is Cdk5 dependent, but migration of subplate neurons seems to be Cdk5 independent in the cerebral cortex (Gilmore and Herrup, 2001). Tangential migration of GABAergic neurons from ganglionic eminence to cerebral cortex is also Cdk5 independent (Gilmore and Herrup, 2001). Migration along radial glial fibers is Cdk5 dependent in many cases including radial migration of cerebral cortical neurons and inward migration of granule cells in the cerebellum (Table 1). The list of examples of Cdk5-depedent migration (Table 1) will be expanded by further analysis of mutant mice.

Table 1 Comparison of migration defects in neuronal types in CNS among mutant mice

Structure or neuronal type in CNS	Cdk5 KO	p35 KO	*Reeler*/Dab1 mutant
Olfactory bulb mitral cell	++	−	−
Cerebral cortex			
(1) Subplate neurons	−	−	++
(2) Cortical neurons	++	+	++
(3) GABAergic neurons	−	−	±
Hippocampus			
(1) Pamidal cell layer	++	+	++
(2) Dentate gyrus	+	+	++
Midbrain dopamin neurons in SN	++	−	+
Cerebellum			
(1) Purkinje cells	++	+	++
(2) Granule cells (inward)	n.d.	+	−
Brain stem			
(1) Facial motor nucleus	++	−	+
(2) Inferior olive	++	−	+

++, severe defect; +, mild defect; n.d., could not be determined because of perinatal lethality. This type of migration of granule cells is Cdk5 dependent (Ohshima et al., 1999). CNS, central nervous system; SN, substantia nigra.

Normal Migration of Cortical Neurons and Alteration by Cdk5/ p35 Deficiency

Neurons settle on six layers in the cerebral cortex of the mammalian brain in an inside-out manner, with the earliest-generated neurons positioning themselves in the deepest layer and the later-generated neurons occupying the more superficial layers (Angevine and Sidman, 1961). The first wave of radially migrating neurons (earliest-generated neurons) splits the pre-existing subpial layer, which is called the preplate zone (PPZ), and "somal translocation" type migration is observed at this stage (Nadarajah et al., 2001). Successive waves of neurons (later-generated neurons) exhibit distinct migratory behaviors. In the subventricular zone (SVZ) and lower intermediate zone (IZ), immature neurons transiently become multipolar, and extend multiple neurites, and just before migrating to the cortical plate (CP), they change shape to become bipolar (Tabata and Nakajima, 2003; Noctor et al., 2004). In the CP, neurons migrate by "locomotion" along radial glial fibers (Nadarajah et al., 2001).

Analyses of Cdk5-null mice and p35-null mice have revealed defects in migratory behavior in Cdk5/p35 deficiency (Gupta et al., 2003; Ohshima et al., 2007b). The PPZ splits into the marginal zone and the subplate in the developing cerebral cortex of Cdk5 KO and p35 KO mice (Chae et al., 1997; Gilmore et al., 1998), in contrast to the impaired PPZ splitting that occurs in *reeler* as discussed below. The migration of Cdk5-null neurons destined for layer II–V stalls below the subplate (Gilmore et al., 1998), indicating that the locomotion type of radial migration is greatly affected by Cdk5/p35 deficiency. A time-lapse imaging study of p35-null brain slices revealed abnormal leading process morphology in migrating p35-null neurons (Gupta et al., 2003). In Cdk5 KO mice, the multipolar-to-bipolar transition is completely blocked, and neurons retain their multipolar morphology (Ohshima et al., 2007b). Introduction of Cdk5-dominant negative (DN) recaptures the phenotype of the migration defect of Cdk5/p35 deficiency in a cell-autonomous manner. When Cdk5-DN was introduced in high levels, the multipolar-to-bipolar transition was impaired, and radial migration was greatly disturbed. When Cdk5-DN was introduced at low levels, branched leading processes were observed, and migration was impaired to a moderate degree (Ohshima et al., 2007b). These observations suggest that the branched leading processes observed in p35-null migrating neurons represent a compensated phenotype in which there is an incomplete transition to the bipolar morphology. The results of an analysis of the migratory behavior of p35-null neurons also suggested reduced association of migrating neurons during locomotion with radial glial fibers (Gupta et al., 2003). Cdk5/p35 deficiency causes migration defects in later-generated cortical neurons and results in inverted layering of the cerebral cortex (Fig. 1).

Fig. 1 Inversion of layer structure of the cerebral cortex in Cdk5–/– mice. mRNA expressions of layer markers, Cux2 for layer II/III, ER81 for layer V, and Foxp2 for layer VI, indicate cortical structure is inverted in Cdk5–/– mice at E18

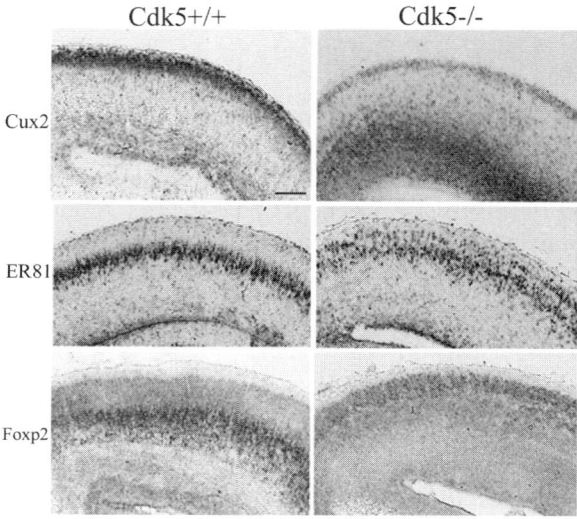

Molecular Mechanism of the Regulatory Function of Cdk5/p35 in Neuronal Migration

How does Cdk5 regulate neuronal migration and positioning? Initially, a possible relation with Reelin signaling was discussed because of the phenotypic similarities between Cdk5 KO mice and *reeler* mice in which Reelin is defective. As summarized in Table 1, there is considerable overlap between neuronal populations affected in *reeler* and Cdk5 KO mice. An unsplit preplate is a characteristic feature (*reeler* phenotype) of deficiency of any component of Reelin signaling pathway (Fig. 2). The relationship between Cdk5/p35 and Reelin signaling was investigated by using double mutant mice with the defects of components of these two signaling pathways (Ohshima et al., 2001; Beffert et al., 2004). Genetic evidence has shown that Reelin and its receptors, VLDLR and ApoER2, and Dab1 belong to the same pathway called "Reelin signaling" (see review by Ohshima and Mikoshiba, 2002). Exaggerated neuronal migration defects were typically observed in the hippocampus of p35/Dab1, p35/Reelin, p35/ApoER2, and p35/VLDLR double KO mice compared with KO mice for the respective components (Ohshima et al., 2001; Beffert et al., 2004). Exaggerated defects in Purkinje cell migration in the cerebellum have also been reported in p35/Reelin and p35/Dab1 double KO mice (Ohshima et al., 2001). These findings indicate that Reelin signaling and Cdk5 function in parallel to effect proper neuronal migration and positioning in the developing brain (Ohshima and Mikoshiba, 2002; Beffert et al., 2004).

Cdk5 modulates actin-cytoskeleton dynamics through phosphorylation of Pak1 (Nikolic et al., 1998; Rashid et al., 2001) and filamin 1 (Fox et al.,

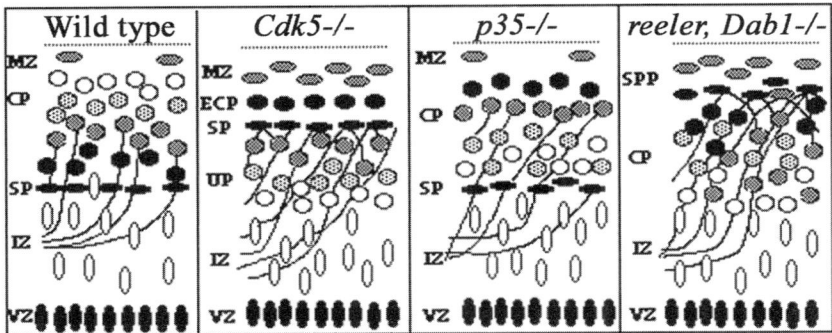

Fig. 2 Schematic representation of the cerebral cortex in the wild-type and mutant mice. In the wild-type mice, migrating neurons split preplate into a marginal zone (MZ) and subplate (SP), to form the cortical plate (CP). In the Cdk5-/- mice, the initial wave of migrating neurons (*indicated by black circles*) splits the preplate to form a narrow ectopic cortical plate (ECP), but later-born neurons (*light gray and white circles*) stack up *reeler*-like under the subplate in an inverted fashion as an underplate (UP). In the p35-/- mice, the later compensatory effects of Cdk5/p39 result in normal positioning of the subplate neurons. In *reeler* and Dab1 mutant mice, the preplate is not split and remains as a superplate (SPP) with mutant neurons stacked up in an inverse order (adapted from Ohshima and Mikoshiba, 2002 with modification)

1998), and it modulates microtubule dynamics through phosphorylation of microtubule-associated proteins, including tau (Kobayashi et al., 1993), MAP1b (Paglini et al., 1998), doublecortin (Tanaka et al., 2004), Nudel (Sasaki et al., 2000; Niethammer et al., 2000), and CRMPs (Uchida et al., 2005). A defect in one of these substrates, filamin 1 in humans, causes periventricular heterotopia (Fox et al., 1998), and defects in two other substrates, Lis1 and doublecortin in humans, cause lissencephaly type 1 (Reiner et al., 1993; des Portes et al., 1998; Gleeson et al., 1998). These findings indicate that Cdk5 regulates cytoskeletal dynamics that determine the speed of migration, extension of the leading processes, and cell-soma propulsion in migrating neurons, in addition to proper transition from multipolar-to-bipolar morphology (Fig. 3). Impaired regulation of cytoskeletal proteins of migrating neurons may produce defective migration in Cdk5/p35 deficiency as described above. Cdk5 may also mediate cellular adhesion in neuronal–glial interactions through phosphorylation of β-catenin to regulate the interaction between β-catenin and N-cadherin (Kwon et al., 2000), and this Cdk5-mediated adhesion may also be important for neuronal migration along radial glial fibers (Fig. 3), as reduced association of p35-null migrating neurons with radial glial fiber was reported (Gupta et al., 2003). Although Cdk5 phosphorylates Dab1 at Ser491 *in vivo* (Keshvara et al., 2002) and at multiple sites *in vitro* (Ohshima et al., 2007a), the functional significance of the Cdk5-mediated Dab1 phosphorylation in Reelin signaling remains to be elucidated.

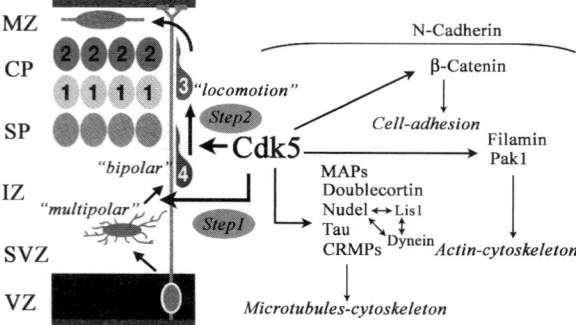

Fig. 3 Cellular and molecular functions of Cdk5 in neuronal migration. In the cerebral cortex, Cdk5 is required for the radial migration of later-generated neurons. Cdk5 is necessary for multipolar-to-bipolar transition (*Step 1*) and locomotion through the regulation of nucleo-kinesis of migrating neurons (*Step 2*). For these steps, Cdk5 regulates the dynamics of actin-cytoskeleton and microtubules-cytoskeleton and cell adhesion through the phosphorylation of its substrate proteins (*See* Color Insert)

Summary

Because Cdk5 phosphorylates a number of proteins in the developing brain, Cdk5 deficiency causes multiple defects in the phosphorylation-mediated regulation of protein interactions required for proper neuronal migration. In addition to the impacts on the dynamics of cytoskeletal proteins, defective phosphorylation of substrate proteins may also alter expression of genes for neuronal differentiation and maturation. Analysis of Cdk5-deficient mouse models may reveal much more about the precise cellular and molecular mechanisms of neuronal migration in the developing brain, and recent studies have begun to demonstrate the involvement of Cdk5 in neuronal migration in the adult brain, and this field remains to be explored.

References

Angevine Jr. J, and Sidman RL. (1961) Autoradiographic study of cell migration during histogenesis of cerebral cortex in the mouse. Nature 192:766–768.

Beffert U, Weeber EJ, Morfini G, Ko J, Brady ST, Tsai L-H, Sweatt D and Herz J. (2004) Reelin and Cyclin-dependent kinase 5-dependent signals cooperate in regulating neuronal migration and synaptic transmission. J. Neurosci. 24:1897–1906.

Chae T, Kwon YT, Bronson R, Dikkes P, Li E and Tsai L-H. (1997) Mice lacking p35, a neuronal specific activator of Cdk5, display cortical lamination defects, seizures and adult lethality. Neuron 18:29–42.

des Portes V, Pinard JM, Billuart P, Vinet MC, Koulakoff A, Carrie A, Gelot A, Dupuis E, Motte J, Berwald-Netter Y, Catala M, Kahn A, Beldjord C and Chelly J. (1998) A novel CNS gene required for neuronal migration and involved in X-linked subcortical laminar heterotopia and lissencephaly syndrome. Cell 92:51–61.

Dhavan R and Tsai L-H. (2002) A decade of Cdk5. Nature Rev. Mol. Cell Biol. 2:749–759.

Fox JW, Lamperti ED, Eksioglu YZ, Hong SE, Feng Y, Graham DA, Scheffer IE, Dobyns WB, Hirsch BA, Radtke RA, Berkovic SF, Huttenlocher PR and Walsh CA. (1998) Mutations in filamin 1 prevent migration of cerebral cortical neurons in human periventricular heterotopia. Neuron 21:1315–1325.

Gilmore EC and Herrup K. (2001) Neocortical cell migration: GABAergic neurons and cells in layers I and VI move in a Cyclin-dependent kinase 5-independent manner. J. Neurosci. 21:9690–9700.

Gilmore EC, Ohshima T, Goffinet AM, Kulkarni AB and Herrup K. (1998) Cyclin-dependent kinase 5-deficient mice demonstrate novel developmental arrest in cerebral cortex. J. Neurosci. 18:6370–6377.

Gleeson JG, Allen KM, Fox JW, Lamperti ED, Berkovic S, Scheffer I, Cooper EC, Dobyns WB, Minnerath SR, Ross ME and Walsh CA. (1998) Doublecortin, a brain-specific gene mutated in human X-linked lissencephaly and double cortex syndrome, encodes a putative signaling protein. Cell 92:63–72.

Gupta A, Sanada K, Miyamoto DT, Rovelstad S, Nadarajah B, Pearlman AL, Brunstrom J and Tsai L-H. (2003) Layering defect in p35 deficiency is linked to improper neuronal-glial interaction in radial migration. Nat. Neurosci. 6:1284–1291.

Hellmich MR, Pant HC, Wada E and Battey JF. (1992) Neuronal cdc2-like kinase: A CDC2-related protein kinase with predominantly neuronal expression. Proc. Natl Acad. Sci. USA 89:10867–10871.

Keshvara L, Magdaleno S, Benhayon D and Curran T. (2002) Cyclin-dependent kinase 5 phosphorylates disabled 1 independently of Reelin signaling. J. Neurosci. 22:4869–4877.

Kobayashi S, Ishiguro K, Omori A, Takamatsu M, Arioka M, Imahori K and Uchida T. (1993) Cdc2-related kinase PSSALRE/Cdk5 is homologous with the 30 kDa subunit of Tau protein kinase II, a proline-directed protein kinase associated with microtubule. FEBS Lett. 335:171–175.

Ko J, Humbert S, Brorson T, Takahashi S, Kulkarni AB, Li E and Tsai L-H. (2001) p35 and p39 are essential for Cdk5 function during neurodevelopment. J. Neurosci. 21:6758–6771.

Kwon YT, Gupta A, Zhou Y, Nikolic M and Tsai L-H. (2000) Regulation of the N-cadherin-mediated adhesion by the p35/Cdk5 kinase. Curr. Biol. 10:363–372.

Lew J, Beaudette CM, Litwin CM and Wang JH. (1992) Purification and characterization of a novel proline-directed protein kinase from bovine brain. J. Biol. Chem. 267:13383–13390.

Lew J, Huang QQ, Qi Z, Winkfein RJ, Aebersold R, Hunt T and Wang JH. (1994) Neuronal cdc2-like kinase is a complex of cyclin-dependent kinase 5 and a novel brain-specific regulatory subunit. Nature 371:423–425.

Meyerson M, Enders GH, Wu CL, Su LK, Gorka, C, Nelson C, Harlow E and Tsai L-H. (1992) A family of human CDC2-related protein kinases. EMBO J. 11:2909–2917.

Nadarajah B, Brunstrom JE., Grutzendler J, Wong ROL and Pearlman AL. (2001) Two modes of radial migration in early development of the cerebral cortex. Nat. Neurosci. 4:143–150.

Niethammer M, Smith DS, Ayala R, Peng J, Ko, J, Lee MS, Morabito M and Tsai L-H. (2000) NUDEL is a novel Cdk5 substrate that associates with LIS1 and cytoplasmic dynein. Neuron 28:697–711.

Nikolic M, Chou MM, Lu W, Mayer BJ and Tsai L-H. (1998) The p35/Cdk5 kinase is a neuron-specific Rac effector that inhibits Pak1 activity. Nature 395:194–198.

Noctor SC, Martinez-Cerdeno V, Ivic L and Kriegstein AR. (2004) Cortical neurons arise in symmetric and asymmetric division zones and migrate through specific phases. Nat. Neurosci. 7:136–144.

Ohshima T, Ward JM, Huh CG, Longenecker G, Veeranna, Pant HC, Brady RO, Martin LJ and Kulkarni AB. (1996) Targeted disruption of the cyclin-dependent kinase 5 gene results in abnormal corticogenesis, neuronal pathology and perinatal death. Proc. Natl Acad. Sci. USA 93:11173–11178.

Ohshima T, Gilmore EC, Longenecker G, Jacobowitz DM, Brady RO, Herrup K and Kulkarni AB. (1999) Migration defect of cdk5−/− neurons in the developing cerebellum is cell autonomous. J. Neurosci. 19:6017–6026.

Ohshima T, Ogawa M, Veeranna, Hirasawa M, Longenecker G, Ishiguro K, Pant HC, Brady RO. Kulkarni AB and Mikoshiba K. (2001) Synergistic contributions of cyclin-dependent kinase 5/p35 and Reelin/Dab1 to the positioning of cortical neurons in the developing mouse brain. Proc. Natl Acad. Sci. USA 98:2764–2769.

Ohshima T, Ogawa M, Takeuchi K, Takahashi S, Kulkarni AB and Mikoshiba K. (2002) Cyclin-dependent kinase 5/p35 contributes synergistically with Reelin/Dab1 to the positioning of facial branchiomoter and inferior olive neurons in the developing mouse hindbrain. J. Neurosci. 22:4036–4044.

Ohshima T and Mikoshiba K. (2002) Reelin signaling and Cdk5 in the control of neuronal positioning. Mol. Neurobiol. 26:153–166.

Ohshima T, Suzuki H, Morimura T, Ogawa M and Mikoshiba K. (2007a) Modulation of Reelin signaling by Cyclin-dependent kinase 5. Brain Res. 1140:84–95.

Ohshima T, Hirasawa M, Tabata H, Mutoh T, Adachi T, Suzuki H, Saruta K, Iwasato T, Itohara S, Hashimoto M, Nakajima K, Ogawa M, Kulkarni AB and Mikoshiba K. (2007b) Cdk5 is required for multipolar-to-bipolar transition during radial neuronal migration and proper dendrite development of pyramidal neurons in the cerebral cortex. Development 134:2273–2282.

Paglini G, Pigino G, Kunda P, Morfini G, Maccioni R, Quiroga S, Ferreira A and Caceres A. (1998) Evidence for the participation of the neuron-specific CDK5 activator P35 during laminin-enhanced axonal growth. J. Neurosci. 18:9858–9869.

Rashid T, Banerjee M and Nikolic M. (2001) Phosphorylation of Pak1 by the p35/Cdk5 kinase affects neuronal morphology. J. Biol. Chem. 276:49043–49052.

Reiner O, Carrozzo R, Shen Y, Wehnert M, Faustinella F, Dobyns W B, Caskey CT and Ledbetter DH. (1993) Isolation of a Miller-Dieker lissencephaly gene containing G- protein -subunit-like repeats. Nature 364:717–721.

Sasaki S, Shionoya A, Ishida M, Gambello MJ, Yingling J, Wynshaw-Boris A and Hirotsune S. (2000) A LIS1/NUDEL/cytoplasmic dynein heavy chain complex in the developing and adult nervous system. Neuron 28:681–696.

Tabata H and Nakajima K. (2003) Multipolar migration: The third mode of radial neuronal migration in the developing cerebral cortex. J. Neurosci. 23:9996–1001.

Tanaka T, Serneo FF, Tseng HC, Kulkarni AB, Tsai L-H and Gleeson JG. (2004) Cdk5 phosphorylation of doublecortin ser297 regulates its effect on neuronal migration. Neuron 41:215–227.

Tang D, Yeung J, Lee KY, Matsushita M, Matsui H, Tomizawa K, Hatase O and Wang JH. (1995) An isoform of the neuronal cyclin-dependent kinase 5 (cdk5) activator. J. Biol. Chem. 270:26897–26903.

Tsai L.-H, Takahashi T, Caviness Jr. VS and Harlow E. (1993) Activity and expression pattern of cyclin-dependent kinase 5 in the embryonic mouse nervous system. Development 119:1029–1040.

Tsai L-H, Delalle I, Caviness Jr. VS, Chae T and Harlow E. (1994) p35 is a neural-specific regulatory subunit of cyclin-dependent kinase 5. Nature 371:419–423.

Uchida Y, Ohshima T, Sasaki Y, Suzuki H, Yanai S, Yamashita N, Nakamura F, Takei K, Ihara Y, Mikoshiba K, Kolattukudy P, Honnorat J and Goshima Y. (2005) Semaphorin3A signaling is mediated via sequential Cdk5 and GSK3β phosphorylation of CRMP2: Implication of common phosphorylating mechanism underlying axon guidance and Alzheimer's disease. Genes Cells 10:165–179.

CRMP Family Protein: Novel Targets for Cdk5 That Regulates Axon Guidance, Synapse Maturation, and Cell Migration

Yoshio Goshima, Yukio Sasaki, Yutaka Uchida, Naoya Yamashita, and Fumio Nakamura

Abstract In the developing nervous system, post-mitotic neurons migrate and extend their neurites and form precise patterns of connections that emerge through the interaction between the growth cone and a myriad of environmental cues such as attractive or repulsive axon guidance molecules. Semaphorin3A (Sema3A) is the prototypical repulsive axon guidance molecule that potently induces growth cone collapse stalling neurite extension. Neuropilin-1 (NRP-1) and Plexin-As are ligand-binding and signal-transducing receptor components for Sema3A, respectively. Collapsin response mediator protein (CRMP) was identified as a signaling molecule of Sema3A. However, its molecular mechanisms have been ill-defined. CRMPs are now known to be composed of five homologous cytosolic proteins CRMP1–5; all of the family proteins are highly phosphorylated in developing brains. By screening pharmaceutical reagents and utilizing gene-deficient mice and through biochemical analysis, we found that Fyn and cyclin-dependent kinase 5 (Cdk5) mediate Sema3A-induced response in dorsal root ganglion (DRG) neurons. Cdk5 was associated with PlexA2 through the active state of Fyn. This raised the possibility that Sema3A induced growth cone collapse response through phosphorylation of CRMPs by Cdk5. The 2-D gel analysis of brain lysate from Cdk5-deficient mice revealed that CRMP2 was a substrate for Cdk5 *in vivo*. *In vitro* kinase assay revealed that Ser522 was the major site of CRMP1 and CRMP2 phosphorylation by Cdk5. Cdk5 primarily phosphorylated CRMP2 at Ser522, and GSK3β secondarily phosphorylates at Thr509. The dual-phosphorylated CRMP2 was recognized by the antibody 3F4, which is highly reactive with the neurofibrillary tangles of Alzheimer's disease. In DRG neurons, Sema3A stimulation enhanced the levels of the phosphorylated form of CRMP2 detected by 3F4. Overexpression of CRMP2 mutant substituting either Ser522 or Thr509 with Ala attenuated Sema3A-induced growth cone collapse. Knockdown of CRMP1 and CRMP2 inhibited Sema3A-induced growth cone collapse. The phosphorylation of

Y. Goshima
Department of Molecular Pharmacology & Neurobiology, Yokohama City University Graduate School of Medicine, 236-0004, Japan
e-mail: goshima@med.yokohama-cu.ac.jp

N.Y. Ip, L.-H. Tsai (eds.), *Cyclin Dependent Kinase 5 (Cdk5)*,
DOI: 10.1007/978-0-387-78887-6_2, © Springer Science+Business Media, LLC 2008

CRMP1 and/or CRMP2 is therefore an essential step for Sema3A signaling. CRMP1 and CRMP2 were also good substrates for Fyn. The phosphorylation of CRMP1 by Cdk5 and Fyn also appears to be involved in Sema3A and Reelin signaling, contributing to spine maturation and the regulation of cell migration during the development of the cerebral cortex.

Introduction

Cyclin-dependent kinase 5 (Cdk5), a member of the serine/threonine kinase Cdk family, has enzymatic activity only in post-mitotic neurons due to a neuron-specific expression of the regulatory subunit p35 (Lew and Wang, 1995). Cdk5 has been implicated in various aspects of neural development, such as neurite outgrowth, axonal path finding, dendritic branching, and neural plasticity. Knockout mice studies reveal that Cdk5 and p35 play critical roles in laminar formation of the cerebral cortex by regulating the migration of neurons (Ohshima et al., 1996; Chae et al., 1997). However, the molecular mechanisms by which Cdk5 regulates the processes of neural development are unclear.

Neurons form precise patterns of connections that emerge through the interaction between the growth cone and extracellular signals in the developing nervous system. Neuronal extension from somas and navigating extensive length to their crucial targets are controlled by four types of axon guidance cues, membrane-bound or soluble attractive and repulsive molecules (Tessier-Lavigne and Goodman, 1996). Morphological changes and motility of neuronal growth cones are closely related to reorganization of actin, tubulin, and other cytoskeletal proteins, and interplay between actin and microtubule cytoskeletons has been shown to be critical for growth cone navigation (Tanaka and Sabry, 1995). Semaphorins, a large family of guidance cues for axonal/dendritic projections, are comprised of both secreting and membrane-bound proteins sharing a Sema domain (Kolodkin, 1998; Raper, 2000). The class 3 subfamily of the eight semaphorin subfamilies, composed of at least seven known members including Sema3A, has been the best-characterized subfamily. Sema3A promotes a variety of cellular responses, including cytoskeletal reorganization, endocytosis, and facilitation of axonal transport (Goshima et al., 1997; Fournier et al., 2000; Raper, 2000; Li et al., 2004). The semaphorin receptor mediating class 3 semaphorin signals has been identified as a complex of neuropilins and plexins (Takahashi et al., 1999; Tamagnone et al., 1999). Neuropilin (NRP)-1 and -2 bind to class 3 semaphorins with high affinities and are necessary for Sema3A-mediated repulsive guidance events. The mammalian plexin (Plex)-A subfamily does not bind directly to class 3 semaphorins. NRPs and Plex-As are ligand-binding and signal-transducing subunits of class 3 semaphorin receptor complexes, respectively (Takahashi et al., 1999; Tamagnone et al., 1999). Recent studies have shown that several intracellular molecules, including the small GTPase Rac1, R-Ras, the collapsin (Sema3A) response mediator protein (CRMP), LIM kinase, and FARP2, are implicated as mediators of Sema3A signaling (Puschel, 2007). However, the whole picture of the mechanisms involved in transducing Sema3A to actin and microtubule cytoskeleton for neuronal guidance remains obscure.

The chicken collapsin (formerly called semaphorin) response mediator protein (CRMP-62) molecule (also known as CRMP2) was originally identified as signaling molecule of Sema3A (Goshima et al., 1995). CRMP2, which has also been independently identified, is one of the five isoforms (CRMP1–5). The interactions among CRMP isomers favor heterophilic oligomerization over homophilic oligomerization (Wang and Strittmatter, 1997). CRMPs are extensively phosphorylated during neuronal development (Byk et al., 1996). This raised the possibility that CRMPs act as downstream components of sema-phorin–PlexA signal transduction pathway through their phosphorylation to regulate cytoskeletal reorganization.

This chapter will discuss the role of CRMPs as substrate of Cdk5 that regulate axon guidance, spine maturation, and cell migration in developing nervous system.

Fyn and Cdk5 Are Involved in Sema3A Signaling

Sema3A is a common guidance cue for nerve projection in the CNS and PNS. The collapse assay system using dorsal root ganglion (DRG) is the most useful method to characterize Sema3A responses. To search protein kinases involved in Sema3A signaling, we first screened various pharmacological reagents against Sema3A-induced growth cone collapse of DRG neurons. Among the more than 20 compounds we tested, lavendustin A ($10\,\mu M$), a tyrosine kinase inhibitor, significantly suppressed Sema3A-induced growth cone collapse. We previously reported that olomoucine, a Cdk inhibitor, blocks growth cone collapse induced by chick caudal tectal membrane containing ephrins (Nakayama et al., 1999). Olomoucine ($10\,\mu M$) also blocks Sema3A-induced growth cone collapse of chick DRG neurons. These data suggest that tyrosine kinase(s) and Cdk(s) are involved in Sema3A-induced growth cone collapse.

Because lavendustin A is thought to inhibit a variety of tyrosine kinases, it was difficult to identify tyrosine kinase(s) responsible for Sema3A signaling by using pharmacological methods. Among nonreceptor tyrosine kinases, Fyn is one of the plausible candidates for Sema3A signaling because Fyn is enriched in growth cones (Helmke and Pfenninger, 1995), and *fyn*-deficient mice show neural phenotypes (Yagi et al., 1993). In embryonic homogygous null mutant mice lacking *fyn* gene, the trajectory of the olfactory nerve displays defasciculation (Morse et al., 1998). Recently, Fyn has been appreciated as a mediator for axon guidance, since phosphorylation of deleted colorectal cancer receptor by Fyn mediates Netrin-1 signaling in growth cone (Meriane et al., 2004), and Netrin requires focal adhesion kinase and Src family kinases for axon outgrowth and attraction (Liu et al., 2004).

To examine whether Fyn is involved in Sema3A signaling, we analyzed Sema3A response in *fyn*-deficient DRG neurons. The rate of growth cone collapse induced by Sema3A in *fyn* $^{-/-}$ DRG neurons at E17 showed a significant decrease compared with that in *fyn*$^{+/-}$ neurons at lower concentrations (Sasaki et al., 2002). Olomoucine is reported to inhibit kinase activities of several kinds of Cdk/cyclin complexes as well as Cdk5/p35. Within the Cdk family, Cdk5 is the most plausible candidate for Sema3A signaling because Cdk5 activity is detected only in neurons, and *cdk5*-deficient mice show neural phenotypes (Ohshima et al., 1996). To assess whether Cdk5 participates in Sema3A-induced growth cone collapse, we used a DRG explant culture of E12 *cdk5*-deficient mice. The collapse rate induced by Sema3A in *cdk5*$^{-/-}$ mice decreased to 30–40%, whereas the rate reached 80% in *cdk5*$^{+/-}$ mice at 0.3 nM Sema3A (Sasaki et al., 2002). These findings demonstrate that Fyn and Cdk5 are involved in Sema3A-induced growth cone collapse.

After discovery of CRMP-62, we investigated whether Sema3A had some effects on axonal transport, because CRMPs are related to UNC-33: *unc-33* mutant shows multiple axon guidance defects, and abnormalities in the form and number of microtubules (Li et al., 1992). UNC-33 has been therefore implicated in the regulation of tubulin dynamics. We found that Sema3A facilitates antero- and retrograde axonal transport (Goshima et al., 1997, 1999). To test the possibility that Fyn and Cdk5 mediate Sema3A-induced axonal transport as well, we examined the effects of lavendustin A and olomoucine on axonal transport (Li et al., 2004). Pretreatment of lavendustin A or olomoucine inhibited Sema3A-induced facilitation of antero- and retrograde axonal transport, without affecting basal levels of the antero- and retrograde axonal transport. Furthermore, Sema3A-induced facilitation of axonal transport was attenuated in *fyn*- and p35-deficient mouse DRG (Li et al., 2004). These findings indicate that Fyn and Cdk5 mediate Sema3A-induced axonal transport as well as growth cone collapse.

Plex-A2 Associates with Cdk5 via Interaction with Active Fyn, and Fyn Activates Cdk5 via Tyr15 Phosphorylation

Plex-A3 and -B1 has been reported to associate with some unknown kinase activity (Tamagnone et al., 1999). We thus hypothesized that Fyn associated with the Plex-A/NRP-1 heterodimer receptor to transduce Sema3A signaling. Coimmunoprecipitation experiments demonstrated that Plex-A2/NRP-1 complex could associate with all Fyn variants, irrespective of their kinase activities (Sasaki et al., 2002). Constitutively, active and wild-type Fyn induced tyrosine phosphorylation of Plex-A2, whereas kinase-negative and dominant-negative mutants did not promote the phosphorylation. Fyn was coimmunoprecipitated with Plex-A2 in the absence of NRP-1. The tyrosine phosphorylation level of Plex-A2 was reduced without NRP-1, suggesting that NRP-1 might have a cooperative effect on Plex-A2 phosphorylation by Fyn. Fyn was not coimmunoprecipitated with Plex-A2 lacking most of the cytosolic region (Plex-A2Δcyto), and no tyrosine phosphorylation of Plex-A2Δcyto was detected, indicating that Fyn phosphorylated and associated with the cytosolic region of Plex-A2. We also examined whether Cdk5 was associated with Plex-A2. Cdk5 was not coimmunoprecipitated with Plex-A2 in the absence of Fyn. When constitutively active Fyn was co-expressed with Cdk5 and Plex-A2, Cdk5 was detected in the immunoprecipitate of Plex-A2. The association of Cdk5 with Plex-A2 was hardly observed in the presence of either wild-type or dominant-negative Fyn. Furthermore, Cdk5 was associated with constitutively active Fyn in the absence of Plex-A2. These data indicate that Cdk5 is associated with PlexA2 through the active state of Src family kinases.

Because Cdk5 cannot associate with Plex-A2 in the absence of Fyn, it was possible that the Src family kinases were necessary to regulate Cdk5 activity downstream of Plex-As. Thus, we examined whether Fyn promoted activation of Cdk5. Transfection of COS-7 cells with Cdk5, p35, and constitutively active Fyn enhanced kinase activity of the Cdk5/p35 complex by about 2-fold compared with Cdk5/p35 alone. Dominant-negative Fyn did not facilitate Cdk5 activity. Wild-type Fyn activated Cdk5 to a lesser extent than did the constitutively active form. It is reported that a nonreceptor tyrosine kinase, Abl, facilitates Cdk5 kinase activity via Tyr15 phosphorylation of Cdk5 (Zukerberg et al., 2000). The kinase activity of a Cdk5 mutant in which Tyr15 was converted to Ala (Cdk5Y15A) was not facilitated by constitutively active Fyn. An anti-phospho-Cdc2 (Tyr15) antibody also recognizes Tyr15-phosphorylated Cdk5. Immunoblot analysis using this antibody showed that constitutively active Fyn phosphorylated Tyr15 of Cdk5. These results clearly show that Fyn facilitates Cdk5 activity mediated through Tyr15 phosphorylation of Cdk5.

Sema3A-Induced Growth Cone Collapse Is Mediated Through Cdk5 Activation via Tyr15 Phosphorylation

To examine whether Sema3A promoted Cdk5 activation via Tyr15 phosphorylation, we performed a Cdk5 kinase assay in COS cells expressing Plex-A2, NRP-1, Cdk5, and p35. Sema3A facilitated kinase activity of Cdk5/p35 within 5 min, and the activity persisted up to 20 min. We examined the effect of the Cdk5Y15A mutant on Sema3A-induced growth cone collapse because Fyn did not facilitate the activity of the Cdk5Y15A mutant. We expressed wild-type Cdk5, Cdk5Y15A, or kinase-negative mutant in DRG neurons. The expression of Cdk5Y15A and kinase-negative mutant suppressed the Sema3A-induced response. In contrast, wild-type Cdk5 in DRG neurons did not block the growth cone collapse. In culture without Sema3A, anti-phospho-Cdc2 (Tyr15) antibody stained only faintly in well-spread growth cones of DRG explants. At 0.5–2 min after Sema3A application, the anti-phospho-Cdc2 (Tyr15) antibody intensely stained partially collapsed, but not fully extended, growth cones. The staining was visible in the central domain of growth cones and in most of the filopodial processes. The staining by the phospho-specific antibody was not detected within 5 min. These data suggest that Sema3A-induced growth cone collapse is mediated through Tyr15 phosphorylation of Cdk5.

Tau is a known substrate of Cdk5 that regulates the dynamics of tubulin (Hosoi et al., 1995). To examine whether phosphorylated tau increased in growth cones after Sema3A stimulation, we used an anti-phosphorylated tau antibody AT-8, which recognized phosphorylation sites by Cdk5. In nonstimulated culture, phosphorylated tau was localized in the axon, but not in the growth cone. At 2–5 min after Sema3A application, phosphorylated tau was mainly localized at the proximal region of the axon. Anti-tau antibody showed that total tau content in growth cones did not appear to change until complete collapse. These results indicate that Sema3A promotes activation of Cdk5 in growth cones, followed by phosphorylation of tau (Sasaki et al., 2002).

CRMP2 Is an *In Vivo* Substrate of Cdk5

These findings suggest that Sema3A regulates growth cone motility through Fyn-Cdk5 cascade. This prompted us to determine whether Sema3A signaling was mediated through phosphorylation of CRMP by Cdk5. To examine whether CRMP was a substrate for Cdk5, we performed *in vitro* kinase assay using purified CRMP2, Cdk5, and p25 or p35. CRMP2 was phosphorylated in the presence of both Cdk5 and p25. Cdk5/p35 enhanced phosphorylation of CRMP2 in HEK293T cells. Phosphorylation of CRMP1, 4, and 5 were also enhanced by Cdk5/p35 in HEK293T cells, a result consistent with the fact that CRMP1, 4, and 5 had the consensus sequence of phosphorylation by Cdk5 near amino acid residue 522 as follows. The consensus sequence of phosphorylation by Cdk5 was amino acid sequence (S/T)PX(K/H/R), where S or T was the phosphorylatable serine or threonine (Songyang et al., 1996). Amino acids 522–525 (SPAK) of CRMP2 matched with the consensus sequence. To test whether Ser522 was the phosphorylation site of CRMP2 by Cdk5, we produced a non-phosphorylated CRMP2 mutant, CRMP2S522A, in which Ser522 was replaced by Ala. Cdk5 phosphorylated wild-type CRMP2 (CRMP2wt), but not CRMP2S522A *in vitro*. We prepared a rabbit polyclonal antibody, anti-pS522-CRMP1/2 antibody, that recognized CRMP1 and CRMP2 phosphorylated at Ser522 *in vitro*, but not CRMP3, CRMP4, and CRMP5. Immunoblot analysis using anti-pS522-CRMP1/2 antibody revealed that the bands were at low intensity or were missing in lysate of embryonic brains from $cdk5^{-/-}$ when compared to wild-type mice.

During a search for *in vivo* substrate of Cdk5 using $cdk5^{-/-}$ mice and a monoclonal antibody, which recognized phospho-threonine followed by proline (pThr-Pro antibody), Ohshima and his colleagues found that the threonine phosphorylation was not seen with brain lysate from $cdk5^{-/-}$ mice. They found that three bands were at low intensity or missing in $cdk5^{-/-}$ brain lysate. These bands corresponded to 65 kDa, 70 kDa, and 120 kDa in SDS-PAGE electrophoresis. 2-D gel electrophoresis followed by Western blot with pThr-Pro antibody showed that five spots were at very low intensity or absent in $cdk5^{-/-}$ brain lysate. Spot 1 and 2 corresponded to 65 kDa, spot 3 and 4 to 70 kDa, and spot 5 to 120 kDa bands. MALDI-TOP mass spectrometry analysis and database search indicated that spot 3 fitted to CRMP2. Immunoblot analysis of 2-D gel with pThr-Pro antibody and C4G, which was phosphorylation-independent anti-CRMP2 monoclonal antibody (Gu et al., 2000), revealed that two of them matched spot 3 and 4. Phosphorylation of CRMP2 was previously reported in Alzheimer's disease (AD) patient brains using monoclonal antibody 3F4 (Yoshida et al., 1998). 3F4 recognized three sites of phosphorylation of CRMP2 at Thr509, Ser518, and Ser522 (Gu et al., 2000). Immunoblot analysis with this monoclonal antibody confirmed that spots 3 and 4 completely matched those sites detected by 3F4 and C4G antibodies. Furthermore, immunoblotting with these antibodies revealed that CRMP2 was not phosphorylated at 3F4 recognition site in $cdk5^{-/-}$ embryonic brain. These results indicated that CRMP2 was phosphorylated by Cdk5 in the embryonic mouse brain, and the phosphorylation site(s) was (were) Thr509 and/or Ser518 and/or Ser522. These findings suggest that CRMP2 is also phosphorylated at the threonine residue(s) *in vivo*. In addition, no band shift was observed with CRMP2S522A co-expressed with Cdk5/p35.

Cdk5-Primed GSK3β Phosphorylation of CRMP2

Based on these findings, we speculated that phosphorylation of Ser522 was required for phosphorylation of other sites, and that other Ser/Thr kinase(s) might be involved in phosphorylation of CRMP2 at threonine residue. Tau is known for its Cdk5-primed phosphorylation by a Ser/Thr kinase GSK3β (Cho and Johnson, 2003). Indeed, when CRMP2 was co-expressed with GSK3β, CRMP2 was phosphorylated and intensity of upper band increased compared with CRMP2 co-expressed with Cdk5/p35 in HEK293T cells. When co-expressed with GSK3β and Cdk5/p35, CRMP2 was more phosphorylated and the intensity of upper band relatively increased, thereby indicating phosphorylation at additional sites. No phosphorylation and band shift were observed in the co-expression of CRMP2S522A with Cdk5/p35, GSK3β, or both. This result suggests that CSK3β as well as Cdk5 phosphorylated CRMP2 at Ser522, or Cdk5 phosphorylation of Ser522 is essential for the additional phosphorylation of CRMP2 by GSK3β. *In vitro* kinase assay using purified protein GSK3β alone did not phosphorylate CRMP2. To examine whether the phosphorylation by Cdk5 enhances additional phosphorylation of CRMP2 by GSK3β, we prepared GST-CRMP2c (aa486 to carboxyl terminus), and then phosphorylated this GST-fusion protein by immobilized Cdk5/p35 bound to anti-p35 antibody and protein-A beads, with non-labeled ATP *in vitro*. After removing the immobilized Cdk5/p35, we mixed the phosphorylated GST-CRMP2c and GSK3β with [γ-^{32}P]ATP for kinase reaction. Incorporation of [^{32}P]phosphate was exclusively observed in Cdk5-phosphorylated GST-CRMP2cwt, but not GST-CRMP2cS522A, confirming the essential role of Cdk5 phosphorylation for the further phosphorylation of CRMP2 by GSK3β.

The Phosphorylation of CRMPs Disrupts Association of CRMPs with Tubulin

It is known that microtubule-associated protein tau, when phosphorylated by Cdk5 and GSK3β, decreases its affinity to microtubules and reduces their stability (Cho and Johnson, 2003). To explore biological significance of the phosphorylation of CRMP2, we examined whether the interaction between CRMP2 and tubulin was altered by CRMP2 phosphorylation by Cdk5 and GSK3β. We first examined localization of CRMPs in the COS cells transfected with GFP-CRMP1, 2, 3, 4, and 5. Each GFP-tagged CRMP family proteins showed distinct subcellular localization in the COS cells. GFP-CRMP2, 3, 4, and 5 were mainly localized in the cytoplasmic region. GFP-CRMP2 and GFP-CRMP5 sometimes showed dot-like pattern, and some of the GFP-CRMP5 signals were localized in the nucleus (data not shown). GFP-CRMP1 showed filamentous structure (Fig. 1a), which was characterized by the presence of microtubule bundles. Colocalization was confirmed by double labeling with the antibody to α-tubulin. This filamentous pattern was altered to diffuse distribution, when GFP-CRMP1 was co-transfected with Cdk5/p35 (Fig. 1b). This alteration was not observed with GFP-CRMP1S522A, suggesting that the phosphorylation of CRMP1 lowered the affinity of CRMP1 for tubulin. On the other hand, CRMP2 was diffusely distributed in the cytoplasm, a result consistent with previous findings (Rosslenbroich et al., 2003). However, in some neuronal cell line cells overexpressing CRMP2, CRMP2 was shown to be colocalized with microtubule (Gu et al., 2000), and to be associated with tubulin heterodimers (Fukata et al., 2002). We therefore examined whether the interaction between CRMP2 and tubulin was altered by CRMP2 phosphorylation by Cdk5 and GSK3β. In HEK293T cells expressing CRMP2 alone, CRMP2 was coimmunoprecipitated with endogenous tubulin. This immunoprecipitation with anti-α-tubulin antibody became barely detectable when CRMP2 was co-expressed with Cdk5/p35 or Cdk5/p35 plus GSK3β. When CRMP2S522A was expressed with Cdk5/p35 and GSK3β, the association of the mutant CRMP2 with tubulin was not altered when compared to CRMP2S522A alone. This indicates that phosphorylation of CRMP2 by Cdk5 and GSK3β lowered the affinity of CRMP2 to tubulin. Furthermore, the association of CRMP2 with tubulin was enhanced by 2-fold in *cdk5*$^{-/-}$

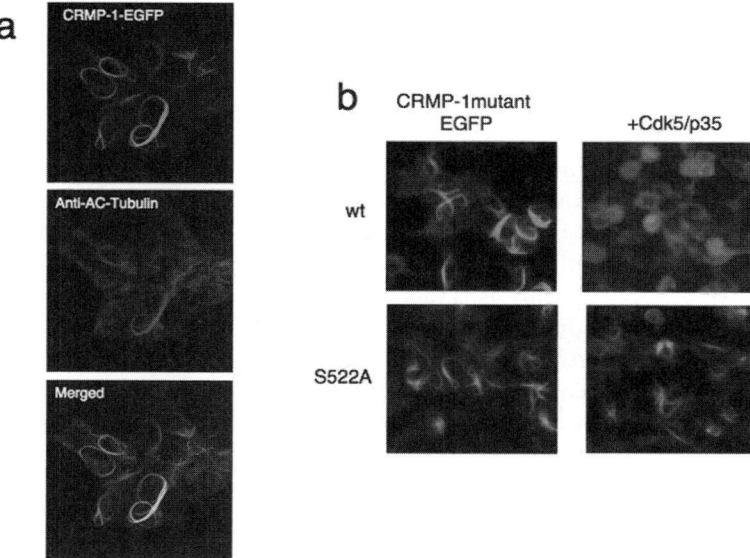

Fig. 1 Distribution of CRMP1 and tubulin in COS-7 cells. CRMP1-EGFP or Cdk5/p35 was ectopically expressed in COS-7. In (**a**), the cells were immunostained with anti-α-tubulin antibody. The localization of wild-type CRMP1-EGFP showed filamentous structure, and this distribution pattern became diffusely distributed within the cytoplasm when co-transfected with Cdk5/p35 (**b**) (*See* Color Insert)

embryonic brain, when compared to wild-type brain, thereby indicating that CRMP2 phosphorylated by Cdk5 reduced its affinity for tubulin in the embryonic mouse brain *in vivo*. To determine the role of GSK3β, we estimated the affinity of CRMP2 to tubulin in cell lysate of cortical neuron primary culture with or without LiCl, a GSK3β inhibitor. LiCl induced band shifting of CRMP2 to the lower molecular weight. The treatment of LiCl increased the amount of CRMP2 coimmunoprecipitated with tubulin when compared to the untreated control. These results suggest that sequential phosphorylation of CRMP2 by Cdk5 and GSK3β is necessary to negatively regulate the association of CRMP2 and tubulin.

Sema3A-Induced Growth Cone Collapse Through Phosphorylation of CRMP2, and Evidence for Both CRMP1 and CRMP2 Are Involved in Sema3A Signaling

We examined whether Sema3A induced growth cone collapse through phosphorylation of CRMP2 by Cdk5 and GSK3β. Immunoblot analysis using anti-pS522-CRMP1/2 antibody revealed that Sema3A increased phosphorylation of endogenous CRMP1 and/or CRMP2 at Ser522 in an olomoucine-sensitive manner. In immunocytochemistry with anti-CRMP1 and -CRMP2 antibodies, CRMP1 and CRMP2 were found to be expressed in the growth (Uchida et al., 2005). These CRMP1 and CRMP2 antibodies recognized both non-phosphorylated and phosphorylated forms of CRMP1 and CRMP2, respectively. Upon Sema3A treatment, the levels of phosphorylated CRMP1 and/or CRMP2 detected by pS522-CRMP1/2 antibody, were increased (Fig. 2).

To examine the roles of CRMP1 and CRMP2 phosphorylation by Cdk5 and GSK3β, we expressed the wild-type and unphosphorylated mutants of CRMP1 and CRMP2 in DRG neurons, and saw the effect of Sema3A on growth cone morphology. Introduction of CRMP2T509A, CRMP2522A, and CRMP2T509A/S522A into DRG neurons significantly suppressed Sema3A-induced growth cone collapse, while that of CRMP2wt does not show any effect (Fig. 3). Introduction of the non-phosphorylated mutant of CRMP1 also suppressed Sema3A-induced growth cone response. Either olomoucine or TDZD-8, a GSK3β inhibitor, alone suppressed Sema3A-induced response, but combined application of olomoucine and TDZD-8 shows no additive effect on Sema3A-induced growth cone collapse. These results suggest that dual phosphorylation of CRMP2 by Cdk5 and GSK3β and the phosphorylation of CRMP1 by Cdk5 are involved in Sema3A-induced response. Immunoblot analysis indicates that Sema3A stimulation enhanced the levels of the dual-phosphorylated form of CRMP2 detected by 3F4 antibody.

Yeast two-hybrid analysis and *in vitro* binding analysis suggest that CRMP1 and CRMP2 prefer hetero-oligomerization to homo-oligomerization (Wang

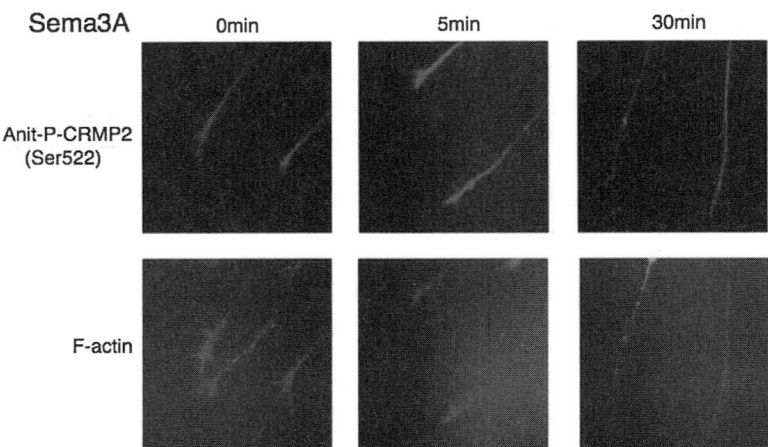

Fig. 2 Sema3A induced growth cone collapse and increased the levels of CRMP2 phosphorylated at Ser522 in the growth cones. Cultured E7 chick DRG neurons were stimulated with Sema3A (1 nM), fixed and double stained with FITC-phalloidin and anti-P-CRMP1/2 antibody. Scale bar, 50 μm (*See* Color Insert)

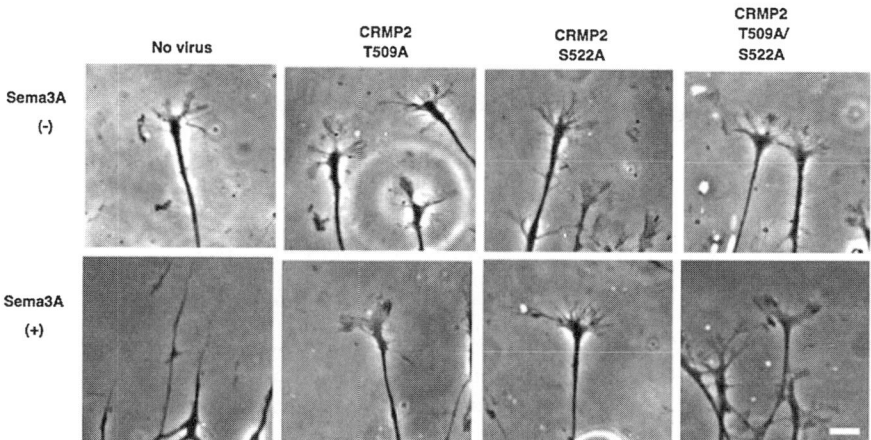

Fig. 3 Suppression of Sema3A-induced response by introduction of non-phosphorylated mutants of CRMP2. Chick DRG neurons were infected with recombinant HSV preparations directing the expression of CRMP2wt, CRMP2T509A, CRMP2S522A, and CRMP2T509A/S522A. Expression of CRMP2T509A, CRMP2S522, and CRMP2T509A/S522A, but not CRMP2wt, suppressed the Sema3A-induced growth cone collapse. Scale bar, 50 μm. Reproduced with permission (Uchida et al., 2005)

and Strittmatter, 1996). CRMP1 and CRMKP2 could therefore function in concert with each other for physiological responses. Indeed, knockdown of CRMP1 and CRMP2 by siRNA suppressed Sema3A-induced growth cone collapse (Uchida et al., 2005).

CRMP1 Regulates Spine Maturation Through Mediating Sema3A Signaling *In Vivo*

To elucidate *in vivo* function of CRMP family proteins, we performed phenotypic analysis of *crmp1*-deficient mice (Charrier et al., 2006). We further investigated possible role of the phosphorylation of CRMP1 by Cdk5 in mediating Sema3A response. We focused on dendritic patterning and spine maturation in the layer V pyramidal neurons, since Sema3A induces spine maturation and dendritic branching *in vivo* (Morita et al., 2006). In cultured cortical neurons from *crmp1*$^{-/-}$ mice, a phenotype of enlarged clusters of synapsin I and PSD-95, which is similar to those from *cdk5*$^{-/-}$ mice, was observed (Yamashita et al., 2007). This phenotype was rescued by wild-type CRMP1 but not by CRMP1T509A/S522A, a mutant of non-phosphorylated Cdk5. In the cultured cortical neurons from *crmp1*$^{+/-}$ mice, Sema3A increased the number of clusters of synapsin I and postsynaptic density-95 (PSD-95), but this increase was attenuated in *crmp1*$^{-/-}$ mice. This phenotype was rescued by introduction of wild-type CRMP1, but not CRMP1-T509A/S522A, into the neurons. Golgi impregnation method showed that $crmp^{-/-}$ layer V cortical neurons showed lowered density of synaptic bouton-like structures and this phenotype had genetic interaction with *sema3A*. These findings suggest that CRMP1 regulates spine maturation through phosphorylation of CRMP1 by Sema3A-Cdk5 cascade.

CRMP1 Mediates Reelin Signaling in Cortical Neuronal Migration

In the course of phenotypic analysis of $crmp1^{-/-}$ cortices, we found that radial migration of cortical neurons was retarded (Yamashita et al., 2006). In the brains of P10 injected at E14.5 $crmp1^{+/-}$ cortex, most labeled neurons were positioned in the deep layers of the cerebral cortex destined to form layers IV and V by BrdU birth-dating analysis. In $crmp^{-/-}$ mice, the majority of labeled neurons were positioned at layers IV and V. The percentage of labeled cells positioned in those regions, however, was decreased, and the percentage in other regions was increased. This phenotype was not observed in the $sema3A^{-/-}$, $crmp1^{+/-}$, and sema3A$^{+/-}$ cortices. On the other hand, CRMP1 was colocalized with disabled-1 (Dab1), an adaptor protein in Reelin (Reln) signaling. In the $Reln^{rl/rl}$ cortex, CRMP1 and Dab1 were expressed at a higher level, yet tyrosine phosphorylated at a lower level. Loss of $crmp1$ in a $dab1$ heterozygous background led to the disruption of hippocampal lamination, a Reeler-like phenotype (Fig. 4). These results suggest that CRMP1 mediates Reln signaling but not Sema3A in cortical layer formation.

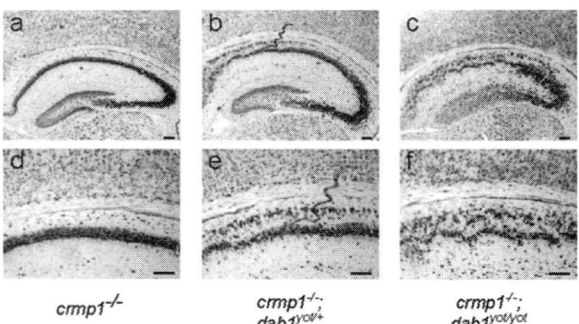

$crmp1^{-/-}$ $crmp1^{-/-};$ $crmp1^{-/-};$
                                 $dab1^{yot/+}$ $dab1^{yot/yot}$

Fig. 4 Genetic interaction between $crmp1$ and $dab1$ in the hippocampus. Nissl staining of the brains at P10 of $crmp1^{-/-}$ (**a,d**); $crmp1^{-/-}$ and dab $^{yot/+}$ (**b,e**); and $crmp1^{-/-}$ and $dab1^{yot/yot}$ (**c,f**) mice. Coronal sections of the hippocampus region are presented. Magnified images of the CA1 region in **a, b,** and **c** are shown in **d, e,** and **f,** respectively. Scale bars, 100 μm. Reproduced with permission (Yamashita et al., 2006)

The Phosphorylation of CRMPs in Mediating Extacellular Signals

We have shown that Cdk5/p35 and Rho/ROCK kinase mediate ephrin-A5-induced signaling in retinal ganglion cells (Chen et al., 2003), and ephrin-A5 induces phosphorylation of CRMP2 via Rho/ROCK kinase during growth cone collapse (Arimura and Kaibuchi, 2005). Lysophosphatidic acid (LPA) is reported to induce growth cone collapse in DRG neurons through Rho/ROCK kinase (Arimura et al., 2000). LPA induces collapse of growth cone of neurons of both $cdk5^{+/-}$ and $cdk5^{-/-}$ with similar extent, indicating that Cdk5 is not involved in LPA signaling (Sasaki et al., 2002). It has been shown that the phosphorylation of CRMP2 at Thr514 by GSK3b induces mobility shift of CRMP2 (Yoshimura et al., 2005). On the other hand, NT-3 and BDNF but not NGF enhance axon elongation and branching in hippocampal neurons. GSK3β is constitutively active, and its activity can be inhibited by treatment with NT-3 and BDNF, probably through phosphorylation of GSK3β at Ser9 by PI3-kinase/Akt pathway (Grimes and Jope, 2001; Huang and Reichardt, 2001; Segal, 2003). NT-3 and BDNF decrease the phosphorylation levels of CRMP2 and increase the phosphorylation levels of Akt at Ser473 and GSK3β at Ser9, whereas NGF has no effects on the phosphorylation levels of CRMP2, GSK3β, and Akt (Yoshimura et al., 2005). Upon stimulation by NT-3, the phosphorylation levels of CRMP2 at Thr514 are decreased in axonal growth cone and in the shaft but not in the cell body. Therefore, these findings suggest that NT-3 and BDNF decrease the phosphorylation levels of CRMP2 at Thr514 via the PI3-kinase/Akt/GSK3β pathway. These findings suggest that various types of protein kinases could be involved in the phosphorylation of CRMPs mediating multiple extracellular guidance signals. The physiological significance of these distinct signaling cascades remains unknown.

Conclusion

Since CRMP-62 was identified as an intracellular molecule that mediates Sema3A-induced response, the molecular mechanism by which CRMPs regulate axon/dendrite projection has long been unclear. Now that CRMPs are good substrates for Cdk5 *in vivo*, CRMPs are emerging as key regulators in a wide variety of developmental processes from cell migration to neural network formation (Fig. 5).

Acknowledgments I am grateful to many colleagues who have participated in developing these ideas, especially my Yokohama collaborators, T. Nakayama, K. Takei, and other laboratory members for instructive inputs. I am grateful also to T. Ohshima for our fruitful collaboration. This work was supported by Grants-in-aid for Scientific Research in a Priority Area and The Yokohama City University Center of Excellence Program form the Ministry of Education, Science, Sports and Culture, Yokohama Medical Foundation, CREST (Core

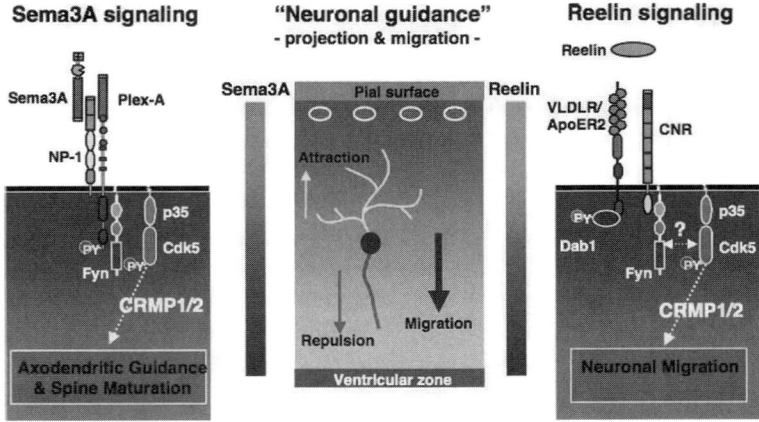

Fig. 5 Role of Sema3A and Reelin in cell migration, axon/dendrite projections, and neural connections. During development, post-mitotic neurons migrate to their correct positions and elaborate their axons and dendrites to establish appropriate neuronal connections. CRMP family proteins may play an integrative role in regulating cytoskeletal organization by mediating various extracellular signals such as Sema3A and Reelin that contribute to neuronal development and maturation

Research for Evolutional Science and Technology) of JST (Japan Science and Technology Corporation), Uehara Memorial Foundation.

References

Arimura N, Inagaki N, Chihara K, Menager C, Nakamura N, Amano M, Iwamatsu A, Goshima Y, Kaibuchi K (2000) Phosphorylation of collapsin response mediator protein-2 by Rho-kinase: evidence for two separate signaling pathways for growth cone collapse. J Biol Chem 275:23973–23980.

Arimura N, Kaibuchi K (2005) Key regulators in neuronal polarity. Neuron 48:881–884.

Byk T, Dobransky T, Cifuentes-Diaz C, Sobel A (1996) Identification and molecular characterization of Unc-33-like phosphoprotein (Ulip), a putative mammalian homolog of the axonal guidance-associated unc-33 gene product. J Neurosci 16:688–701.

Chae T, Kwon YT, Bronson R, Dikkes P, Li E, Tsai LH (1997) Mice lacking p35, a neuronal specific activator of Cdk5, display cortical lamination defects, seizures, and adult lethality. Neuron 18:29–42.

Charrier E, Mosinger Jr B, Meissirel C, Aguera M, Rogemond V, Reibel S, Salin P, Chounlamountri N, Perrot V, Belin M, Goshima Y, Honnorat J, Thomasset N, Kolattukudy P (2006) Transient Alterations in granule cell proliferation, apoptosis and migration in post-natal developing cerebellum of CRMP1$^{-/-}$ mice. Genes Cells 10:165–179.

Cheng Q, Sasaki Y, Shoji M, Sugiyama Y, Tanaka H, Nakayama T, Mizuki N, Nakamura F, Takei K, Goshima Y (2003) Cdk5/p35 and Rho-kinase mediate ephrin-A5-induced signaling in retinal ganglion cells. Mol Cell Neurosci 24:632–645.

Cho JH, Johnson GV (2003) Glycogen synthase kinase 3beta phosphorylates tau at both primed and unprimed sites: differential impact on microtubule binding. J Biol Chem 278:187–193.

Fournier AE, Nakamura F, Kawamoto S, Goshima Y, Kalb RG, Strittmatter SM (2000) Semaphorin3A enhances endocytosis at sites of receptor-F-actin colocalization during growth cone collapse. J Cell Biol 149:411–422.

Fukata Y, Itoh TJ, Kimura T, Menager C, Nishimura T, Shiromizu T, Watanabe H, Inagaki N, Iwamatsu A, Hotani H, Kaibuchi K (2002) CRMP-2 binds to tubulin heterodimers to promote microtubule assembly. Nat Cell Biol 4:583–591.

Goshima Y, Nakamura F, Strittmatter P, Strittmatter SM (1995) Collapsin-induced growth cone collapse mediated by an intracellular protein related to UNC-33. Nature 376:509–514.

Goshima Y, Kawakami T, Hori H, Sugiyama Y, Takasawa S, Hashimoto Y, Kagoshima-Maezono M, Takenaka T, Misu Y, Strittmatter SM (1997) A novel action of collapsin: collapsin-1 increases antero- and retrograde axoplasmic transport independently of growth cone collapse. J Neurobiol 33:316–328.

Goshima Y, Hori H, Sasaki Y, Yang T, Kagoshima-Maezono M, Li C, Takenaka T, Nakamura F, Takahashi T, Strittmatter SM, Misu Y, Kawakami T (1999) Growth cone neuropilin-1 mediates collapsin-1/Sema III facilitation of antero- and retrograde axoplasmic transport. J Neurobiol 39:579–589.

Grimes CA, Jope RS (2001) The multifaceted roles of glycogen synthase kinase 3beta in cellular signaling. Prog Neurobiol 65:391–426.

Gu Y, Hamajima N, Ihara Y (2000) Neurofibrillary tangle-associated collapsin response mediator protein-2 (CRMP-2) is highly phosphorylated on Thr-509, Ser-518, and Ser-522. Biochemistry 39:426–4275.

Helmke S, Pfenninger KH (1995) Growth cone enrichment and cytoskeletal association of non-receptor tyrosine kinases. Cell Motil Cytoskel 30:194–207.

Hosoi T, Uchiyama M, Okumura E, Saito T, Ishiguro K, Uchida T, Okuyama A, Kishimoto T, Hisanaga S (1995) Evidence for cdk5 as a major activity phosphorylating tau protein in porcine brain extract. J Biochem (Tokyo) 117:741–749.

Huang EJ, Reichardt LF (2001) Neurotrophins: roles in neuronal development and function. Annu Rev Neurosci 24:677–736.

Kolodkin AL (1998) Semaphorin-mediated neuronal growth cone guidance. Prog Brain Res 117:115–132.

Lew J, Wang JH (1995) Neuronal cdc2-like kinase. Trends Biochem Sci 20:33–37.

Li C, Sasaki Y, Takei K, Yamamoto H, Shouji M, Sugiyama Y, Kawakami T, Nakamura F, Yagi T, Ohshima T, Goshima Y (2004) Correlation between semaphorin3A-induced facilitation of axonal transport and local activation of a translation initiation factor eukaryotic translation initiation factor 4E. J Neurosci 24:6161–6170.

Li W, Herman RK, Shaw JE (1992) Analysis of the Caenorhabditis elegans axonal guidance and outgrowth gene unc-33. Genetics 132:675–689.

Liu G, Beggs H, Jurgensen C, Park HT, Tang H, Gorski J, Jones KR, Reichardt LF, Wu J, Rao Y (2004) Netrin requires focal adhesion kinase and Src family kinases for axon outgrowth and attraction. Nat Neurosci 7:1222–1232.

Meriane M, Tcherkezian J, Webber CA, Danek EI, Triki I, McFarlane S, Bloch-Gallego E, Lamarche-Vane N (2004) Phosphorylation of DCC by Fyn mediates Netrin-1 signaling in growth cone guidance. J Cell Biol 167:687–698.

Morita A, Yamashita N, Sasaki Y, Uchida Y, Nakajima O, Nakamura F, Yagi T, Taniguchi M, Usui H, Katoh-Semba R, Takei K, Goshima Y (2006) Regulation of dendritic branching and spine maturation by semaphorin3A-Fyn signaling. J Neurosci 26:2971–2980.

Morse WR, Whitesides III JG, LaMantia AS, Maness PF (1998) p59fyn and pp60c-src modulate axonal guidance in the developing mouse olfactory pathway. J Neurobiol 36:53–63.

Nakayama T, Goshima Y, Misu Y, Kato T (1999) Role of cdk5 and tau phosphorylation in heterotrimeric G protein-mediated retinal growth cone collapse. J Neurobiol 41:326–339.

Ohshima T, Ward JM, Huh CG, Longenecker G, Veeranna, Pant HC, Brady RO, Martin LJ, Kulkarni AB (1996) Targeted disruption of the cyclin-dependent kinase 5 gene results in abnormal corticogenesis, neuronal pathology and perinatal death. Proc Natl Acad Sci USA 93:11173–11178.

Puschel AW (2007) GTPases in semaphorin signaling. Adv Exp Med Biol 600:12–23.

Raper JA (2000) Semaphorins and their receptors in vertebrates and invertebrates. Curr Opin Neurobiol 10:88–94.

Rosslenbroich V, Dai L, Franken S, Gehrke M, Junghans U, Gieselmann V, Kappler J (2003) Subcellular localization of collapsin response mediator proteins to lipid rafts. Biochem Biophys Res Commun 305:392–329.

Sasaki Y, Cheng C, Uchida Y, Nakajima O, Ohshima T, Yagi T, Taniguchi M, Nakayama T, Kishida R, Kudo Y, Ohno S, Nakamura F, Goshima Y (2002) Fyn and Cdk5 mediate semaphorin-3A signaling, which is involved in regulation of dendrite orientation in cerebral cortex. Neuron 35:907–920.

Segal RA (2003) Selectivity in neurotrophin signaling: theme and variations. Annu Rev Neurosci 26:299–330.

Songyang Z, Lu KP, Kwon YT, Tsai LH, Filhol O, Cochet C, Brickey DA, Soderling TR, Bartleson C, Graves DJ, DeMaggio AJ, Hoekstra MF, Blenis J, Hunter T, Cantley LC (1996) A structural basis for substrate specificities of protein Ser/Thr kinases: primary sequence preference of casein kinases I and II, NIMA, phosphorylase kinase, calmodulin-dependent kinase II, CDK5, and Erk1. Mol Cell Biol 16:6486–6493.

Takahashi T, Fournier A, Nakamura F, Wang LH, Murakami Y, Kalb RG, Fujisawa H, Strittmatter SM (1999) Plexin-neuropilin-1 complexes form functional semaphorin-3A receptors. Cell 99:59–69.

Tamagnone L, Artigiani S, Chen H, He Z, Ming GI, Song H, Chedotal A, Winberg ML, Goodman CS, Poo M, Tessier-Lavigne M, Comoglio PM (1999) Plexins are a large family of receptors for transmembrane, secreted and GPI-anchored semaphorins in vertebrates. Cell 99:71–80.

Tanaka E, Sabry J (1995) Making the connection: cytoskeletal rearrangements during growth cone guidance. Cell 83:171–176.

Tessier-Lavigne M, Goodman CS (1996) The molecular biology of axon guidance. Science 274:1123–1133.

Uchida Y, Ohshima T, Sasaki Y, Suzuki H, Yanai S, Yamashita N, Nakamura F, Takei K, Ihara Y, Mikoshiba K, Kolattukudy P, Honnorat J, Goshima Y (2005) Semaphorin3A signalling is mediated via sequential Cdk5 and GSK3beta phosphorylation of CRMP2: implication of common phosphorylating mechanism underlying axon guidance and Alzheimer's disease. Genes Cells 10:165–179.

Wang LH, Strittmatter SM (1996) A family of rat CRMP genes is differentially expressed in the nervous system. J Neurosci 16:6197–6207.

Wang LH, Strittmatter SM (1997) Brain CRMP forms heterotetramers similar to liver dihydropyrimidinase. J Neurochem 69:2261–2269.

Yagi T, Aizawa S, Tokunaga T, Shigetani Y, Takeda N, Ikawa Y (1993) A role for Fyn tyrosine kinase in the suckling behaviour of neonatal mice. Nature 366:742–745.

Yamashita N, Uchida Y, Ohshima T, Hirai S, Nakamura F, Taniguchi M, Mikoshiba K, Honnorat J, Kolattukudy P, Thomasset N, Takei K, Takahashi T, Goshima Y (2006) Collapsin response mediator protein 1 mediates reelin signaling in cortical neuronal migration. J Neurosci 26:13357–13362.

Yamashita N, Morita A, Uchida Y, Nakamura F, Usui H, Ohshima T, Taniguchi M, Honnorat J, Kolattukudy P, Thomasset N, Takei K, Takahashi T, Goshima Y (2007) Regulation of spine development by Semaphorin3A through cyclin-dependent kinase 5 phosphorylation of collapsin response mediator protein 1. J Neurosci 27:12546–12554.

Yoshida H, Watanabe A, Ihara Y (1998) Collapsin response mediator protein-2 is associated with neurofibrillary tangles in Alzheimer's disease. J Biol Chem 273:9761–9768.

Yoshimura T, Kawano Y, Arimura N, Kawabata S, Kikuchi A, Kaibuchi K (2005) GSK-3beta regulates phosphorylation of CRMP-2 and neuronal polarity. Cell 120:137–149.

Zukerberg LR, Patrick GN, Nikolic M, Humbert S, Wu CL, Lanier LM, Gertler FB, Vidal M, Van Etten RA, Tsai LH (2000) Cables links Cdk5 and c-Abl and facilitates Cdk5 tyrosine phosphorylation, kinase upregulation, and neurite outgrowth. Neuron 26:633–646.

Cdk5 in Presynapses

Fan-Yan Wei and Kazuhito Tomizawa

Abstract Recent studies have explored the indispensable roles of Cdk5 in presynapse. Presynapse is the structure in which neurotransmitter-containing synaptic vesicles are fused to the synaptic membrane and recycled to internal compartments via exocytosis and endocytosis, respectively. Cdk5 is involved in the regulation of the exocytosis and endocytosis of synaptic vesicles.

Cdk5 in Presynapses

The role of Cdk5 in neurons has been mainly studied in the postsynaptic region of neurons. However, recent studies have also explored the indispensable roles of Cdk5 in presynapse. Presynapse is the structure in which neurotransmitter-containing synaptic vesicles are fused to the synaptic membrane and recycled to internal compartments via exocytosis and endocytosis, respectively. In this chapter, we will describe the presynaptic substrates of Cdk5 and present insights into the physiological role of Cdk5 during exocytosis and endocytosis.

Cdk5 in Exocytosis

Exocytosis involves sequential intracellular movements including trafficking of synaptic vesicles from the reserve pool to the ready releasable pool and release of the neurotransmitter in the ready releasable pool by fusion of the vesicle membrane with the presynaptic terminal. It has been shown that Cdk5 phosphorylates certain presynaptic proteins which are essential for exocytosis.

K. Tomizawa
Department of Physiology, Okayama University Graduate School of Medicine,
Dentistry and Pharmaceutical Sciences, 2-5-1 Shikata-cho, Okayama 700-8558, Japan
e-mail: tomikt@md.okayama-u.ac.jp

N.Y. Ip, L.-H. Tsai (eds.), *Cyclin Dependent Kinase 5 (Cdk5)*,
DOI: 10.1007/978-0-387-78887-6_3, © Springer Science+Business Media, LLC 2008

Cdk5 and Synapsin

In the presynapse, neurotransmitter-containing vesicles can be classified into two pools based on their localization (Greengard et al., 1993). The ready releasable pool is docked at the plasma membrane, while the reserve pool, which is not docked at the membrane, contains a large proportion of vesicles which are tethered to each other and associated with the actin cytoskeleton. Exocytosis consists of sequential events, which include the transport of synaptic vesicles from the reserve pool to the ready releasable pool in the presynaptic plasma membrane, and membrane fusion between synaptic vesicles and the plasma membrane to release neurotransmitters. Synapsin I is a neuron-specific protein which associates with the cytoplasmic surface of synaptic vesicles (De et al., 1983a,b; Huttner et al., 1983). Interestingly, the immunoreactivity of synapsin is mainly found in the reserve pool of synaptic vesicles, not in the ready releasable pool (De et al., 1983a). Therefore, synapsin has been considered to be a modulator of synaptic vesicle trafficking in the presynaptic terminal.

Synapsin is a prominent phosphoprotein regulated in response to various physiological stimuli (Johnson et al., 1971; Krueger et al., 1977). Protein kinase A (PKA) and Ca^{2+}/calmodulin-dependent kinase I (CaMKI) phosphorylate Ser 9 (P-site 1), while CaMKII phosphorylates Ser 566 and Ser 603 (P-site 2 and P-site 3). Ser 62 and Ser 67 (P-site 4 and P-site 5) can be phosphorylated by mitogen-activated protein kinase (MAP kinase) ERK1/2 (Czernik et al., 1987; Huttner and Greengard, 1979; Jovanovic et al., 2001; Jovanovic et al., 1996; Matsubara et al., 1996; Sihra et al., 1989). Cdk5 phosphorylates Ser 549 (P-site 6), which is also phosphorylated by MAP kinase *in vitro*, while Ser 551 (P-site 7) is only phosphory-lated by Cdk5 (Czernik et al., 1987; Jovanovic et al., 1996; Matsubara et al., 1996). Synapsin has been shown to interact with synaptic vesicles and actin. Phosphor-ylation of synapsin at P-sites 1, 2, and 3 decreases its binding with synaptic vesicles as well as its interaction with the actin cytoskeleton (Bahler and Greengard, 1987; Benfenati et al., 1990; Schiebler et al., 1986; Valtorta et al., 1992). These three P-sites have been shown to be efficiently phosphorylated in a Ca^{2+}-dependent manner after stimulation by depolarization of synatosomes, indicating that the phosphorylation at these sites is important for the release of synaptic vesicles (Jovanovic et al., 2001; Hata et al., 1993a). On the other hand, P-sites 4, 5, and 6 are phosphorylated under basal conditions. Phosphorylation at these sites has been shown to decrease the ability of synapsin to bind to actin but not to synaptic vesicles. After the depolarization of synaptosomes, these sites are dephosphory-lated by calcineurin (PP2B). The inhibition of calcineurin can potentiate glutamate release after stimulation, suggesting that phosphorylation at P-sites 4, 5, and 6 by MAPK or Cdk5 facilitates trafficking of synaptic vesicles.

Cdk5 and Munc-18

Munc-18 was isolated as a syntaxin-binding protein, which is a mammalian homologue of UNC-18 in *C. elegans* and Sec-1 in yeast (Novick et al., 1980).

A study using Munc-18 knockout mice clearly showed the indispensable role of Munc-18 in neurotransmitter release (Verhage et al., 2000; Hata et al., 1993b). In these mice, synaptic transmission was completely abolished, whereas synaptogenesis was normal and postsynaptic receptors were functional. Since synaptic vesicles could still dock to the presynaptic membrane in Munc-18 knockout mice, the function of Munc-18 in synaptic vesicle release is most likely downstream of docking. The fusion between synaptic vesicles and the synaptic membrane requires the formation of core complexes containing syntaxin, synaptobrevin, and synaptosomal-associated protein of 25 kDa (SNAP25) (Yang et al., 2000). Based on the evidence that Munc-18 is not associated with the core complex, Munc-18 must be released from the Munc-18–syntaxin complex. It has been shown that PKC phosphorylates Munc-18 at Ser 306 and Ser 313, and Cdk5 phosphorylates Munc-18 at Thr 574 *in vitro* (Fletcher et al., 1999; Fujita et al., 1996). Previous studies showed that phosphorylation of Munc-18 by both PKC and Cdk5 could cause dissociation of the Munc-18–syntaxin complex and promote the release of synaptic vesicles. However, analysis of the 3D structure of Munc-18 revealed that the Thr 574 site would be sterically inaccessible to Cdk5 in the closed conformation (Misura et al., 2000). Although it is still unclear how Cdk5 phosphorylates Munc-18 at Thr 574 under physiological conditions, some biochemical evidence has shown Cdk5 could bind to Munc-18 and be co-purified with Munc-18 from rat brain, suggesting that Cdk5 might have some roles in the regulation of Munc-18 function (Bhaskar et al., 2004; Misura et al., 2000).

Cdk5 and Voltage-Gated Calcium Channels

Activation of voltage-gated calcium channels is the crucial step in the release of synaptic vesicles. In the presynaptic terminal, N-type and P/Q-type calcium channels are the predominant voltage-gated calcium channels. Previous studies showed that Cdk5 could phosphorylate P/Q-type calcium channels at the intracellular loop domain (LII–III) (Tomizawa et al., 2002). Phosphorylation of P/Q-type calcium channels prevents the interaction between the loop domain and SNARE proteins such as SNAP25 and synaptotagmin. Since it has been thought that the interaction between the loop domain in calcium channels and SNARE proteins is important for the channel activity and subsequent synaptic vesicle release, phosphorylation of P/Q-type calcium channels by Cdk5 would have an inhibitory effect on exocytosis. Supporting this hypothesis, pharmacological inhibition of Cdk5 by roscovitine was found to enhance neurotransmitter release in both the synaptosome fraction and cultured neurons. Although it has been reported that roscovitine itself has the ability to activate calcium channels and enhance exocytosis through a Cdk5-independent mechanism (Cho and Meriney, 2006; Yan et al., 2002), studies employing genetic alteration of Cdk5 activity revealed that Cdk5 directly regulated calcium channels and neurotransmitter release. Furthermore, transgenic mice overexpressing p35, in which Cdk5 activity was enhanced, showed attenuation of dopaminergic signaling through both postsynaptic and presynaptic mechanisms (Takahashi et al., 2005). The findings strongly suggest that phosphorylation of presynaptic terminal components, including P/Q-type calcium channels, by Cdk5 is inhibitory for synaptic vesicle release.

Role of Cdk5 in Exocytosis

Accumulating evidence shows that Cdk5-dependent phosphorylation is important in both trafficking and release of neurotransmitters. Interestingly, Cdk5 might play opposite roles in trafficking and release. Phosphorylation of synapsin by Cdk5 facilitates vesicle trafficking from the reserve pool to the ready releasable pool. On the other hand, Cdk5 in the presynaptic terminal phosphorylates voltage-gated calcium channels to inhibit the release of neurotransmitters. The reason why Cdk5 has such complicated regulatory effects on exocytosis is unclear at present. It is worth noting that under some experimental conditions, inhibition of Cdk5 seems to inhibit vesicle release suggesting that Cdk5 might have different roles in cells which have different exocytosis machinery (Rosales et al., 2004; Xin et al., 2004).

Cdk5 in Endocytosis

After synaptic vesicles undergo exocytosis to release neurotransmitters, empty vesicles are pinched off from presynaptic membrane by endocytosis and then transported to early endosomes. Various proteins have been found to participate in the endocytosis process. Among these proteins, three essential proteins, dynamin 1, amphiphysin 1, and synaptojanin, are known to be phosphorylated by Cdk5.

Cdk5 and Dynamin 1

Dynamin 1 is a GTPase, which is indispensable for synaptic vesicle endocytosis (Marks et al., 2001). It has an N-terminal GTPase domain and a C-terminal proline-rich domain which can bind to SH3 domain-containing endocytotic proteins, including amphiphysin and endophilin (Grabs et al., 1997; Bauerfeind et al., 1997). Dynamin 1 can be phosphorylated by various protein kinases *in vitro*, such as PKC, casein kinase, and MAPK (Powell et al., 2000; Robinson et al., 1993; Hosoya et al., 1994; Earnest et al., 1996). However, the physiological role of phosphorylation by these kinases remains unclear. Recently, two independent studies found that Cdk5 phsophorylates dynamin 1 at various sites in the proline-rich domain (Tomizawa et al., 2003; Tan et al., 2003). One of the phsophorylation sites is located at Thr 780 (Tomizawa et al., 2003). Phosphorylation of Thr 780 results in the dissociation of dynamin 1 from amphiphysin as a result of disruption of the interaction of the proline-rich domain of dynamin 1 with the SH3 domain in amphiphysin. A recent study also showed that phosphorylation of the proline-rich domain in dynamin 1 decreased its ability to bind to the SH3 domain in endophilin (Solomaha et al., 2005). These biochemical data strongly suggest that Cdk5-dependent phosphorylation at Thr 780 in dynamin 1 is inhibitory for endocytosis, since interaction between dynamin 1 and amphiphysin is necessary for dynamin 1 GTPase activity and subsequent vesicle fission. The inhibitory effect of Cdk5 on dynamin 1–mediated endocytosis is further supported by the fact that endocytosis in neurons from p35 knockout mice was enhanced compared to that in wild-type mice. These results strongly suggest that Cdk5 acts as an inhibitory kinase, which can suppress endocytosis. In another study (Tomizawa et al., 2003), Cdk5 was found to phosphorylate dynamin 1 at Ser 774 and Ser 778 in the proline-rich domain (Tan et al., 2003). In contrast to the findings of the earlier study (Tomizawa et al., 2003), phosphorylation at Ser 774 and Ser 778 has no influence on the binding of dynamin 1 with the amphiphysin SH3 domain. Furthermore, phosphorylation at these two serine sites even enhanced dynamin 1 GTPase activity and subsequent endocytosis in cultured neurons. The apparent discrepancy about the role of Cdk5 in dynamin 1–dependent endocytosis remains unresolved at present. However, different roles of Cdk5 could exist in endocytosis (Nguyen and Bibb, 2003), as suggested above for the role of Cdk5 in exocytosis. In the basal condition, dynamin 1 might be phosphorylated at Thr 780 to prevent the binding with amphiphysin or endophilin. After depolarization of the presynaptic membrane, dynamin 1 is dephosphorylated by calcineurin and thus initiates the endocytosis process with amphiphysin. After endocytosis, dynamin 1 dissociates from amphiphysin to prepare for the next endocytosis. During the preparation, dynamin 1 might be rephosphorylated by Cdk5 at Ser 774 and Ser 778 through a calcineurin-insensitive pathway, which is important to maintain the steady state in endocytosis. Future studies using highly specific antibody to each phosphorylation site would reveal the details of Cdk5-dependent endocytosis.

Cdk5 and Amphiphysin

It was first shown that amphiphysin could be phosphorylated *in vitro* at Ser 272, 276, and 285 in the proline-rich domain by Cdk5 (Rosales et al., 2000; Floyd et al., 2001). Further studies have identified two new sites, Ser 261 and Thr 310 (Tomizawa et al., 2003; Liang et al., 2007), which could also be phosphorylated by Cdk5 *in vitro*. Binding assays revealed that phosphorylation of amphiphysin by Cdk5 had no effect on its ability to bind with dynamin 1, but could abolish the interaction with other proteins such as adaptin. In a cell-free liposome assay, it was shown that phosphorylation of amphiphysin by Cdk5 could partially inhibit the formation of small liposomes. These results suggest that Cdk5 might also have an inhibitory effect on endocytosis through regulation of amphiphysin. Using mass spectrometry, a very recent study showed that Ser 276 and Ser 285 are the predominant phosphorylation sites in amphiphysin. Moreover, phosphorylation of these sites could reduce the binding ability of amphiphysin to liposomes.

Cdk5 and Synaptojanin

Synaptojanin is a polyphosphoinositide phosphatase enriched in presynaptic terminals. Synaptojanin catalyzes the dephosphorylation of phosphatidylinositol-4,5-biphosphate [PI(4,5)P2], which is an important regulator of both exocytosis and endocytosis (McPherson et al., 1996; Lackner et al., 1999; De et al., 1996). In synaptojanin knockout mice, which have elevated PI(4,5)P2 concentrations, an accumulation of coated vesicles was observed in endocytotic synaptic terminals (Cremona et al., 1999). These findings suggest that dephosphorylation of PI(4,5)P2 by synaptojanin is important in uncoating of clathrin-coated synaptic vesicles. Like other endocytotic proteins such as amphiphysin and dynamin, synaptojanin is phosphorylated under basal conditions, although the kinase responsible for the phosphorylation is unknown. A recent study demonstrated that Cdk5 is the endogenous kinase which phosphorylates synaptojanin at Ser 1144 (Lee et al., 2004). Phosphorylation of synaptojanin directly inhibits its phosphatase activity. Moreover, phosphorylation of synaptojanin could disrupt its interaction with endophilin and amphiphysin. Since binding of endophilin to synaptojanin could enhance its phosphatase activity, phosphorylation of synaptojanin by Cdk5 would further suppress its activity, resulting in the inhibition of synaptic vesicle endocytosis.

Concluding Remarks

In presynaptic terminals, a number of proteins are phosphorylated under basal conditions, and dephosphorylated during depolarization stimulation. Accumulating evidence has shown that Cdk5 is a major protein kinase which is responsible for the phosphorylation under basal conditions at presynaptic terminals. Although it is well accepted that Cdk5 is indispensable for the physiological functions of presynapse, the mechanism by which Cdk5 regulates both exocytosis and endocytosis is still unclear. Further studies such as identifying novel presynaptic substrates of Cdk5 are needed to clarify the precise function of Cdk5 in the presynapse.

References

Bahler, M. and Greengard, P. (1987). Synapsin I bundles F-actin in a phosphorylation-dependent manner. Nature *326*, 704–707.
Bauerfeind, R., Takei, K., and De, C.P. (1997). Amphiphysin I is associated with coated endocytic intermediates and undergoes stimulation-dependent dephosphorylation in nerve terminals. J. Biol. Chem. *272*, 30984–30992.

Benfenati, F., Neyroz, P., Bahler, M., Masotti, L., and Greengard, P. (1990). Time-resolved fluorescence study of the neuron-specific phosphoprotein synapsin I. Evidence for phosphorylation-dependent conformational changes. J. Biol. Chem. *265*, 12584–12595.

Bhaskar, K., Shareef, M.M., Sharma, V.M., Shetty, A.P., Ramamohan, Y., Pant, H.C., Raju, T.R., and Shetty, K.T. (2004). Co-purification and localization of Munc18-1 (p67) and Cdk5 with neuronal cytoskeletal proteins. Neurochem. Int. *44*, 35–44.

Cho, S. and Meriney, S.D. (2006). The effects of presynaptic calcium channel modulation by roscovitine on transmitter release at the adult frog neuromuscular junction. Eur. J. Neurosci. *23*, 3200–3208.

Cremona, O., Di, P.G., Wenk, M.R., Luthi, A., Kim, W.T., Takei, K., Daniell, L., Nemoto, Y., Shears, S.B., Flavell, R.A., McCormick, D.A., and De, C.P. (1999). Essential role of phosphoinositide metabolism in synaptic vesicle recycling. Cell *99*, 179–188.

Czernik, A.J., Pang, D.T., and Greengard, P. (1987). Amino acid sequences surrounding the cAMP-dependent and calcium/calmodulin-dependent phosphorylation sites in rat and bovine synapsin I. Proc. Natl. Acad. Sci. U.S.A. *84*, 7518–7522.

De, C.P., Cameron, R., and Greengard, P. (1983a). Synapsin I (protein I), a nerve terminal-specific phosphoprotein. I. Its general distribution in synapses of the central and peripheral nervous system demonstrated by immunofluorescence in frozen and plastic sections. J. Cell Biol. *96*, 1337–1354.

De, C.P., Emr, S.D., McPherson, P.S., and Novick, P. (1996). Phosphoinositides as regulators in membrane traffic. Science *271*, 1533–1539.

De, C.P., Harris, S.M., Jr., Huttner, W.B., and Greengard, P. (1983b). Synapsin I (Protein I), a nerve terminal-specific phosphoprotein. II. Its specific association with synaptic vesicles demonstrated by immunocytochemistry in agarose-embedded synaptosomes. J. Cell Biol. *96*, 1355–1373.

Earnest, S., Khokhlatchev, A., Albanesi, J.P., and Barylko, B. (1996). Phosphorylation of dynamin by ERK2 inhibits the dynamin-microtubule interaction. FEBS Lett. *396*, 62–66.

Fletcher, A.I., Shuang, R., Giovannucci, D.R., Zhang, L., Bittner, M.A., and Stuenkel, E.L. (1999). Regulation of exocytosis by cyclin-dependent kinase 5 via phosphorylation of Munc18. J. Biol. Chem. *274*, 4027–4035.

Floyd, S.R., Porro, E.B., Slepnev, V.I., Ochoa, G.C., Tsai, L.H., and De, C.P. (2001). Amphiphysin 1 binds the cyclin-dependent kinase (cdk) 5 regulatory subunit p35 and is phosphorylated by cdk5 and cdc2. J. Biol. Chem. *276*, 8104–8110.

Fujita, Y., Sasaki, T., Fukui, K., Kotani, H., Kimura, T., Hata, Y., Sudhof, T.C., Scheller, R.H., and Takai, Y. (1996). Phosphorylation of Munc-18/n-Sec1/rbSec1 by protein kinase C: its implication in regulating the interaction of Munc-18/n-Sec1/rbSec1 with syntaxin. J. Biol. Chem. *271*, 7265–7268.

Grabs, D., Slepnev, V.I., Songyang, Z., David, C., Lynch, M., Cantley, L.C., and De, C.P. (1997). The SH3 domain of amphiphysin binds the proline-rich domain of dynamin at a single site that defines a new SH3 binding consensus sequence. J. Biol. Chem. *272*, 13419–13425.

Greengard, P., Valtorta, F., Czernik, A.J., and Benfenati, F. (1993). Synaptic vesicle phosphoproteins and regulation of synaptic function. Science *259*, 780–785.

Hata, Y., Slaughter, C.A., and Sudhof, T.C. (1993b). Synaptic vesicle fusion complex contains unc-18 homologue bound to syntaxin. Nature *366*, 347–351.

Hata, Y., Slaughter, C.A., and Sudhof, T.C. (1993a). Synaptic vesicle fusion complex contains unc-18 homologue bound to syntaxin. Nature *366*, 347–351.

Hosoya, H., Komatsu, S., Shimizu, T., Inagaki, M., Ikegami, M., and Yazaki, K. (1994). Phosphorylation of dynamin by cdc2 kinase. Biochem. Biophys. Res. Commun. *202*, 1127–1133.

Huttner, W.B. and Greengard, P. (1979). Multiple phosphorylation sites in protein I and their differential regulation by cyclic AMP and calcium. Proc. Natl. Acad. Sci. U.S.A. *76*, 5402–5406.

32																																																																					F.-Y. Wei, K. Tomizawa

Huttner, W.B., Schiebler, W., Greengard, P., and De, C.P. (1983). Synapsin I (protein I), a nerve terminal-specific phosphoprotein. III. Its association with synaptic vesicles studied in a highly purified synaptic vesicle preparation. J. Cell Biol. *96*, 1374–1388.

Johnson, E.M., Maeno, H., and Greengard, P. (1971). Phosphorylation of endogenous protein of rat brain by cyclic adenosine 3′,5′-monophosphate-dependent protein kinase. J. Biol. Chem. *246*, 7731–7739.

Jovanovic, J.N., Benfenati, F., Siow, Y.L., Sihra, T.S., Sanghera, J.S., Pelech, S.L., Greengard, P., and Czernik, A.J. (1996). Neurotrophins stimulate phosphorylation of synapsin I by MAP kinase and regulate synapsin I-actin interactions. Proc. Natl. Acad. Sci. U.S.A. *93*, 3679–3683.

Jovanovic, J.N., Sihra, T.S., Nairn, A.C., Hemmings, H.C., Jr., Greengard, P., and Czernik, A.J. (2001). Opposing changes in phosphorylation of specific sites in synapsin I during Ca2+-dependent glutamate release in isolated nerve terminals. J. Neurosci. *21*, 7944–7953.

Krueger, B.K., Forn, J., and Greengard, P. (1977). Depolarization-induced phosphorylation of specific proteins, mediated by calcium ion influx, in rat brain synaptosomes. J. Biol. Chem. *252*, 2764–2773.

Lackner, M.R., Nurrish, S.J., and Kaplan, J.M. (1999). Facilitation of synaptic transmission by EGL-30 Gqalpha and EGL-8 PLCbeta: DAG binding to UNC-13 is required to stimulate acetylcholine release. Neuron *24*, 335–346.

Lee, S.Y., Wenk, M.R., Kim, Y., Nairn, A.C., and De, C.P. (2004). Regulation of synaptojanin 1 by cyclin-dependent kinase 5 at synapses. Proc. Natl. Acad. Sci. U.S.A. *101*, 546–551.

Liang, S., Wei, F.Y., Wu, Y.M., Tanabe, K., Abe, T., Oda, Y., Yoshida, Y., Yamada, H., Matsui, H., Tomizawa, K., and Takei, K. (2007). Major Cdk5-dependent phosphorylation sites of amphiphysin 1 are implicated in the regulation of the membrane binding and endocytosis. J. Neurochem *102*,1466–1476.

Marks, B., Stowell, M.H., Vallis, Y., Mills, I.G., Gibson, A., Hopkins, C.R., and McMahon, H.T. (2001). GTPase activity of dynamin and resulting conformation change are essential for endocytosis. Nature *410*, 231–235.

Matsubara, M., Kusubata, M., Ishiguro, K., Uchida, T., Titani, K., and Taniguchi, H. (1996). Site-specific phosphorylation of synapsin I by mitogen-activated protein kinase and Cdk5 and its effects on physiological functions. J. Biol. Chem. *271*, 21108–21113.

McPherson, P.S., Garcia, E.P., Slepnev, V.I., David, C., Zhang, X., Grabs, D., Sossin, W.S., Bauerfeind, R., Nemoto, Y., and De, C.P. (1996). A presynaptic inositol-5-phosphatase. Nature *379*, 353–357.

Misura, K.M., Scheller, R.H., and Weis, W.I. (2000). Three-dimensional structure of the neuronal-Sec1-syntaxin 1a complex. Nature *404*, 355–362.

Nguyen, C. and Bibb, J.A. (2003). Cdk5 and the mystery of synaptic vesicle endocytosis. J. Cell Biol. *163*, 697–699.

Novick, P., Field, C., and Schekman, R. (1980). Identification of 23 complementation groups required for post-translational events in the yeast secretory pathway. Cell *21*, 205–215.

Powell, K.A., Valova, V.A., Malladi, C.S., Jensen, O.N., Larsen, M.R., and Robinson, P.J. (2000). Phosphorylation of dynamin I on Ser-795 by protein kinase C blocks its association with phospholipids. J. Biol. Chem. *275*, 11610–11617.

Robinson, P.J., Sontag, J.M., Liu, J.P., Fykse, E.M., Slaughter, C., McMahon, H., and Sudhof, T.C. (1993). Dynamin GTPase regulated by protein kinase C phosphorylation in nerve terminals. Nature *365*, 163–166.

Rosales, J.L., Ernst, J.D., Hallows, J., and Lee, K.Y. (2004). GTP-dependent secretion from neutrophils is regulated by Cdk5. J. Biol. Chem. *279*, 53932–53936.

Rosales, J.L., Nodwell, M.J., Johnston, R.N., and Lee, K.Y. (2000). Cdk5/p25(nck5a) interaction with synaptic proteins in bovine brain. J. Cell Biochem. *78*, 151–159.

Schiebler, W., Jahn, R., Doucet, J.P., Rothlein, J., and Greengard, P. (1986). Characterization of synapsin I binding to small synaptic vesicles. J. Biol. Chem. *261*, 8383–8390.

Sihra, T.S., Wang, J.K., Gorelick, F.S., and Greengard, P. (1989). Translocation of synapsin I in response to depolarization of isolated nerve terminals. Proc. Natl. Acad. Sci. U.S.A. *86*, 8108–8112.

Solomaha, E., Szeto, F.L., Yousef, M.A., and Palfrey, H.C. (2005). Kinetics of Src homology 3 domain association with the proline-rich domain of dynamins: specificity, occlusion, and the effects of phosphorylation. J. Biol. Chem. *280*, 23147–23156.

Takahashi, S., Ohshima, T., Cho, A., Sreenath, T., Iadarola, M.J., Pant, H.C., Kim, Y., Nairn, A.C., Brady, R.O., Greengard, P., and Kulkarni, A.B. (2005). Increased activity of cyclin-dependent kinase 5 leads to attenuation of cocaine-mediated dopamine signaling. Proc. Natl. Acad. Sci. U.S.A. *102*, 1737–1742.

Tan, T.C., Valova, V.A., Malladi, C.S., Graham, M.E., Berven, L.A., Jupp, O.J., Hansra, G., McClure, S.J., Sarcevic, B., Boadle, R.A., Larsen, M.R., Cousin, M.A., and Robinson, P.J. (2003). Cdk5 is essential for synaptic vesicle endocytosis. Nat. Cell Biol. *5*, 701–710.

Tomizawa, K., Ohta, J., Matsushita, M., Moriwaki, A., Li, S.T., Takei, K., and Matsui, H. (2002). Cdk5/p35 regulates neurotransmitter release through phosphorylation and down-regulation of P/Q-type voltage-dependent calcium channel activity. J. Neurosci. *22*, 2590–2597.

Tomizawa, K., Sunada, S., Lu, Y.F., Oda, Y., Kinuta, M., Ohshima, T., Saito, T., Wei, F.Y., Matsushita, M., Li, S.T., Tsutsui, K., Hisanaga, S., Mikoshiba, K., Takei, K., and Matsui, H. (2003). Cophosphorylation of amphiphysin I and dynamin I by Cdk5 regulates clathrin-mediated endocytosis of synaptic vesicles. J. Cell Biol. *163*, 813–824.

Valtorta, F., Greengard, P., Fesce, R., Chieregatti, E., and Benfenati, F. (1992). Effects of the neuronal phosphoprotein synapsin I on actin polymerization. I. Evidence for a phosphorylation-dependent nucleating effect. J. Biol. Chem. *267*, 11281–11288.

Verhage, M., Maia, A.S., Plomp, J.J., Brussaard, A.B., Heeroma, J.H., Vermeer, H., Toonen, R.F., Hammer, R.E., van den Berg, T.K., Missler, M., Geuze, H.J., and Sudhof, T.C. (2000). Synaptic assembly of the brain in the absence of neurotransmitter secretion. Science *287*, 864–869.

Xin, X., Ferraro, F., Back, N., Eipper, B.A., and Mains, R.E. (2004). Cdk5 and Trio modulate endocrine cell exocytosis. J. Cell Sci. *117*, 4739–4748.

Yan, Z., Chi, P., Bibb, J.A., Ryan, T.A., and Greengard, P. (2002). Roscovitine: a novel regulator of P/Q-type calcium channels and transmitter release in central neurons. J. Physiol *540*, 761–770.

Yang, B., Steegmaier, M., Gonzalez, L.C., Jr., and Scheller, R.H. (2000). nSec1 binds a closed conformation of syntaxin1A. J. Cell Biol. *148*, 247–252.

Cyclin-Dependent Kinase 5: A Critical Regulator of Neurotransmitter Release

Sul-Hee Chung

Abstract Neurotransmitter release is tightly regulated through the specific control of the synaptic vesicle cycle, which is composed of Ca^{++}-triggered exocytosis, endocytosis, and recycling. Various protein kinases have been implicated in the regulation of neurotransmitter release. Accumulating evidence indicates that cyclin-dependent kinase 5 (Cdk5) controls the multiple steps of neurotransmitter release through phosphorylation of the various substrates involved in synaptic vesicle exocytosis, endocytosis, neurotransmitter synthesis, Ca^{++} influx, and lipid signaling at presynaptic terminals—the distal ends of axons that specialize in neurotransmitter release. This study is an overview of the most recent information available concerning Cdk5-mediated phosphorylation as a critical regulatory mechanism for neurotransmitter release at presynaptic terminals.

Introduction

All of the processes through which we coordinate our movements, sense our environment, act in response to stimuli, and learn and memorize information rely on complicated networks of neurons that communicate with each other and their targets. Communication between a neuron and its target cell (another neuron, a muscle cell, or an endocrine cell) occurs at the synapses—specialized sites that release chemical signals called neurotransmitters. Neurotransmitters are stored in synaptic vesicles housed in presynaptic axon terminals and are released when excitation of the presynaptic terminal triggers vesicle exocytosis—the secretion of molecules following fusion of the vesicles with the presynaptic membrane. The released neurotransmitters then bind to and stimulate their receptors on postsynaptic target cells, thus altering the function of these cells. This process is referred to as synaptic transmission. Synaptic vesicles

S.-H. Chung
Graduate Program in Neuroscience, Institute for Brain Science and Technology, Inje University, 633-146 Gaegeum 2-dong, Busanjin-gu, Busan, South Korea, 614-735
e-mail: sulchung@inje.ac.kr

N.Y. Ip, L.-H. Tsai (eds.), *Cyclin Dependent Kinase 5 (Cdk5)*,
DOI: 10.1007/978-0-387-78887-6_4, © Springer Science+Business Media, LLC 2008

undergo a trafficking cycle composed of Ca^{++}-triggered exocytosis, endocytosis, and recycling to support rapid and repeated rounds of neurotransmitter release. The synaptic vesicle cycle can be further divided into several steps that include attachment to the presynaptic membrane (docking), prefusion with the presynaptic membrane (priming), triggering of the neurotransmitter release, recycling of the emptied vesicles, and reloading of the vesicles with neurotransmitter (Bellen, 1999; Cowan et al., 2001; Sudhof, 2004).

Table 1 Cdk5 target proteins involved in neurotransmitter release from presynaptic terminals

Protein	Phosphorylation site	Effect of phosphorylation	References
Exocytosis			
Munc18-1	Thr574	Reduces binding to syntaxin 1	Shuang et al. (1998); and Fletcher et al. (1999)
		Increases quantal size and kinetics of exocytosis	Barclay et al. (2004)
Sept5	Ser17	Reduces binding to syntaxin 1	Taniguchi et al. (2007)
Synapsin 1	Ser551 and Ser553	No effect on actin bundling activity	Matsubara et al. (1996)
Pctaire1	Ser95	Increases kinase activity	Cheng et al. (2002)
		Phosphorylates D2 domain (Ser569) of NSF → reduces the hexamerization of NSF	Liu et al. (2006)
Endocytosis			
Dynamin I	Ser774 and Ser778	No effect on the binding to amphiphysin 1	Tan et al. (2003)
	Thr780	Inhibits binding to amphiphysin I	Tomizawa et al. (2003)
Amphiphysin 1	Ser276 and Ser285	No effect on the interaction with dynamin 1 but inhibits binding to beta-adaptin	Floyd et al. (2001); and Tomizawa et al. (2003)
		Reduces binding to lipid membrane	Liang et al. (2007)
Synaptojanin 1	Ser1144	Inhibits phosphatase activity	Lee et al. (2004)
Neurotransmitter synthesis			
Tyrosine hydroxylase	Ser31	Stabilizes and increases activity	Kansy et al. (2004); and Moy and Tsai, (2004)
Presynaptic Ca^{++} and lipid signaling			
P/Q-type Ca^{++} channels	loop between domains II and III	Inhibits binding to SNAP-25 and synaptotagmin	Tomizawa et al. (2002)
PIPKIgamma	Ser650	Inhibits binding to talin	Lee et al. (2005)

Neurotransmitter release is tightly regulated through the precise control of the synaptic vesicle cycle by multiple mechanisms. Many of the proteins associated with the synaptic vesicle cycle are substrates for protein kinases, including protein kinase A (PKA), protein kinase C (PKC), Ca^{++}- and calmodulin-dependent protein kinase II (CaMKII), and cyclin-dependent kinase 5 (Cdk5) (Leenders and Sheng, 2005). Numerous studies point to a role for Cdk5 as a crucial regulator of neurotransmitter release through its functional effects on exocytosis, endocytosis, Ca^{++} influx, and quantal size (Angelo et al., 2006; Cheung et al., 2006). Cdk5 is a proline-directed serine/threonine kinase that is activated by two proteins, p35 and p39. As a predominantly neuron-specific kinase, Cdk5 has been implicated in multiple neuronal functions, including neuronal survival, axon guidance, neuronal migration, synaptic plasticity, learning, and memory (Dhavan and Tsai, 2001; Angelo et al., 2006; Cheung et al., 2006). Under neurotoxic conditions, p35 is cleaved to generate p25, an aberrant activator, a finding that has implicated p25/Cdk5 in neurodegenerative disease (Lee et al., 2000). In this review, I summarize the presynaptic roles of Cdk5 in the regulation of neurotransmitter release through phosphorylation of multiple presynaptic target proteins involved in exocytosis, endocytosis, neurotransmitter synthesis, and Ca^{++} and lipid signaling (Table 1 and Fig. 1).

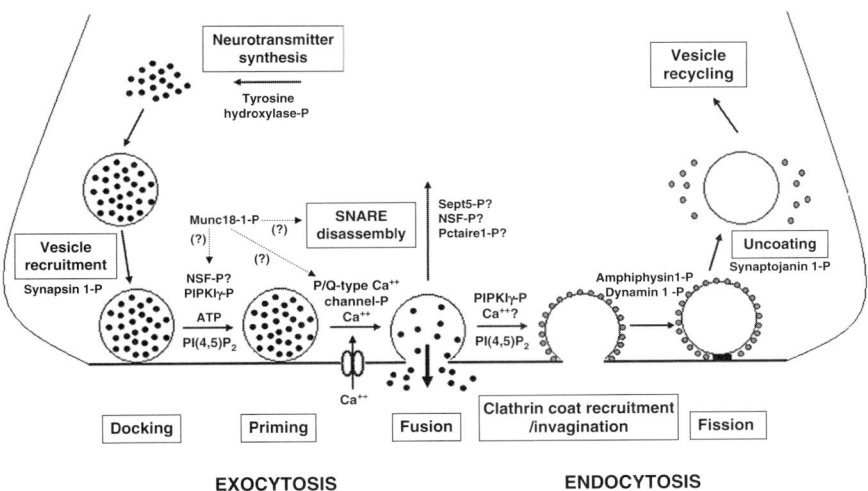

Fig. 1 The Cdk5 target proteins implicated in neurotransmitter release and vesicle recycling in presynaptic terminals. Phosphorylation by Cdk5 contributes to the modulation of a variety of steps (denoted by rectangles) of exocytosis, clathrin-mediated endocytosis, and vesicle recycling. P indicates that the protein has been phosphorylated by Cdk5. For a detailed description of each step of the synaptic vesicle cycle, reviews on exocytosis and endocytosis are useful (Jarousse and Kelly, 2001; Sudhof, 2004)

Effect of Cdk5 on the Exocytosis of Synaptic Vesicles in Presynaptic Terminals

A remarkable convergence of yeast genetics, mammalian biochemistry, and the classical neurophysiological effects of clostridial neurotoxins has revealed a number of proteins involved in the process of exocytosis in presynaptic terminals. The fusion of synaptic vesicle membranes with the presynaptic membranes is mediated by the soluble N-ethylmaleimide-sensitive factor (NSF) attachment protein receptor (SNARE) complex. This complex is a closely packed helical bundle formed by three membrane proteins: syntaxin, vesicle-associated membrane protein (VAMP; also known as synaptobrevin), and 25-kDa synaptosome-associated protein (SNAP-25) (Sollner et al., 1993; Jahn and Scheller, 2006). The SNARE proteins interact with each other through conserved sequences of approximately 60–70 amino acids known as the SNARE motif. The assembly and disassembly of the SNARE complexes are under the control of NSF and the soluble NSF attachment protein (SNAP), as well as the Rab GTPases and Munc18 family proteins (Gonzalez and Scheller, 1999; Sudhof, 2004). Syntaxin is known to play a key role in exocytosis and in the formation of SNARE complex because it interacts with several proteins that modulate these processes, such as Munc18, Munc13, Ca^{++} channel proteins, and others (Sudhof, 2004; Bhalla et al., 2006). The fusion of synaptic vesicle membranes with the presynaptic membranes is a process that is tightly regulated by Ca^{++}. Synaptotagmin I appears to be one of the key molecules that transduces the Ca^{++} signal to the molecular machinery involved in the exocytotic process (Chapman, 2002; Rizo et al., 2006).

There is also a growing realization that phosphorylation may play a key role in regulating SNARE complex interactions and neurotransmitter release. SNAREs and their interacting proteins have been shown to be substrates of purified and recombinant protein kinases *in vitro* and of certain endogenous kinases *in vivo* (Takahashi et al., 2003; Leenders and Sheng, 2005). However, a clear picture has yet to emerge regarding the functional significance and the control of SNARE phosphorylation. Cdk5 has been implicated in mechanisms of synaptic vesicle exocytosis, although controversy exists as to whether Cdk5 has a positive or negative regulatory role in neurotransmitter release at the presynaptic terminal. Cdk5–/– mice show an increase in the frequency of miniature endplate potential (MEPP)—small depolarization in the resting potential of postsynaptic cells—relative to control mice, suggesting that the spontaneous release of neurotransmitters from presynaptic terminals and synaptic transmission is enhanced in the mutant mice (Fu et al., 2005). This result demonstrates that the absence of Cdk5 promotes transmitter release at peripheral synapses *in vivo,* suggesting an inhibitory role of Cdk5 in neurotransmitter release. In support of this view, Cdk5 has been implicated in the inhibition of dopamine release in the striatum when measured by amperometry experiments with mouse brain slices (Chergui et al., 2004). There is also evidence that Cdk5 decreases neurotransmitter release through the reduction of the vesicle fusion pore conductance and quantal size without altering the pore opening times (Barclay et al., 2004). In contrast, other data point to a role of Cdk5 as a positive regulator of exocytosis through the phosphorylation of substrates at nerve terminals and in other secretory cells, including adrenal chromaffin cells, which secrete catecholamines, and pancreatic β–cells, which secrete insulin (Fletcher et al., 1999; Lilja et al., 2001, 2004; Rosales et al., 2004; Xin et al., 2004). Munc18-1, Sept5, synapsin 1, and P/Q-subtype voltage-dependent calcium channels (VDCCs) are the known Cdk5 substrates that have been associated with synaptic vesicle exocytosis (Fig. 1).

Phosphorylation of Munc18-1 by Cdk5

Although there is a consensus in the field that Munc18-1 is essential for proper neurotransmitter release, the precise function of this protein in exocytosis remains elusive due to a collection of contradictory results. Deletion of the *munc18* gene in yeast, *C. elegans, Drosophila*, and mouse causes a complete block of synaptic transmission, supporting a positive regulatory role for Munc18 in neurotransmitter secretion (Gallwitz and Jahn, 2003). However, overexpression of Munc18 inhibits exocytosis in *Drosophila*, supporting a negative role of Munc18 in the secretion of the neurotransmitter (Wu et al., 1998). Moreover, Munc18 overexpression was shown by capacitance measurements to increase flash photolysis-induced secretion in intact chromaffin cells (Voets et al., 2001) and to have no effect on exocytosis when measured by single-spike amperometry in permeabilized chromaffin cells (Fisher et al., 2001). These discrepancies suggest that Munc18 is a versatile regulator of secretion at multiple points in the pathway. Munc18-1 is known to fold syntaxin 1 into a closed conformation that cannot associate with other SNARE proteins and thus block SNARE assembly (Dulubova et al., 1999; Yang et al., 2000). However, recent studies have demonstrated a role of the Munc18 protein family as a stimulatory subunit of its cognate SNARE vesicle fusion machinery (Zilly et al., 2006; Shen et al., 2007). Therefore, the discrepancies among previous data related to the function of Munc18 in exocytosis may be explained in part by the existence of dual roles for Munc18-1 in exocytosis; for example, activation of SNARE complex formation for membrane fusion and inhibition of the SNARE complex by sequestering syntaxin 1 in addition to the differences in experimental techniques and cell systems.

Cdk5 phosphorylation of Thr574 in Munc18-1 induces dissociation of the Munc18-1–syntaxin 1 complex. This dissociation is believed to allow syntaxin 1 to participate in the formation of the SNARE complex, which is required for membrane fusion during exocytosis (Shuang et al., 1998; Fletcher et al., 1999).

Therefore, Cdk5 activity can enhance neurotransmitter release through modulation of the Munc18-1–syntaxin 1 interaction. Studies concerning pancreatic β-cells also support a positive role of Cdk5 in insulin exocytosis through the Cdk5/p39-mediated phosphorylation of Munc18 (Lilja et al., 2001, 2004). In contrast, both a dominant negative mutant version of Cdk5 and a form of Munc18 that houses a Thr574Ala mutation increase the quantal size and broaden the kinetics of vesicle fusion, as measured by amperometry in adrenal chromaffin cells (Barclay et al., 2004), supporting a role of Cdk5 as a negative regulator of exocytosis. On the basis of the various recent findings regarding Cdk5 and Munc18 (Shuang et al., 1998; Zilly et al., 2006; Shen et al., 2007), it is possible to refine the hypothesis that Munc18 functions in several distinct steps in the secretory pathway (Fig. 1). First, it has been speculated that Thr574 phosphorylation of Munc18-1 in a preformed Munc18-1–syntaxin 1 complex disassembles the complex, which in turn stimulates formation of the SNARE complex and neurotransmitter release. However, it is also possible that Thr574 phosphorylation of Munc18-1 inhibits binding to the SNARE complex and subsequently inhibits membrane fusion and neurotransmitter release. Alternatively, Thr574 phosphorylation of Munc18-1 in the context of a preformed Munc18-1–SNARE complex may dissociate the SNARE complex. As Thr574 of Munc18-1 in the Munc18-1–syntaxin 1 complex is sterically inaccessible for protein interaction (Misura et al., 2000; Rosales and Lee, 2006), it is probable that Cdk5 phosphorylates Thr574 of Munc18-1 preferentially in the free Munc18-1 or Munc18-1–SNARE complex rather than in the Munc18-1–syntaxin1 complex.

Phosphorylation of Sept5 by Cdk5

Sept5 (also termed CDCrel-1) is a member of the septin family of proteins, which is predominantly expressed in the brain and localized mainly to presynaptic axon terminals (Dong et al., 2003). Overexpression of Sept5 inhibits dopamine release from cultured PC12 cells (Dong et al., 2003) and inhibits regulated secretion in the HIT-T15 cell line (Beites et al., 1999), supporting the notion that Sept5 exerts an inhibitory role in exocytosis. Sept5 binds to syntaxin 1 alone and to syntaxin 1 in the SNARE complex, but not to syntaxin 1 in the Munc18–syntaxin 1 complex or the α-SNAP–SNARE complex, suggesting that Sept5 regulates the availability of SNARE proteins after exocytosis (Beites et al., 2005). Ser17 of Sept5 is phosphorylated by Cdk5, and this phosphorylation reduces Sept5 binding to syntaxin 1 (Taniguchi et al., 2007). It is interesting that phosphorylation of both Sept5 and Munc18 by Cdk5 suppresses the binding of these proteins to syntaxin 1. Phosphorylation of Sept5 by Cdk5 may be involved in the disassembly of the SNARE complex (Fig. 1), although the functional significance of this phosphorylation event *in vivo* requires further investigation.

Phosphorylation of Synapsin 1 by Cdk5

Synapsin 1 is a neuron-specific phosphoprotein that is associated with the membranes of synaptic vesicles and is involved in the regulation of neurotransmitter release. Synapsin 1 tethers synaptic vesicles to actin filaments in a phosphorylation-dependent manner, controlling the number of vesicles available for release at the nerve terminus. It has been reported that Cdk5 phosphorylates Ser551 and Ser553 of synapsin 1 *in vitro* and that this phosphorylation has no effect on *in vitro* actin bundling activity (Matsubara et al., 1996). However, the physiological significance of the Cdk5-mediated phosphorylation of synapsin 1 has yet to be established. Recently, synapsin 1 was identified as a novel effector for Rab3A, a synaptic vesicle–specific small GTP-binding protein that has been implicated in multiple steps of exocytosis (Giovedi et al., 2004a, b). Synapsin 1 stimulates the GTP binding and GTPase activity of Rab3A, and Rab3A inhibits synapsin 1 binding to actin as well as synapsin-induced actin bundling and vesicle clustering (Giovedi et al., 2004a). Therefore, it would be worthwhile to investigate whether the Cdk5-mediated phosphorylation of synapsin 1 has an effect on the interaction between synapsin 1 and Rab3A to regulate synaptic vesicle exocytosis.

Phosphorylation of Pctaire1, a Kinase to Phosphorylate NSF by Cdk5

Pctaire1, a member of the Cdk5 family, is abundantly expressed in terminally differentiated cells such as postmitotic neurons (Besset et al., 1999). To date, the precise function of Pctaire1 remains elusive. Overexpression of Pctaire1 significantly reduces both basal and high [K$^+$]-stimulated growth hormone release in PC12 cells (Liu et al., 2006). Supporting this result is the observation that the presence of a mutated form of Pctaire1 that lacks kinase activity enhances basal and high [K$^+$]-stimulated growth hormone secretion in PC12 cells (Liu et al., 2006). This study suggests a role of Pctaire1 in modulating exocytosis. Pctaire1 interacts with the Cdk5 activator p35, and phosphorylation of Pctaire1 at Ser95 by Cdk5 enhances the kinase activity of Pctaire1 (Cheng et al., 2002).

NSF is an ATPase required for intracellular trafficking and membrane fusion events. As shown in *Drosophila* temperature-sensitive paralytic mutations in NSF (comatose), NSF is essential for synaptic exocytosis *in vivo* (Littleton et al., 1998). Binding of SNAPs to the SNARE complex allows binding of NSF, and upon ATP hydrolysis, NSF hexamers disassemble the SNARE complex, which resets the SNARE proteins for another round of membrane docking and fusion (Sollner et al., 1993; Brunger, 2005). The D2 domain of NSF, which is required for the oligomerization of NSF subunits, binds directly to and is phosphorylated by Pctaire1 on Ser569 (Liu et al., 2006). Phosphorylation of NSF by Pctaire1 reduces the hexamerization of NSF. An

NSF phosphorylation deficient mutant (S569A) was shown to increase the oligomerization of NSF, as well as the basal and high $[K^+]$-stimulated growth hormone secretion in PC12 cells (Liu et al., 2006). Interestingly, both NSF and Pctaire1 have effects on basal secretion, a relatively uncharacterized regulated secretory pathway that is distinct from constitutive secretion (Matsuuchi and Kelly, 1991). These findings support the hypothesis that the phosphorylation of NSF by Pctaire1 regulates NSF oligomerization and modulates the process of neurotransmitter release. It is notable that Pctaire1 was shown to interact with a number of synaptic vesicle–associated proteins, such as syntaxin 1, synaptotagmin, synapsin 1, and p35, in co-immunoprecipitation (IP) assays (Liu et al., 2006). Furthermore, a majority of the synaptic proteins that interact with Pctaire1, including NSF, were shown to be co-immunoprecipitated with FLAG-tagged Cdk5 from brain extracts, although a direct interaction of purified Cdk5 with NSF *in vitro* was not observed (Liu et al., 2006). It would therefore be valuable to ascertain whether Pctaire1 and Cdk5 exist as components of a large protein complex and together play a key role in phosphorylating multiple proteins to regulate neurotransmitter release. It is probable that Cdk5-mediated activation of Pctaire1 inhibits exocytosis by suppressing the oligomerization of NSF (Fig. 1).

Effect of Cdk5 on Endocytosis in Presynaptic Terminals

At presynaptic terminals, exocytosis is followed by the endocytosis of synaptic vesicles through a rapid pathway (also known as fast or kiss-and-run endocytosis) and a slow pathway (also known as clathrin-mediated endocytosis) (Evans and Cousin, 2007). Nerve terminals are enriched with several proteins known to be involved in clathrin-mediated endocytosis, such as clathrin, AP2, epsin, dynamin, amphiphysin, and synaptojanin (Cowan et al., 2001; Jarousse and Kelly, 2001). Endocytosis of synaptic vesicles can be further subdivided into several distinct steps (coat assembly on membranes, invagination, fission, and uncoating) (Jarousse and Kelly, 2001), suggesting that clathrin-mediated endocytosis is a tightly regulated process. Synaptic vesicle endocytosis is triggered by a coordinated dephosphorylation of a group of phosphoproteins called the dephosphins, which include dynamin 1, amphiphysin, synaptojanin, AP-180, Epsin, and Eps15 (Cousin and Robinson, 2001). The stimulated dephosphorylation of the dephosphins by the Ca^{++}-dependent protein phosphatase calcineurin and their subsequent rephosphorylation by Cdk5 are required for maintaining synaptic vesicle recycling and thus synaptic transmission (Tan et al., 2003; Tomizawa et al., 2003; Evans and Cousin, 2007). Cdk5 inhibitors enhance electrical stimulation–induced endocytosis in hippocampal neurons, and endocytosis is also enhanced in the neurons of p35-deficient mice (Tomizawa et al., 2003). However, as with exocytosis, researchers disagree on whether Cdk5 positively or negatively regulates endocytosis (Nguyen and Bibb,

2003; Tan et al., 2003; Tomizawa et al., 2003). It is possible that Cdk5 has the ability to activate and inhibit synaptic vesicle endocytosis, depending on certain physiological conditions, as this important regulatory protein is involved in multiple stages of this process (for example, fission and the clathrin uncoating of vesicles). However, it may also be possible that Cdk5 inhibits the first round of endocytosis, but is necessary for subsequent rounds, as was suggested previously (Nguyen and Bibb, 2003).

Phosphorylation of Dynamin 1 by Cdk5

Dynamin 1 is a large GTPase that polymerizes into oligomeric rings at the neck of invaginating buds and catalyzes vesicle fission upon GTP hydroysis (Takei and Haucke, 2001). During synaptic vesicle endocytosis, dynamin 1 is dephosphorylated by calcineurin and is subsequently rephosphorylated by Cdk5 on Ser774, Ser778 (Tan et al., 2003), and Thr780 (Tomizawa et al., 2003), although a controversy exists regarding the precise *in vivo* phosphorylation sites and the functional effects of phosphorylation (Tan et al., 2003; Tomizawa et al., 2003; Graham et al., 2007). The phosphorylation of dynamin 1 on Thr780 inhibits dynamin 1 binding to amphiphysin 1 (see below), and Cdk5 phosphorylation of both amphiphysin 1 and dynamin 1 completely disrupts co-polymerization of these two proteins into a ring formation (Tomizawa et al., 2003), thus pointing to Cdk5 as a negative regulator of endocytosis.

Phosphorylation of Amphiphysin I by Cdk5

Amphiphysin 1 plays a key role in clathrin-mediated synaptic vesicle endocytosis, acting as a membrane curvature sensor, a linker of clathrin coat proteins, and an enhancer of dynamin GTPase activity through multiple interactions with other endocytic proteins (Liang et al., 2007). Amphiphysin 1 interacts with p35 and can be phosphorylated by Cdk5 at Ser276 and Ser285 (Floyd et al., 2001; Tomizawa et al., 2003; Liang et al., 2007). Phosphorylation of amphiphysin 1 has no effect on the interaction with dynamin 1 but significantly inhibits its binding to β-adaptin (Tomizawa et al., 2003). The binding of amphiphysin 1 to a lipid membrane *in vitro* is reduced by the Cdk5-dependent phosphorylation of wild-type amphiphysin I, but not of the mutated form of the protein of both sites (S276, 285A) (Liang et al., 2007). Interestingly, endocytosis is increased in PC12 cells that express a form of amphiphysin 1 that carries an S276, 285A mutation, relative to the wild-type version (Liang et al., 2007). These results suggest that Cdk5-dependent phosphorylation at Ser276 and Ser285 is important for the membrane binding of amphiphysin and thus the regulation of endocytosis.

Phosphorylation of Synaptojanin 1 by Cdk5

Synaptojanin 1 is a polyphosphoinositide phosphatase. Concentrated in axon terminals, it dephosphorylates the inositol ring of various phosphoinositides, including phosphatidylinositol-4,5-bisphosphate [PI(4,5)P$_2$], and has been implicated in the uncoating of endocytosed synaptic vesicles during clathrin-mediated endocytosis (Fig. 1). PI(4,5)P$_2$ has been shown to play a role in the plasma membrane recruitment of proteins involved in synaptic vesicle endocytosis. Similar to other dephosphins such as dynamin 1 and amphiphysin 1, synaptojanin 1 undergoes constitutive phosphorylation in resting nerve terminals and Ca^{++}-dependent dephosphorylation upon nerve terminal depolarization. In general, phosphorylation inhibits, whereas dephosphorylation promotes, the assembly of endocytic complexes (Cousin and Robinson, 2001). Cdk5 phosphorylates synaptojanin 1 at Ser1144 in resting synapses, and this phosphorylation inhibits the phosphatase activity of synaptojanin 1 (Lee et al., 2004). Ser1144 is adjacent to a binding site for endophilin 1, a lysophosphatidic acid acyl transferase that catalyzes the conversion of lysophosphatidic acid to phosphatidic acid and is known to play a role in clathrin-mediated endocytosis (Schmidt et al., 1999). Phosphorylation of Ser1144 inhibits the membrane recruitment of synaptojanin 1 during endocytosis, perhaps by modulating its physical and regulatory interactions with endophilin 1 and amphiphysin 1 (Lee et al., 2004; Sahin and Bibb, 2004).

Effect of Cdk5 on Neurotransmitter Synthesis

Efficient neurotransmitter release from the presynaptic terminal is regulated by the neurotransmitter cycle as well as by the synaptic vesicle cycle. The neurotransmitter cycle involves the biosynthesis, storage, reuptake, and degradation of neurotransmitters and is regulated by phosphorylation. Cdk5 has been linked to neurotransmitter synthesis via the phosphorylation of tyrosine hydroxylase, which catalyzes the rate-limiting step in the synthesis of catecholamine neurotransmitters, such as dopamine, norepinephrine, and epinephrine (Kansy et al., 2004; Moy and Tsai, 2004).

Phosphorylation of Tyrosine Hydroxylase by Cdk5

Tyrosine hydroxylase exists in cells as a tetramer of four identical subunits, each of which is composed of an N-terminal regulatory domain and a C-terminal catalytic domain. The activity of tyrosine hydroxylase is regulated by phosphorylation, especially of Ser8, Ser19, Ser31, and Ser40, which are housed in the regulatory domain of the rat protein. Tyrosine hydroxylase serves as a substrate for several protein kinases, including Cdk5, PKA, PKC, CaMKII, and mitogen-activated protein kinase (MAP kinase). Cdk5 phosphorylates tyrosine hydroxylase at Ser31 *in vitro* and *in vivo* (Kansy et al., 2004; Moy and Tsai, 2004); this modification not only stimulates tyrosine hydroxylase activity but also markedly elevates the stability of the protein (Moy and Tsai, 2004). These results are supportive of an important role of Cdk5 in modulating the production and release of catecholamines.

Control of Presynaptic Ca^{++} and Lipid Signaling by Cdk5

Ca^{++} is a critical element in the regulation of neurotransmitter release because it triggers both exocytosis and endocytosis (Fig. 1). Several studies point to synaptotagmin as a Ca^{++} sensor that triggers both exocytosis and endocytosis (Chapman, 2002; Rizo et al., 2006). When a presynaptic action potential invades nerve terminals, Ca^{++} influx occurs through P/Q- or N-type VDCCs. An electrophysiological study revealed that the Cdk5 inhibitor roscovitine enhances Ca^{++} influx through the P/Q-type VDCC, implicating Cdk5 in presynaptic Ca^{++} signaling (Tomizawa et al., 2003).

Phosphoinositides have emerged as major regulators of the binding of cytosolic proteins to the bilayer lipid membrane. Accumulating evidence points to important regulatory roles of PI(4,5)P$_2$ in the synaptic vesicle exocytosis and endocytosis beyond its classical role as a precursor of second messengers (Wenk and De Camilli, 2004). Phosphatidylinositol(4)phosphate 5-kinase type I (PIPKIγ), the kinase involved in the generation of PI(4,5)P$_2$, in concert with the polyphosphoinositide phosphatase synaptojanin, plays an important regulatory function in neurotransmitter release through regulation of a plasma membrane pool of PI(4,5)P$_2$ (Wenk et al., 2001; Di Paolo et al., 2004). Interestingly, PIPKIγ and synaptojanin 1 were both recently identified as substrates of Cdk5 (Lee et al., 2004, 2005), suggesting that Cdk5 regulates neurotransmitter release through the modulation of PI(4,5)P$_2$ metabolism.

Phosphorylation of P/Q-Type Calcium Channels by Cdk5

The influx of Ca^{++} through VDCCs is the central stimulus for neurotransmitter release at the presynaptic terminals, and P/Q-type VDCCs are highly concentrated at central synapses. Cdk5 phosphorylates intracellular loop-connecting domains II and III of P/Q-type VDCCs, and this phosphorylation inhibits the interaction of Ca^{++} channels with SNAP-25 and synaptotagmin (Tomizawa et al., 2002). Furthermore, roscovitine potentiates Ca^{++} currents through P/Q-type VDCCs, thereby enhancing the release of glutamate, the principal neurotransmitter in the brain, known to be involved in learning and memory as well as in neurotoxicity (Tomizawa et al., 2002; Yan et al., 2002; Monaco and Vallano, 2005). These results support the contention that Cdk5 inhibits neurotransmitter release through the phosphorylation of P/Q-type Ca^{++} channels and downregulation of Ca^{++} channel activity. Due to the key role of Ca^{++} in endocytosis, it will be important to determine whether Cdk5-mediated phosphorylation of VDCCs also has effects on endocytosis.

Phosphorylation of PIPKIγ by Cdk5

PIPKIγ is expressed at high amounts in the brain and is involved in the generation of PI(4,5)P$_2$. Interaction of the protein talin with PIPKIγ is essential for clathrin-mediated synaptic vesicle endocytosis (Morgan et al., 2004). Talin serves as an adaptor between integrin and the actin cytoskeleton at sites of cell adhesion and was recently found to be present at neuronal synapses. In resting synapses, Cdk5 phosphorylates PIPKIγ at Ser650, which is within the talin-binding sequence, and this phosphorylation disrupts the association between PIPKIγ and talin (Lee et al., 2005). It is interesting that Cdk5 functions in the regulation of PI(4,5)P$_2$ synthesis via phosphorylation of both the kinase PIPKIγ and the phosphatase synaptojanin (Lee et al., 2004, 2005). These studies of PI(4,5)P$_2$ synthesis are consistent with data that suggest that Cdk5 promotes endocytosis (Tan et al., 2003). Given that PIPKIγ is also a critical priming factor for exocytosis (Wang et al., 2005), additional experiments are required to determine whether Cdk5-mediated phosphorylation of PIPKIγ modulates synaptic vesicle exocytosis.

Conclusion

There is increasing evidence that Cdk5 is involved in neurotransmitter release through its regulation of multiple steps in the exocytosis and endocytosis pathways. The controversy that exists as to whether Cdk5 acts as a positive or negative regulator of neurotransmitter release may arise from the fact that Cdk5 participates at a number of points in synaptic vesicle cycling. It is certainly expected that the list of Cdk5 substrates will continue to grow and that an increasing amount of evidence will support roles of Cdk5 in various stages of exocytosis and endocytosis, Ca^{++} signaling, and lipid signaling in presynaptic terminals. However, a clear picture of the functions of Cdk5 will depend most heavily on progress made by researchers who study neurotransmitter release. Abnormal Cdk5 activity as well as altered neurotransmission has been implicated in the progression of neurodegenerative disorders, such as Alzheimer's disease. Considering the critical role of Cdk5 in neurotransmitter release, it is important that researchers decipher how aberrant activity of this pivotal protein contributes to neurodegeneration through abnormal regulation of neurotransmitter release in presynaptic terminals. Such lines of investigation may then lead to insights into how one might slow the progression of devastating neurodegenerative diseases.

Acknowledgments This work was supported by KOSEF (R01-2007-000-11910-0) funded by the Korean government (MOST). This work was also supported by a Korea Research Foundation grant funded by the Korean Government (MOEHRD) (KRF-2007-331-E00198) and by the IBST Grant 2006 and 2007 from Inje University. Due to space limitation, the author regrets not being able to cite all relevant publications in this review.

References

Angelo M, Plattner F, Giese KP (2006) Cyclin-dependent kinase 5 in synaptic plasticity, learning and memory. J Neurochem 99:353–370.

Barclay JW, Aldea M, Craig TJ, Morgan A, Burgoyne RD (2004) Regulation of the fusion pore conductance during exocytosis by cyclin-dependent kinase 5. J Biol Chem 279:41495–41503.

Beites CL, Campbell KA, Trimble WS (2005) The septin Sept5/CDCrel-1 competes with alpha-SNAP for binding to the SNARE complex. Biochem J 385:347–353.

Beites CL, Xie H, Bowser R, Trimble WS (1999) The septin CDCrel-1 binds syntaxin and inhibits exocytosis. Nat Neurosci 2:434–439.

Bellen H (1999) Neurotransmitter release: Oxford university press, Oxford.

Besset V, Rhee K, Wolgemuth DJ (1999) The cellular distribution and kinase activity of the Cdk family member Pctaire1 in the adult mouse brain and testis suggest functions in differentiation. Cell Growth Differ 10:173–181.

Bhalla A, Chicka MC, Tucker WC, Chapman ER (2006) Ca(2+)-synaptotagmin directly regulates t-SNARE function during reconstituted membrane fusion. Nat Struct Mol Biol 13:323–330.

Brunger AT (2005) Structure and function of SNARE and SNARE-interacting proteins. Q Rev Biophys 38:1–47.

Chapman ER (2002) Synaptotagmin: a Ca(2+) sensor that triggers exocytosis? Nat Rev Mol Cell Biol 3:498–508.

Cheng K, Li Z, Fu WY, Wang JH, Fu AK, Ip NY (2002) Pctaire1 interacts with p35 and is a novel substrate for Cdk5/p35. J Biol Chem 277:31988–31993.

Chergui K, Svenningsson P, Greengard P (2004) Cyclin-dependent kinase 5 regulates dopaminergic and glutamatergic transmission in the striatum. Proc Natl Acad Sci U S A 101:2191–2196.

Cheung ZH, Fu AK, Ip NY (2006) Synaptic roles of Cdk5: implications in higher cognitive functions and neurodegenerative diseases. Neuron 50:13–18.

Cousin MA, Robinson PJ (2001) The dephosphins: dephosphorylation by calcineurin triggers synaptic vesicle endocytosis. Trends Neurosci 24:659–665.

Cowan WM, Sudhof TC, Stevens CF, Davies K (2001) Synapses: The Johns Hopkins University Press, Maryland.

Dhavan R, Tsai LH (2001) A decade of CDK5. Nat Rev Mol Cell Biol 2:749–759.

Di Paolo G, Moskowitz HS, Gipson K, Wenk MR, Voronov S, Obayashi M, Flavell R, Fitzsimonds RM, Ryan TA, De Camilli P (2004) Impaired PtdIns(4,5)P2 synthesis in nerve terminals produces defects in synaptic vesicle trafficking. Nature 431:415–422.

Dong Z, Ferger B, Paterna JC, Vogel D, Furler S, Osinde M, Feldon J, Bueler H (2003) Dopamine-dependent neurodegeneration in rats induced by viral vector-mediated overexpression of the parkin target protein, CDCrel-1. Proc Natl Acad Sci U S A 100:12438–12443.

Dulubova I, Sugita S, Hill S, Hosaka M, Fernandez I, Sudhof TC, Rizo J (1999) A conformational switch in syntaxin during exocytosis: role of munc18. EMBO J 18:4372–4382.

Evans GJ, Cousin MA (2007) Activity-dependent control of slow synaptic vesicle endocytosis by cyclin-dependent kinase 5. J Neurosci 27:401–411.

Fisher RJ, Pevsner J, Burgoyne RD (2001) Control of fusion pore dynamics during exocytosis by Munc18. Science 291:875–878.

Fletcher AI, Shuang R, Giovannucci DR, Zhang L, Bittner MA, Stuenkel EL (1999) Regulation of exocytosis by cyclin-dependent kinase 5 via phosphorylation of Munc18. J Biol Chem 274:4027–4035.

Floyd SR, Porro EB, Slepnev VI, Ochoa GC, Tsai LH, De Camilli P (2001) Amphiphysin 1 binds the cyclin-dependent kinase (cdk) 5 regulatory subunit p35 and is phosphorylated by cdk5 and cdc2. J Biol Chem 276:8104–8110.

Fu AK, Ip FC, Fu WY, Cheung J, Wang JH, Yung WH, Ip NY (2005) Aberrant motor axon projection, acetylcholine receptor clustering, and neurotransmission in cyclin-dependent kinase 5 null mice. Proc Natl Acad Sci U S A 102:15224–15229.

Gallwitz D, Jahn R (2003) The riddle of the Sec1/Munc-18 proteins - new twists added to their interactions with SNAREs. Trends Biochem Sci 28:113–116.

Giovedi S, Darchen F, Valtorta F, Greengard P, Benfenati F (2004a) Synapsin is a novel Rab3 effector protein on small synaptic vesicles. II. Functional effects of the Rab3A-synapsin I interaction. J Biol Chem 279:43769–43779.

Giovedi S, Vaccaro P, Valtorta F, Darchen F, Greengard P, Cesareni G, Benfenati F (2004b) Synapsin is a novel Rab3 effector protein on small synaptic vesicles. I. Identification and characterization of the synapsin I-Rab3 interactions in vitro and in intact nerve terminals. J Biol Chem 279:43760–43768.

Gonzalez L, Jr., Scheller RH (1999) Regulation of membrane trafficking: structural insights from a Rab/effector complex. Cell 96:755–758.

Graham ME, Anggono V, Bache N, Larsen MR, Craft GE, Robinson PJ (2007) The in vivo phosphorylation sites of rat brain dynamin I. J Biol Chem 282: 14695–14707.

Jahn R, Scheller RH (2006) SNAREs–engines for membrane fusion. Nat Rev Mol Cell Biol 7:631–643.

Jarousse N, Kelly RB (2001) Endocytotic mechanisms in synapses. Curr Opin Cell Biol 13:461–469.

Kansy JW, Daubner SC, Nishi A, Sotogaku N, Lloyd MD, Nguyen C, Lu L, Haycock JW, Hope BT, Fitzpatrick PF, Bibb JA (2004) Identification of tyrosine hydroxylase as a physiological substrate for Cdk5. J Neurochem 91:374–384.

Lee MS, Kwon YT, Li M, Peng J, Friedlander RM, Tsai LH (2000) Neurotoxicity induces cleavage of p35 to p25 by calpain. Nature 405:360–364.

Lee SY, Wenk MR, Kim Y, Nairn AC, De Camilli P (2004) Regulation of synaptojanin 1 by cyclin-dependent kinase 5 at synapses. Proc Natl Acad Sci U S A 101:546–551.

Lee SY, Voronov S, Letinic K, Nairn AC, Di Paolo G, De Camilli P (2005) Regulation of the interaction between PIPKI gamma and talin by proline-directed protein kinases. J Cell Biol 168:789–799.

Leenders AG, Sheng ZH (2005) Modulation of neurotransmitter release by the second messenger-activated protein kinases: implications for presynaptic plasticity. Pharmacol Ther 105:69–84.

Liang S, Wei FY, Wu YM, Tanabe K, Abe T, Oda Y, Yoshida Y, Yamada H, Matsui H, Tomizawa K, Takei K (2007) Major Cdk5-dependent phosphorylation sites of amphiphysin 1 are implicated in the regulation of the membrane binding and endocytosis. J Neurochem 102: 1466-1476.

Lilja L, Yang SN, Webb DL, Juntti-Berggren L, Berggren PO, Bark C (2001) Cyclin-dependent kinase 5 promotes insulin exocytosis. J Biol Chem 276:34199–34205.

Lilja L, Johansson JU, Gromada J, Mandic SA, Fried G, Berggren PO, Bark C (2004) Cyclin-dependent kinase 5 associated with p39 promotes Munc18-1 phosphorylation and Ca(2+)-dependent exocytosis. J Biol Chem 279:29534–29541.

Littleton JT, Chapman ER, Kreber R, Garment MB, Carlson SD, Ganetzky B (1998) Temperature-sensitive paralytic mutations demonstrate that synaptic exocytosis requires SNARE complex assembly and disassembly. Neuron 21:401–413.

Liu Y, Cheng K, Gong K, Fu AK, Ip NY (2006) Pctaire1 phosphorylates N-ethylmaleimide-sensitive fusion protein: implications in the regulation of its hexamerization and exocytosis. J Biol Chem 281:9852–9858.

Matsubara M, Kusubata M, Ishiguro K, Uchida T, Titani K, Taniguchi H (1996) Site-specific phosphorylation of synapsin I by mitogen-activated protein kinase and Cdk5 and its effects on physiological functions. J Biol Chem 271:21108–21113.

Matsuuchi L, Kelly RB (1991) Constitutive and basal secretion from the endocrine cell line, AtT-20. J Cell Biol 112:843–852.

Misura KM, Scheller RH, Weis WI (2000) Three-dimensional structure of the neuronal-Sec1-syntaxin 1a complex. Nature 404:355–362.

Monaco EA, 3rd, Vallano ML (2005) Roscovitine triggers excitotoxicity in cultured granule neurons by enhancing glutamate release. Mol Pharmacol 68:1331–1342.

Morgan JR, Di Paolo G, Werner H, Shchedrina VA, Pypaert M, Pieribone VA, De Camilli P (2004) A role for talin in presynaptic function. J Cell Biol 167:43–50.

Moy LY, Tsai LH (2004) Cyclin-dependent kinase 5 phosphorylates serine 31 of tyrosine hydroxylase and regulates its stability. J Biol Chem 279:54487–54493.

Nguyen C, Bibb JA (2003) Cdk5 and the mystery of synaptic vesicle endocytosis. J Cell Biol 163:697–699.

Rizo J, Chen X, Arac D (2006) Unraveling the mechanisms of synaptotagmin and SNARE function in neurotransmitter release. Trends Cell Biol 16:339–350.

Rosales JL, Lee KY (2006) Extraneuronal roles of cyclin-dependent kinase 5. Bioessays 28:1023–1034.

Rosales JL, Ernst JD, Hallows J, Lee KY (2004) GTP-dependent secretion from neutrophils is regulated by Cdk5. J Biol Chem 279:53932–53936.

Sahin B, Bibb JA (2004) Protein kinases talk to lipid phosphatases at the synapse. Proc Natl Acad Sci U S A 101:1112–1113.

Schmidt A, Wolde M, Thiele C, Fest W, Kratzin H, Podtelejnikov AV, Witke W, Huttner WB, Soling HD (1999) Endophilin I mediates synaptic vesicle formation by transfer of arachidonate to lysophosphatidic acid. Nature 401:133–141.

Shen J, Tareste DC, Paumet F, Rothman JE, Melia TJ (2007) Selective activation of cognate SNAREpins by Sec1/Munc18 proteins. Cell 128:183–195.

Shuang R, Zhang L, Fletcher A, Groblewski GE, Pevsner J, Stuenkel EL (1998) Regulation of Munc-18/syntaxin 1A interaction by cyclin-dependent kinase 5 in nerve endings. J Biol Chem 273:4957–4966.

Sollner T, Bennett MK, Whiteheart SW, Scheller RH, Rothman JE (1993) A protein assembly-disassembly pathway in vitro that may correspond to sequential steps of synaptic vesicle docking, activation, and fusion. Cell 75:409–418.

Sudhof TC (2004) The synaptic vesicle cycle. Annu Rev Neurosci 27:509–547.

Takahashi M, Itakura M, Kataoka M (2003) New aspects of neurotransmitter release and exocytosis: regulation of neurotransmitter release by phosphorylation. J Pharmacol Sci 93:41–45.

Takei K, Haucke V (2001) Clathrin-mediated endocytosis: membrane factors pull the trigger. Trends Cell Biol 11:385–391.

Tan TC, Valova VA, Malladi CS, Graham ME, Berven LA, Jupp OJ, Hansra G, McClure SJ, Sarcevic B, Boadle RA, Larsen MR, Cousin MA, Robinson PJ (2003) Cdk5 is essential for synaptic vesicle endocytosis. Nat Cell Biol 5:701–710.

Taniguchi M, Taoka M, Itakura M, Asada A, Saito T, Kinoshita M, Takahashi M, Isobe T, Hisanaga S (2007) Phosphorylation of adult type Sept5 (CDCrel-1) by cyclin-dependent kinase 5 inhibits interaction with syntaxin-1. J Biol Chem 282:7869–7876.

Tomizawa K, Ohta J, Matsushita M, Moriwaki A, Li ST, Takei K, Matsui H (2002) Cdk5/p35 regulates neurotransmitter release through phosphorylation and downregulation of P/Q-type voltage-dependent calcium channel activity. J Neurosci 22:2590–2597.

Tomizawa K, Sunada S, Lu YF, Oda Y, Kinuta M, Ohshima T, Saito T, Wei FY, Matsushita M, Li ST, Tsutsui K, Hisanaga S, Mikoshiba K, Takei K, Matsui H (2003) Cophosphorylation of amphiphysin I and dynamin I by Cdk5 regulates clathrin-mediated endocytosis of synaptic vesicles. J Cell Biol 163:813–824.

Voets T, Toonen RF, Brian EC, de Wit H, Moser T, Rettig J, Sudhof TC, Neher E, Verhage M (2001) Munc18-1 promotes large dense-core vesicle docking. Neuron 31:581–591.

Wang L, Li G, Sugita S (2005) A central kinase domain of type I phosphatidylinositol phosphate kinases is sufficient to prime exocytosis: isoform specificity and its underlying mechanism. J Biol Chem 280:16522–16527.

Wenk MR, De Camilli P (2004) Protein-lipid interactions and phosphoinositide metabolism in membrane traffic: insights from vesicle recycling in nerve terminals. Proc Natl Acad Sci U S A 101:8262–8269.

Wenk MR, Pellegrini L, Klenchin VA, Di Paolo G, Chang S, Daniell L, Arioka M, Martin TF, De Camilli P (2001) PIP kinase Igamma is the major PI(4,5)P(2) synthesizing enzyme at the synapse. Neuron 32:79–88.

Wu MN, Littleton JT, Bhat MA, Prokop A, Bellen HJ (1998) ROP, the Drosophila Sec1 homolog, interacts with syntaxin and regulates neurotransmitter release in a dosage-dependent manner. EMBO J 17:127–139.

Xin X, Ferraro F, Back N, Eipper BA, Mains RE (2004) Cdk5 and Trio modulate endocrine cell exocytosis. J Cell Sci 117:4739–4748.

Yan Z, Chi P, Bibb JA, Ryan TA, Greengard P (2002) Roscovitine: a novel regulator of P/Q-type calcium channels and transmitter release in central neurons. J Physiol 540:761–770.

Yang B, Steegmaier M, Gonzalez LC, Jr., Scheller RH (2000) nSec1 binds a closed conformation of syntaxin1A. J Cell Biol 148:247–252.

Zilly FE, Sorensen JB, Jahn R, Lang T (2006) Munc18-bound syntaxin readily forms SNARE complexes with synaptobrevin in native plasma membranes. PLoS Biol 4:e330.

Cdk5 in Dendrite and Synapse Development: Emerging Role as a Modulator of Receptor Tyrosine Kinase Signaling

Zelda H. Cheung and Nancy Y. Ip

Abstract Proper functioning of the nervous system relies on the precise wiring of neuronal connections. Regulation of neuronal architecture and synapse formation is critical in shaping the neural circuitry. Recently, accumulating evidence indicates that cyclin-dependent kinase 5 (Cdk5), a kinase implicated in the control of neuronal migration and survival, regulates multiple aspects of dendrite development and synaptic function. Here we summarize the existing findings on the mechanisms by which Cdk5 regulates dendrite, spine, and synapse development.

Dendrite, Spine, and Synapse Development

When the first neuron was illustrated by Ramón Y Cajal more than a hundred years ago, it was immediately obvious that neurons are unique. As neurons are polarized in nature, information is transmitted in the form of currents from dendrites toward the cell body, which may lead to the initiation of an action potential, propagating the signal along axons. Initiation of an action potential results in the release of neurotransmitter from the axon terminal. These chemicals diffuse across a specialized junction known as synapse to bind to neurotransmitter receptors on dendritic protrusions known as spines, or at cell body of neighboring neurons. Since neuronal architecture directly affects the sites of synapse formation, hence complexity of the neuronal network, growth and arborization of dendrites and axons are under tight control throughout neuronal differentiation. Strength of the synapses, on the other hand, is regulated by the composition of the scaffold proteins and neurotransmitter receptors at the post-synaptic densities (PSDs). Recently, increasing evidence suggests that spine

N.Y. Ip
Department of Biochemistry, Biotechnology Research Institute and Molecular Neuroscience Center, Hong Kong University of Science and Technology, Clear Water Bay, Hong Kong, China
e-mail: boip@ust.hk

N.Y. Ip, L.-H. Tsai (eds.), *Cyclin Dependent Kinase 5 (Cdk5)*,
DOI: 10.1007/978-0-387-78887-6_5, © Springer Science+Business Media, LLC 2008

51

morphology also heavily impacts on the efficiency of synaptic transmission. More strikingly, spine morphology appears to be highly plastic. *In vivo* observation of spines using time-lapse imaging technology reveals that shapes of spines are rapidly modified by a myriad of stimuli ranging from extracellular factors to neural activity. More importantly, changes in spine morphology correlates with modification of synaptic strength, thus spawning the idea that rapid changes in spine morphology may represent the physical changes in neural circuitry and synaptic transmission that underlie learning and memory formation. These realizations have rekindled interest in the molecular pathways involved in dendrite, spine, and synapse development (Whitford et al., 2002; Kummer et al., 2006; Tada and Sheng, 2006).

Although dendrite and synapse development are regulated by distinct signaling pathways, there is similarity in the mechanisms by which they are controlled and fine tuned during development. For example, neuronal architecture, synapse formation, and spine morphogenesis are all modulated by extracellular cues. Dendrite growth and extension are guided by extracellular cues such as semaphorins and neurotrophins, while factors such as neuregulin (NRG) and agrin are major players in synapse formation at the neuromuscular junction (NMJ) (Lai and Ip, 2003). Ephrins, on the other hand, have emerged as important players in spine morphogenesis (Klein, 2004). Secondly, neural activity also contributes to dendrite and spine growth. This activity has been demonstrated to enhance dendrite growth and arborization (Dijkhuizen and Ghosh, 2005). Similarly, long-term potentiation (LTP) is associated with increase in spine number and spine enlargement, while long-term depression (LTD) is predominantly linked to spine retraction (Dillon and Goda, 2005). The mechanisms by which neural stimulation and exposure to extracellular factors translate into actual changes in dendrite and spine shapes also share certain similarities. Both dendrites and spines are abundant in actin filament, although microtubules are only present in dendrites. Rho GTPases, a family of small GTPase serving as essential regulators of actin dynamics, frequently function as the effectors for mediating extracellular cues and neural activity to changes in dendrite and spine shapes. Therefore, precise control of Rho GTPase activity will also contribute to dendrite and synapse development.

Remarkably, increasing studies reveal that cyclin-dependent kinase 5 (Cdk5), a proline-directed serine/threonine kinase, is involved in multiple aspects of dendrite, synapse, and spine development through regulating neural activity, signaling downstream of various extracellular factors, and Rho GTPase activity (Cheung et al., 2006). In this chapter, we will summarize the existing findings in support of an emerging role of Cdk5 in dendrite and synapse development.

Cdk5: an Atypical Cyclin-Dependent Kinase

In an attempt to search for homologues of the cell cycle player Cdc2 in the brain, Cdk5 was purified as a Cdc2-related protein (Lew et al., 1992; Meyerson et al., 1992). Concurrently, Cdk5 was also identified as a Tau kinase (Kobayashi et al., 1993). Although Cdk5 is similar to other Cdks in that it associates with cyclins such as cyclin D, it is interestingly not activated by cyclins. Rather, Cdk5 was purified with a p25 regulatory subunit, which was later found to be a truncated fragment of p35, a neural-specific activator of Cdk5. p39, an isoform of p35, was later also identified as a Cdk5 activator. The importance of p35 and p39 as activators of Cdk5 is underscored by two observations. Similar to other Cdks, Cdk5 is ubiquitously expressed. Nonetheless, its activity is restricted largely to the nervous system in accordance with the expression pattern of p35 and p39, which is predominantly neural specific. This suggests that p35 and p39 are essential activators of Cdk5. This notion is further supported by transgenic animal studies. Mice lacking Cdk5 are characterized by cortical lamination defects and perinatal death. Remarkably, mice lacking both p35 and p39 displayed a phenotype that is essentially indistinguishable from that of Cdk5 knockout animals, thus accentuating the importance of p35 and p39 in Cdk5 activation (Dhavan and Tsai, 2001).

In addition to being activated by non-cyclin activators, Cdk5 is also unique among its family members as its functions are not restricted to the control of cell cycle progression. Rather, Cdk5 is implicated in almost all aspects of neuronal development, including the regulation of neuronal migration, cytoskeletal dynamics, modulation of synaptic transmission, and its involvement in neuronal death in neurodegenerative disease models (Dhavan and Tsai, 2001). Furthermore, Cdk5 is also increasingly implicated in dendrite and synapse formation (Cheung et al., 2006; Cheung and Ip, 2007). Here we discuss the signaling mechanisms by which Cdk5 regulates dendrite, spine, and synapse development.

Regulation of Dendrite Development by Cdk5

The first hint about a potential role of Cdk5 in dendrite development came from the observation that inhibition of Cdk5 activity attenuates dendrite outgrowth in cortical neurons (Nikolic et al., 1996). In support of this possibility, Cdk5, p35, and p39 are all localized to the growth cones of extending neurites (Humbert et al., 2000; Fu et al., 2002). Moreover, a recent study using a transgenic mice with cortex-specific deletion of Cdk5 reveals that dendrite morphology in layer V pyramidal neurons was abnormal, consistent with an important role of Cdk5 in dendrite development (Ohshima et al., 2007). Considerable progress has been made in recent years to understand the mechanisms by which Cdk5 affect dendrite development. While the list of Cdk5 substrates continues to increase, it is clear that Cdk5 modulates dendrite development through multiple pathways (Fig. 1).

Cdk5 as a Modulator of Extracellular Cues

Soon after the neurons reach their final destination during neuronal migration, the cells initiate the formation of new processes. Among the few processes extended, one differentiates into an axon, while the other ones develop into dendrites. Extension and arborization of dendrites are guided heavily by extracellular factors such as neurotrophins, semaphorins, and ephrins. Thus, regulation of the downstream signaling of these guidance cues will have monumental impact on dendrite development.

Neurotrophins

Neurotrophins are a family of trophic factors whose members include nerve growth factor (NGF), brain-derived neurotrophic factor (BDNF), neurotrophin-3 (NT-3), and NT-4/5. Actions of neurotrophins are mediated by two classes of receptors: the low-affinity p75 neurotrophin receptor and the receptor tyrosine kinases (RTKs) Trk receptors (TrkA, TrkB, and TrkC). While all neurotrophins bind to p75 with comparable affinity, NGF associates preferentially with TrkA; BDNF and NT-4/5 bind preferentially to TrkB; and NT-3 binds strongly to TrkC but also weakly to TrkA and TrkB (Huang and Reichardt, 2003).

  Neurotrophins typically act as chemoattractants for extending neurites and have been demonstrated to stimulate dendrite growth and arborization in cortical and hippocampal neurons (Whitford et al., 2002). Cdk5 was initially implicated in neurotrophin-stimulated neurite growth when inhibition of Cdk5 activity was observed to attenuate NGF-induced neurite outgrowth in PC12 cells (Harada et al., 2001). In particular, NGF stimulation results in enhanced p35 levels and Cdk5 activity through sustained Erk activation and the activation of transcription factor Egr-1. Inhibiting this increase in Cdk5 activity attenuates neurite outgrowth, revealing a pivotal role of Cdk5 in NGF-induced neuronal differentiation in PC12 cells (Harada et al., 2001). On the other hand, Cdk5 may also contribute to NGF-induced neurite outgrowth through modulating the activity of transcription factor STAT3. NGF treatment has been observed to induce STAT3 activation in PC12 cells. Interestingly, inhibition of STAT3 activation downstream of neurotrophin stimulation attenuated neurite growth in hippocampal neurons (Ng et al., 2006). Since Cdk5-mediated phosphorylation of STAT3 is crucial for the maximal activation of STAT3 (Fu et al., 2004), it is plausible that Cdk5 may also modulate NGF-induced neurite outgrowth through affecting STAT3 activity, although further experiments will be required to verify this possibility.

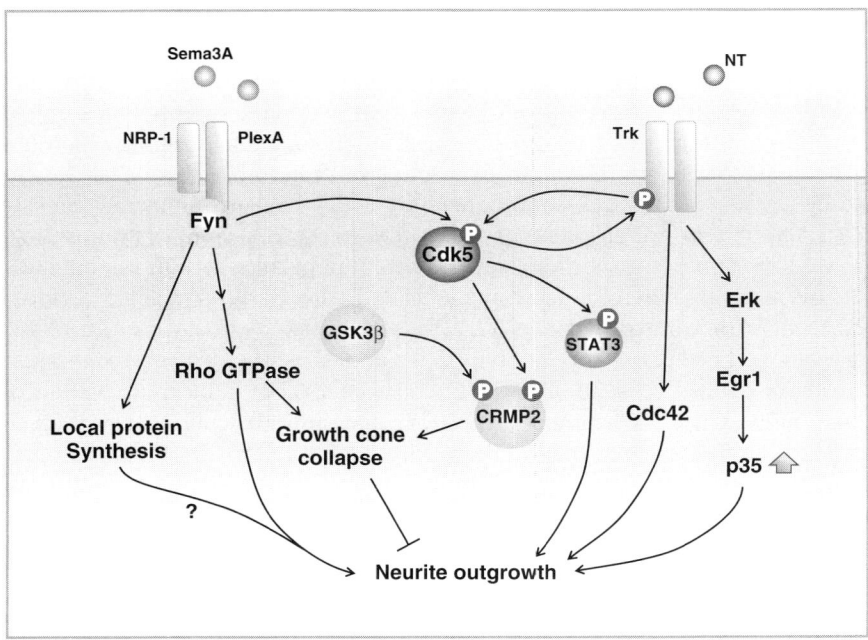

Fig. 1 Regulation of neurite growth by Cdk5. Cdk5 was observed to regulate neurite development through modulating the downstream signaling of extracellular cues including semaphorins 3A (Sema 3A) and neurotrophins. Cdk5 was found to be required for Sema 3A–induced growth cone collapse via two mechanisms. Phosphorylation of Cdk5 by Fyn at Tyr15 leads to enhanced Cdk5 activity and recruitment of Cdk5 to the Sema 3A receptor complex of neuropilin 1 (NRP-1) and plexinA (PlexA). Both Fyn and Cdk5 are required for Sema 3A–induced growth cone collapse and induction of local protein synthesis, but whether they play a role in the regulation of Rho GTPase activity downstream of Sema 3A treatment remains to be explored. On the other hand, Cdk5 also modulates Sema 3A–induced growth cone collapse through phosphorylation of CRMP2. Dual phosphorylation of CRMP2 by GSK3β and Cdk5 is required for Sema 3A–triggered growth cone collapse. In addition to Sema 3A, Cdk5 also contributes to NGF-induced neurite outgrowth in PC12 cells. NGF treatment leads to sustained Erk1/2 activation, induction of transcription factor Egr-1, and elevated transcription of p35. This enhances Cdk5 activity, and reversal of this increase inhibits NGF-induced neurite outgrowth. Furthermore, Cdk5 may also affect neurite outgrowth through phosphorylation of STAT3, since STAT3 activation following NGF stimulation is required for neurite outgrowth. A recent study reveals that Cdk5 also contributes to BDNF-induced dendrite growth in hippocampal neurons. BDNF stimulation triggers phosphorylation of Cdk5 at Tyr15, resulting in enhanced Cdk5 activity. Cdk5 in turn phosphorylates TrkB, which is required for BDNF-stimulated increase in primary dendrites and activation of Rho GTPase Cdc42. Finally, Cdk5 also modulates Rho GTPase activity through phosphorylation of GEFs or their effectors. Tyrosine phosphorylation is depicted by blue circles, while red circles represent serine/threonine phosphorylation (*See* Color Insert)

Aside from NGF, BDNF has also been observed to enhance Cdk5 activity (Tokuoka et al., 2000; Cheung et al., 2007). Interestingly, BDNF was observed to increase Cdk5 activity prior to any changes in p35 levels. Rather, activation of TrkB results in the recruitment of Cdk5 to the activated receptors, leading to phosphorylation of Cdk5 by TrkB at Tyr15 (Cheung et al., 2007), which has been demonstrated to enhance Cdk5 activity (Zukerberg et al., 2000). This finding reveals an alternative mechanism by which neurotrophins may elevate Cdk5 activity, although whether other Trk receptors similarly phosphorylate Cdk5 remains to be explored. More importantly, BDNF-stimulated increase in Cdk5 activity is recently associated with BDNF-triggered dendrite growth in hippocampal neurons. TrkB was identified as a substrate of Cdk5. Inhibition of TrkB phosphorylation by Cdk5 abolishes BDNF-triggered increase in primary dendrites, in addition to attenuating BDNF-induced activation of Rho GTPase Cdc42 (Cheung et al., 2007), suggesting that phosphorylation of TrkB by Cdk5 is required for BDNF-induced dendrite growth. It will be interesting to further examine if Cdk5 is similarly required for the induction of dendrite arborization by neurotrophins.

Semaphorin 3A

Unlike neurotrophins, which predominantly function as chemoattractants, the role of semaphorin 3A as a guidance cue is more complex. While semaphorin 3A has long been established as a repulsive guidance cue for developing axon, it was recently demonstrated as a chemoattractant for apical dendrites of cortical neurons (Polleux et al., 2000). In addition, semaphorin 3A was observed to induce neurite outgrowth and dendritic branching in cortical neurons (Morita et al., 2006), further supporting a chemoattractant role of semaphorin 3A in dendrites. Actions of semaphorin 3A are mediated by a receptor complex comprising neuropilins-1/2 and plexin-As. Binding of semaphorin 3A to the neuropilin/plexin complex leads to the recruitment of downstream mediators such as Rho GTPase Rac, Rho GTPase effector LIM kinase, and collapsing response mediator protein (CRMP) to regulate dendrite growth.

Cdk5 was first observed to modulate semaphorin 3A signaling when inhibition of Cdk5 activity was found to attenuate semaphorin 3A–induced growth cone collapse in Cdk5$^{-/-}$ dorsal root ganglion neurons. Cdk5 is recruited to plexinA2 through binding to an activated Src kinase known as Fyn. This leads to phosphorylation of Cdk5 at Tyr15 by Fyn, which is required for semaphorin 3A–triggered growth cone collapse (Sasaki et al., 2002). In addition, Cdk5 and Fyn are also involved in semaphorin 3A–stimulated local protein synthesis at growth cones via the translation initiation factor eIF-4E. Enhanced activation of eIF-4E was markedly suppressed in p35$^{-/-}$ neurons, revealing the important role of Cdk5/p35 in semaphorin 3A–induced local protein synthesis (Li et al., 2004a). Interestingly, Cdk5 also modulates semaphorin 3A signaling through the phosphorylation of CRMP2, an intracellular mediator for semaphorin 3A–induced growth cone collapse. Sequential phosphorylation of CRMP2 by

Cdk5 and GSK3β leads to the dissociation of CRMP2 from tubulin. Furthermore, phosphorylation of CRMP2 by both Cdk5 and GSK3β are required for semaphorin 3A–triggered growth cone collapse. These observations collectively reveal an essential role of Cdk5 in semaphorin 3A signaling. Nonetheless, it should be noted that majority of these studies were carried out in axons. Whether Cdk5 also contributes to regulation of dendrite growth by semaphorin 3A awaits further investigations.

Cdk5: Regulator of Cytoskeletal Components

As mentioned earlier, exposure to extracellular guidance cues is often translated to actual changes in dendrite growth through engaging key regulators of the cytoskeletal components. Rho GTPases, for example, are important regulators of dendrite growth through modulation of actin dynamics. Since its isolation, Cdk5 has long been implicated in the regulation of actin and microtubule dynamics. Indeed, Cdk5 was initially identified as a kinase for Tau, a microtubule-binding protein. Through the years, a plethora of microtubule-binding proteins and Rho GTPase effectors or guanine nucleotide exchange factors (GEFs) were identified as Cdk5 substrates. Nonetheless, how these phosphorylation events directly contribute to dendrite growth remains rather ambiguous. In light of the essential roles of actin and microtubule dynamics in the regulation of dendrite growth, however, it is important to discuss the involvement of Cdk5 as a regulator of cytoskeletal components.

Regulation of Actin Dynamics Through Modulation of Rho GTPases

Rho GTPases are a family of small GTPase that directly modulate actin dynamics through the mobilization of downstream effectors such as actin-binding proteins. Members of the Rho GTPase family include RhoA, Rac1, and Cdc42. Activated via binding to the appropriate GEFs, thereby facilitating recruitment of a GTP, these GTPases are inactivated upon hydrolysis of the GTP to GDP. Binding to GTPase activating proteins (GAPs) accelerates hydrolysis of GTP, thus promoting inactivation of the Rho GTPase. Upon activation, Rho GTPases are often recruited to the plasma membrane where they interact with their respective downstream effectors. Interestingly, while members of the Rho GTPase are all involved in the regulation of actin dynamics, they exhibit opposing roles. While RhoA is mostly associated with neurite retraction, Rac1 and Cdc42 favor neurite outgrowth and extension (Dillon and Goda, 2005).

A number of GEFs and effectors for Rho GTPases have been demonstrated as substrates of Cdk5. Among various GEFs, RasGRF1 (Kesavapany et al., 2006), RasGRF2 (Kesavapany et al., 2004), and ephexin1 (Fu et al., 2007) were identified as Cdk5 substrates. Cdk5 also phosphorylates effectors of Rho

GTPase such as Pak1 (Nikolic et al., 1998; Rashid et al., 2001) and WAVE1 (Kim et al., 2006). On the other hand, Cdk5 interacts with α-chimaerin, a GAP for Rac and Cdc42 (Brown et al., 2004; Qi et al., 2004). Interestingly, some of these phosphorylation events are not directly associated with the regulation of Rho GTPase activity. For example, although Cdk5/p35 inhibits Pak1 activity, it was later delineated to be independent of direct phosphorylation by Cdk5 (Nikolic et al., 1998; Rashid et al., 2001). In addition, phosphorylation of RasGRF1 results in the degradation of RasGRF1 by calpain (Kesavapany et al., 2006), but how this affects Rho GTPase activity remains unknown. Nonetheless, through the phosphorylation of ephexin1, Cdk5 has been observed to facilitate ephexin1-mediated RhoA activity (Fu et al., 2007). In addition, phosphorylation of RasGRF2 by Cdk5 reduces Rac activity (Kesavapany et al., 2004). These observations indicate that Cdk5 could modulate Rho GTPase activity.

Regulation of Microtubule Dynamics

In addition to the regulation of actin dynamics, Cdk5 has also been demonstrated to phosphorylate a myriad of microtubule-binding proteins, including Tau, MAP2, and MAP1b (Kobayashi et al., 1993; Pigino et al., 1997; Tseng et al., 2005). Indeed, hyperphosphorylation of Tau was evident in the neurons of a transgenic mice overexpressing p25 (Ahlijanian et al., 2000). Interestingly, reduction of MAP1b phosphorylation was accompanied by inhibition of laminin-induced axon growth when Cdk5 activity was inhibited, suggesting that Cdk5 may also contribute to axon growth through phosphorylation of MAP1b (Paglini et al., 1998). Nonetheless, whether phosphorylation of these microtubule-binding proteins by Cdk5 affects microtubule dynamics and whether this contributes to dendrite growth remain enigmatic. Further investigation will be required to delineate the precise role of these phosphorylation events in dendrite outgrowth.

Regulation of Synapse Formation by Cdk5

Synapse formation occurs when the extending axon of a developing neuron comes into contact with the dendrite or cell body of a neighboring neuron. Being a specialized junction for efficient transmission of information, the part of the post-synaptic neurons that are directly juxtaposed against the axon terminal need to differentiate and mature into the PSD, which is characterized by the abundant presence of neurotransmitter receptors and various scaffold proteins. Furthermore, synapses are to be maintained subsequent to formation as lack of activity has been observed to lead to dispersal of neurotransmitter receptors at the synapse. Strength of the synaptic connection is often determined by the amount of surface neurotransmitter receptors expressed, and the way they are clustered (Kummer et al., 2006). The expression of Cdk5 and its activators at the synapse suggests a potential role of this kinase in synapse development (Humbert et al., 2000; Fu et al., 2002). Indeed, despite the predominantly neural-specific nature of Cdk5 activity, we found that Cdk5 and its activators are also expressed in the post-synaptic muscle fiber at the NMJ (Fu et al., 2001). Using the NMJ as a model, it was found that Cdk5 is involved in multiple aspects of synapse development (Cheng and Ip, 2003; Fig. 2).

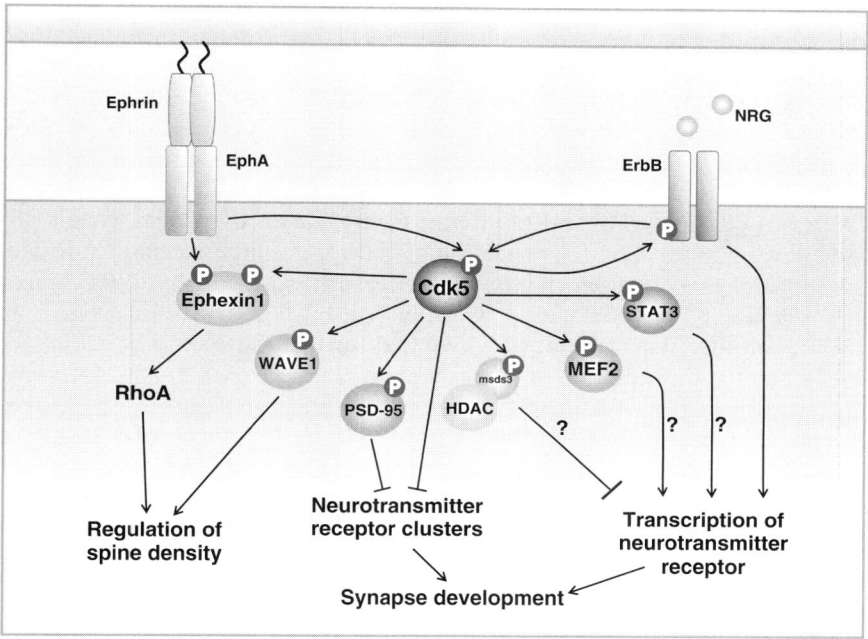

Fig. 2 Regulation of spine morphogenesis and synapse formation by Cdk5. Cdk5 is recently implicated in spine morphogenesis through modulation of Rho GTPase activity. Through phosphorylation of Rho GEF ephexin, Cdk5 is required for ephrinA1-induced RhoA activation and spine retraction in hippocampal neurons. Cdk5 also inhibits activity of WAVE1, thereby affecting spine morphology. On the other hand, Cdk5 modulates transcription of neurotransmitter receptors at the synapse through regulation of neuregulin (NRG)/ErbB signaling. NRG-triggered increase in Cdk5 activity is required for NRG-induced transcription of acetylcholine receptors and GABA$_A$ receptors at the NMJ and central synapses, respectively. Cdk5-mediated phosphorylation of STAT3 also contributes to NRG-stimulated gene transcription. In addition, Cdk5 regulates gene transcription through inhibition of MEF2 and facilitation of mSds3-HDAC-dependent gene repression, although whether these events are implicated in the regulation of neurotransmitter receptor transcription remains to be determined. Finally, Cdk5 is an important regulator of neurotransmitter receptor clustering. Acetylcholine receptor endplate is broadened in Cdk5-deficient diaphragm. Cdk5 also phosphorylates PSD-95 to inhibit clustering of neurotransmitter receptors in the CNS. Tyrosine phosphorylation is depicted by blue circles, while red circles represent serine/threonine phosphorylation (*See* Color Insert)

Regulation of Neurotransmitter Receptor Expression

Cdk5 as a Modulator of Extracellular Cues Signaling

Earlier studies have revealed that NRG, an extracellular factor secreted by axon terminal, contributes to the regulation of acetylcholine receptor expression in cultured myotubes. Unexpectedly, similar to the pivotal role of Cdk5 in the regulation of dendrite growth by extracellular cues, we found that Cdk5 activity is required for the induction of acetylcholine receptor expression by NRG in myotubes (Fu et al., 2001). NRG receptor ErbB, a RTK, was identified as a Cdk5 substrate. Phosphorylation of ErbB by Cdk5 is required for the initiation of ErbB downstream signaling and NRG-induced transcription of acetylcholine receptors, thus revealing an important role of Cdk5 in the regulation of NRG-mediated transcription of neurotransmitter receptor (Fu et al., 2001). In addition to the NMJ, Cdk5-mediated phosphorylation of ErbB is also demonstrated in neurons (Li et al., 2003). Furthermore, Cdk5 is required for the induction of $GABA_A$ transcription by NRG in the CNS (Xie et al., 2004). While the mechanisms implicated in the regulation of neurotransmitter receptor transcription by NRG remain largely unknown, we recently observed that NRG stimulation triggers STAT3 activation. Cdk5 was identified as a serine kinase of STAT3, and this phosphorylation is required for the transcriptional activity of STAT3 and the transcription of downstream target gene such as *c-fos* following NRG treatment (Fu et al., 2004). These observations collectively reveal that through affecting downstream signaling of NRG, Cdk5 contributes to the regulation of neurotransmitter receptor expression in both the CNS and PNS.

Regulation of Gene Transcription Machinery by Cdk5

In addition to affecting neurotransmitter receptor transcription by modulating signaling of extracellular cues, identification of novel Cdk5 substrates in recent years indicates that Cdk5 may also control gene transcription by directly regulating the gene transcription machinery. For example, phosphorylation of the transcription factor MEF2 by Cdk5 inhibits the transcriptional activity of MEF2, and triggers degradation of the transcription factor (Gong et al., 2003; Tang et al., 2005). Furthermore, it was recently observed that regulation of MEF2 levels by neural activity affects the number of excitatory synapses (Flavell et al., 2006), suggesting that regulation of MEF2 activity by Cdk5 may directly impact on synaptic transcription.

Aside from the phosphorylation of transcription factors, mSds3, a component of the histone deacetylase 1 (HDAC1) complex, was also identified as a Cdk5 substrate (Li et al., 2004b). Structural remodeling of the chromatin has emerged as a key mechanism by which gene transcription is regulated. Acetylation of histones, mediated mostly by histone acetyltransferase (HATs), is associated with activation of gene transcription. Deacetylation of histones, on the other hand, is mediated by HDACs and triggers silencing of gene transcription.

Phosphorylation of mSds3 by Cdk5 reduces histone acetylation and enhances mSds3-mediated transcriptional repression, indicating that the phosphorylation enhances the deacetylase activity of the mSds3–HDAC1 complex (Li et al., 2004b). It will be interesting to further examine if Cdk5 may affect neurotransmitter receptor expression at the synapse through modulation of HDAC activity.

Regulation of Neurotransmitter Receptor Clustering

Clustering of neurotransmitter receptors at the PSD is an important determining factor of synaptic strength. Examination of the NMJ phenotype in Cdk5$^{-/-}$ animals reveals a critical role of Cdk5 in the clustering of acetylcholine receptor. The central band of acetylcholine receptor cluster is significantly broader in Cdk5 knockout diaphragm. Furthermore, agrin-induced acetylcholine receptor clusters are significantly larger in Cdk5$^{-/-}$ myotubes (Fu et al., 2005; Lin et al., 2005). These observations collectively demonstrated an indispensable role of Cdk5 in the regulation of acetylcholine receptor clustering.

Formation of neurotransmitter clusters requires the coordinated expression or trafficking of scaffold proteins at the PSD. A recent study suggests that Cdk5 may affect clustering of neurotransmitter receptors by phosphorylating PSD-95, an important synaptic scaffold protein in the CNS. Phosphorylation of PSD-95 by Cdk5 reduces the multimerization of PSD-95, and leads to enlarged PSD-95 clusters, potentially affecting the clustering of NMDA receptors at synapses (Morabito et al., 2004). In addition, Aβ was observed to reduce PSD-95 levels in cortical neurons, a process that requires Cdk5 activity (Roselli et al., 2005). This suggests that Cdk5 may regulate both the protein level and clustering of PSD-95, thereby affecting clustering of neurotransmitter receptors.

Regulation of Spine Development

Regulation of spine development has attracted massive interests, as spine morphogenesis is believed to underlie the formation of new memory in different learning paradigms due to the rapid dynamics of spine motility and the correlation of spine shapes with synaptic strength. Indeed, significant progress has been made in the past few years to enhance our understanding of the formation and maturation of spines. Spine development begins when a filopodia extends from a dendritic shaft toward the approaching axon terminal. Maturation of these newly formed spines will lead to the generation of mushroom-shaped spines. Similar to the extension of developing dendrites, changes in actin cytoskeleton is responsible for effecting the actual modulation of spine shape (Dillon and Goda, 2005). Given the regulation of Rho GTPase activity by Cdk5, it is not surprising that Cdk5 may also affect spine morphogenesis.

Indeed, spine density is enhanced in p25 inducible transgenic mice characterized by transient elevation of Cdk5 activity in their forebrain (Fischer et al., 2005). Nonetheless, further evidence in support of this possibility, and the mechanisms implicated, has only been recently demonstrated. Figure 2 summarizes our existing knowledge on the role of Cdk5 in spine morphogenesis.

Regulation of Spine Morphology Through Modulation of Extracellular Cues and Rho GTPase

Traditionally recognized as a repulsive axon guidance cue, ephrins are recently revealed to play important roles in the regulation of spine morphogenesis. Ephrins are subdivided into ephrin As and ephrin Bs, which bind specifically to their cognate receptors EphAs and EphBs. Eph receptors are currently the largest family of RTKs. While EphB signaling is predominantly associated with stimulation of spine growth, EphA signaling was recently demonstrated to lead to spine retraction (Klein, 2004). Interestingly, Cdk5 was observed to be required for EphA4-mediated spine retraction. Activation of EphA4 triggers RhoA activation, which has been demonstrated to underlie EphA4-triggered spine retraction. Upon stimulation with ligand ephrin A1, Cdk5 is recruited to EphA4 and phosphorylated at Tyr15, leading to an enhancement in Cdk5 activity. The activated Cdk5 then phosphorylates Rho GEF ephexin1, thereby regulating RhoA activity to mediate EphA4-stimulated spine retraction (Fu et al., 2007). Our observations reveal that Cdk5, through modulating EphA4 signaling and direct phosphorylation of Rho GEF, regulates spines morphology in hippocampal neurons.

A recent study indicates that Cdk5 may also directly phosphorylate effectors of Rho GTPase to affect spine morphogenesis. WAVE1, an effector of Rac, was identified as a Cdk5 substrate (Kim et al., 2006). Phosphorylation of WAVE1 by Cdk5 inhibits WAVE1 activity and reduces Arp2/3 complex-dependent actin polymerization. Furthermore, knockdown of WAVE1 expression in neurons reduces the number of dendritic spines. Interestingly, while WAVE1 is highly phosphorylate in the brain, dephosphorylation of WAVE1 at the Cdk5 sites by cAMP enhances spine density (Kim et al., 2006). These observations collectively suggest that Cdk5 may affect spine morphogenesis through the phosphorylation of WAVE1, although the precise mechanism implicated awaits further investigation.

Regulation of Spine Morphology Through Modulating Synaptic Activity

Although alteration in spine shape is associated with changes in synaptic strength, it is interesting to note that synaptic activity can also in turn regulate spine morphology. In particular, while LTP triggers spine growth and maturation, LTD induces spine regression (Dillon and Goda, 2005). Recent animal studies have revealed an important role of Cdk5 in LTP and LTD induction in various learning paradigms. For example, Cdk5 activity is enhanced during associative learning and fear conditioning, and is required for effective associative learning (Fischer et al., 2002). Moreover, LTP induction is facilitated in transgenic mice transiently overexpressing p25, and this is accompanied by an increase in spine density (Fischer et al., 2005). On the other hand, spatial learning and LTD induction are both impaired in p35-deficient mice (Ohshima et al., 2005). Furthermore, inhibition of Cdk5 activity was found to enhance LTP- and NMDA-mediated currents in hippocampal neurons (Hawasli et al., 2007). Interestingly, histone acetylation is recently associated with reversal of learning impairment due to neuronal loss in a transgenic mice with inducible p25 expression (Fischer et al., 2007). In light of the phosphorylation of mSds3 by Cdk5, it is plausible that Cdk5 may also affect learning and memory through controlling mSds3-HDAC activity. These observations indicate that Cdk5 may affect spine morphogenesis through regulating synaptic strength and plasticity.

Cdk5: Modulator of RTK Signaling

Since its isolation, the multifaceted nature of Cdk5 is increasingly accentuated by the ever-growing list of Cdk5 substrates. As summarized in this chapter, Cdk5 substrates are found in almost all subcellular compartments, and are involved in the regulation of dendrite and synapse development through multiple mechanisms. Nonetheless, these observations also reveal an emerging importance of Cdk5 as a regulator of RTK signaling. A number of receptors for the extracellular factors involved in the regulation of dendrite and synapse development are RTKs, including Trk receptors, ErbB, and Eph receptors. In systems as diverse as the NMJ and synapses in the hippocampus, Cdk5 was observed to be required for the downstream signaling and functions of these RTKs (Fig. 3). More importantly, activation of these RTKs are often associated with enhancement of Cdk5 activity, possibly through direct phosphorylation of Cdk5 at Tyr15, thus providing a mechanism for ligand-induced activation of Cdk5. Given the importance of RTK signaling in almost all aspects of neuronal development, the identification of Cdk5 as a potential regulator of RTK signaling will have far-reaching implications. Further characterization of Cdk5 as a modulator of RTK will likely provide novel insights into the regulation of RTK signaling, and knowledge on how Cdk5 may impact on the downstream functions of these extracellular factors.

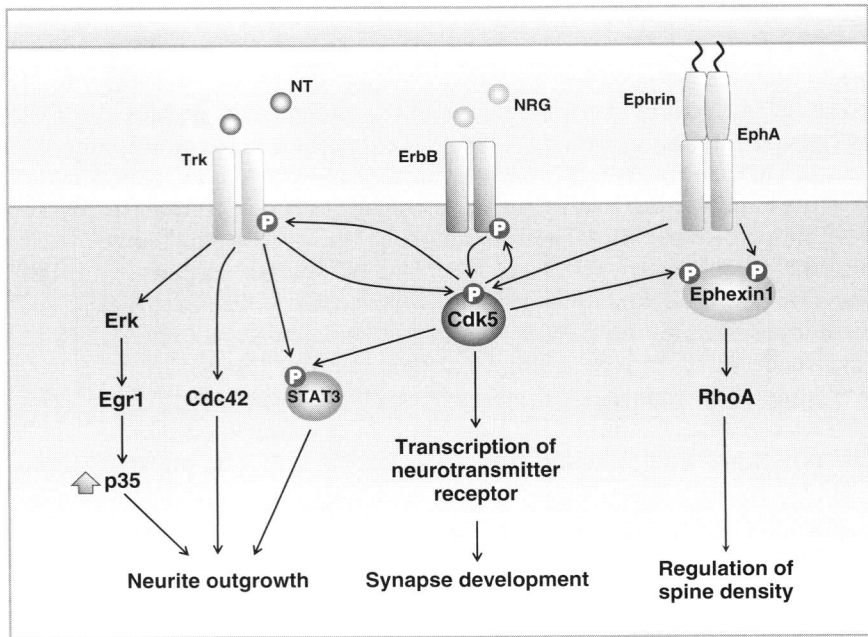

Fig. 3 The emerging role of Cdk5 as a regulator of RTK signaling. Cdk5 was observed to modulate signaling and functions downstream of RTK activation, thereby affecting dendrite outgrowth, spine morphogenesis, and synapse formation. Cdk5 is required for BDNF-stimulated dendrite growth and Cdc42 activation in hippocampal neurons. On the other hand, Cdk5 is also required for the NGF-induced neurite outgrowth in PC12 cells. NGF stimulation triggers sustained Erk activation and induction of transcription factor Egr-1, leading to p35 expression. Inhibition of this increase attenuates NGF-induced neuronal differentiation. In addition, STAT3 activation downstream of NGF stimulation also contributes to neurite outgrowth in PC12 cells. Since Cdk5 phosphorylates STAT3 to enhance its transcriptional activity, Cdk5 may also affect neurite outgrowth through phosphorylation of STAT3. Cdk5 also regulates synapse formation by modulating the signaling of NRG. Cdk5-mediated phosphorylation of NRG receptor ErbB is required for NRG-stimulated upregulation of acetylcholine receptor expression. NRG also triggers phosphorylation of STAT3 by Cdk5, thereby regulating gene transcription at synapses. Finally, Cdk5 affects spine morphogenesis by modulating signaling downstream of EphA4 activation. Cdk5-mediated phosphorylation of Rho GEF ephexin1 is required for the activation of RhoA and the induction of spine retraction following ephrinA1 treatment. Tyrosine phosphorylation is depicted by blue circles, while red circles represent serine/threonine phosphorylation (*See* Color Insert)

Acknowledgments We apologize to the many researchers whose works were not cited due to limited space. We would like to thank Ka-Chun Lok for his excellent help on preparing the figures. The studies by N.Y. Ip were supported in part by the Research Grants Council of Hong Kong (HKUST3/03C, 6130/03 M, 6119/04 M, 6421/05 M, 6444/06 M and 6431/06 M), the Area of Excellence Scheme of the University Grants Committee (AoE/B-15/01), and Hong Kong Jockey Club. N.Y. Ip and Z.H. Cheung are Croucher Foundation Senior Research Fellow and Croucher Foundation Fellow, respectively.

References

Ahlijanian MK, Barrezueta NX, Williams RD, Jakowski A, Kowsz KP, McCarthy S, Coskran T, Carlo A, Seymour PA, Burkhardt JE, Nelson RB, McNeish JD (2000) Hyperphosphorylated tau and neurofilament and cytoskeletal disruptions in mice overexpressing human p25, an activator of cdk5. Proc Natl Acad Sci U S A 97: 2910–2915.

Brown M, Jacobs T, Eickholt B, Ferrari G, Teo M, Monfries C, Qi RZ, Leung T, Lim L, Hall C (2004) Alpha2-chimaerin, cyclin-dependent Kinase 5/p35, and its target collapsin response mediator protein-2 are essential components in semaphorin 3A-induced growth-cone collapse. J Neurosci. 24:8994–9004.

Cheng K, Ip NY (2003) Cdk5: a new player at synapses. Neurosignals. 12:180–190.

Cheung ZH, Ip NY (2007) The roles of cyclin-dependent kinase 5 in dendrite and synapse development. Biotechnol J.

Cheung ZH, Fu AK, Ip NY (2006) Synaptic roles of Cdk5: implications in higher cognitive functions and neurodegenerative diseases. Neuron. 50:13–18.

Cheung ZH, Chin WH, Chen Y, Ng YP, Ip NY (2007) Cdk5 is involved in BDNF-stimulated dendritic growth in hippocampal neurons. PLoS Biol. 5:e63.

Dhavan R, Tsai LH (2001) A decade of CDK5. Nat Rev Mol Cell Biol. 2:749–759.

Dijkhuizen PA, Ghosh A (2005) Regulation of dendritic growth by calcium and neurotrophin signaling. Prog Brain Res. 147:17–27.

Dillon C, Goda Y (2005) The actin cytoskeleton: integrating form and function at the synapse. Annu Rev Neurosci. 28:25–55.

Fischer A, Sananbenesi F, Schrick C, Spiess J, Radulovic J (2002) Cyclin-dependent kinase 5 is required for associative learning. J Neurosci. 22:3700–3707.

Fischer A, Sananbenesi F, Pang PT, Lu B, Tsai LH (2005) Opposing roles of transient and prolonged expression of p25 in synaptic plasticity and hippocampus-dependent memory. Neuron. 48:825–838.

Fischer A, Sananbenesi F, Wang X, Dobbin M, Tsai LH (2007) Recovery of learning and memory is associated with chromatin remodelling. Nature. 447:178–182.

Flavell SW, Cowan CW, Kim TK, Greer PL, Lin Y, Paradis S, Griffith EC, Hu LS, Chen C, Greenberg ME (2006) Activity-dependent regulation of MEF2 transcription factors suppresses excitatory synapse number. Science. 311:1008–1012.

Fu AK, Fu WY, Cheung J, Tsim KW, Ip FC, Wang JH, Ip NY (2001) Cdk5 is involved in neuregulin-induced AChR expression at the neuromuscular junction. Nat Neurosci. 4:374–381.

Fu WY, Wang JH, Ip NY (2002) Expression of Cdk5 and its activators in NT2 cells during neuronal differentiation. J Neurochem. 81:646–654.

Fu AK, Fu WY, Ng AK, Chien WW, Ng YP, Wang JH, Ip NY (2004) Cyclin-dependent kinase 5 phosphorylates signal transducer and activator of transcription 3 and regulates its transcriptional activity. Proc Natl Acad Sci U S A. 101:6728–6733.

Fu AK, Ip FC, Fu WY, Cheung J, Wang JH, Yung WH, Ip NY (2005) Aberrant motor axon projection, acetylcholine receptor clustering, and neurotransmission in cyclin-dependent kinase 5 null mice. Proc Natl Acad Sci U S A 102:15224–15229.

Fu WY, Chen Y, Sahin M, Zhao XS, Shi L, Bikoff JB, Lai KO, Yung WH, Fu AK, Greenberg ME, Ip NY (2007) Cdk5 regulates EphA4-mediated dendritic spine retraction through an ephexin1-dependent mechanism. Nat Neurosci 10:67–76.

Gong X, Tang X, Wiedmann M, Wang X, Peng J, Zheng D, Blair LA, Marshall J, Mao Z (2003) Cdk5-mediated inhibition of the protective effects of transcription factor MEF2 in neurotoxicity-induced apoptosis. Neuron 38:33–46.

Harada T, Morooka T, Ogawa S, Nishida E (2001) ERK induces p35, a neuron-specific activator of Cdk5, through induction of Egr1. Nat Cell Biol 3:453–459.

Hawasli AH, Benavides DR, Nguyen C, Kansy JW, Hayashi K, Chambon P, Greengard P, Powell CM, Cooper DC, Bibb JA (2007) Cyclin-dependent kinase 5 governs learning and synaptic plasticity via control of NMDAR degradation. Nat Neurosci 10:880–886.

Huang EJ, Reichardt LF (2003) Trk receptors: roles in neuronal signal transduction. Annu Rev Biochem 72:609–642.

Humbert S, Lanier LM, Tsai LH (2000) Synaptic localization of p39, a neuronal activator of cdk5. Neuroreport 11:2213–2216.

Kesavapany S, Amin N, Zheng YL, Nijhara R, Jaffe H, Sihag R, Gutkind JS, Takahashi S, Kulkarni A, Grant P, Pant HC (2004) p35/cyclin-dependent kinase 5 phosphorylation of ras guanine nucleotide releasing factor 2 (RasGRF2) mediates Rac-dependent Extracellular Signal-regulated kinase 1/2 activity, altering RasGRF2 and microtubule-associated protein 1b distribution in neurons. J Neurosci 24:4421–4431.

Kesavapany S, Pareek TK, Zheng YL, Amin N, Gutkind JS, Ma W, Kulkarni AB, Grant P, Pant HC (2006) Neuronal nuclear organization is controlled by cyclin-dependent kinase 5 phosphorylation of Ras guanine nucleotide releasing factor-1. Neurosignals 15:157–173.

Kim Y, Sung JY, Ceglia I, Lee KW, Ahn JH, Halford JM, Kim AM, Kwak SP, Park JB, Ho Ryu S, Schenck A, Bardoni B, Scott JD, Nairn AC, Greengard P (2006) Phosphorylation of WAVE1 regulates actin polymerization and dendritic spine morphology. Nature 442:814–817.

Klein R (2004) Eph/ephrin signaling in morphogenesis, neural development and plasticity. Curr Opin Cell Biol 16:580–589.

Kobayashi S, Ishiguro K, Omori A, Takamatsu M, Arioka M, Imahori K, Uchida T (1993) A cdc2-related kinase PSSALRE/cdk5 is homologous with the 30 kDa subunit of tau protein kinase II, a proline-directed protein kinase associated with microtubule. FEBS Lett 335:171–175.

Kummer TT, Misgeld T, Sanes JR (2006) Assembly of the postsynaptic membrane at the neuromuscular junction: paradigm lost. Curr Opin Neurobiol 16:74–82.

Lai KO, Ip NY (2003) Central synapse and neuromuscular junction: same players, different roles. Trends Genet 19:395–402.

Lew J, Winkfein RJ, Paudel HK, Wang JH (1992) Brain proline-directed protein kinase is a neurofilament kinase which displays high sequence homology to p34cdc2. J Biol Chem 267:25922–25926.

Li BS, Ma W, Jaffe H, Zheng Y, Takahashi S, Zhang L, Kulkarni AB, Pant HC (2003) Cyclin-dependent kinase-5 is involved in neuregulin-dependent activation of phosphatidylinositol 3-kinase and Akt activity mediating neuronal survival. J Biol Chem 278:35702–35709.

Li C, Sasaki Y, Takei K, Yamamoto H, Shouji M, Sugiyama Y, Kawakami T, Nakamura F, Yagi T, Ohshima T, Goshima Y (2004a) Correlation between semaphorin3A-induced facilitation of axonal transport and local activation of a translation initiation factor eukaryotic translation initiation factor 4E. J Neurosci 24:6161–6170.

Li Z, David G, Hung KW, DePinho RA, Fu AK, Ip NY (2004b) Cdk5/p35 phosphorylates mSds3 and regulates mSds3-mediated repression of transcription. J Biol Chem 279:54438–54444.

Lin W, Dominguez B, Yang J, Aryal P, Brandon EP, Gage FH, Lee KF (2005) Neurotransmitter acetylcholine negatively regulates neuromuscular synapse formation by a cdk5-dependent mechanism. Neuron 46:569–579.

Meyerson M, Enders GH, Wu CL, Su LK, Gorka C, Nelson C, Harlow E, Tsai LH (1992) A family of human cdc2-related protein kinases. EMBO J 11:2909–2917.

Morabito MA, Sheng M, Tsai LH (2004) Cyclin-dependent kinase 5 phosphorylates the N-terminal domain of the postsynaptic density protein PSD-95 in neurons. J Neurosci 24:865–876.

Morita A, Yamashita N, Sasaki Y, Uchida Y, Nakajima O, Nakamura F, Yagi T, Taniguchi M, Usui H, Katoh-Semba R, Takei K, Goshima Y (2006) Regulation of dendritic branching and spine maturation by semaphorin3A-Fyn signaling. J Neurosci 26:2971–2980.

Ng YP, Cheung ZH, Ip NY (2006) STAT3 as a downstream mediator of Trk signaling and functions. J Biol Chem 281:15636–15644.

Nikolic M, Dudek H, Kwon YT, Ramos YF, Tsai LH (1996) The cdk5/p35 kinase is essential for neurite outgrowth during neuronal differentiation. Genes Dev 10:816–825.

Nikolic M, Chou MM, Lu W, Mayer BJ, Tsai LH (1998) The p35/Cdk5 kinase is a neuron-specific Rac effector that inhibits Pak1 activity. Nature 395:194–198.

Ohshima T, Ogura H, Tomizawa K, Hayashi K, Suzuki H, Saito T, Kamei H, Nishi A, Bibb JA, Hisanaga S, Matsui H, Mikoshiba K (2005) Impairment of hippocampal long-term depression and defective spatial learning and memory in p35 mice. J Neurochem 94:917–925.

Ohshima T, Hirasawa M, Tabata H, Mutoh T, Adachi T, Suzuki H, Saruta K, Iwasato T, Itohara S, Hashimoto M, Nakajima K, Ogawa M, Kulkarni AB, Mikoshiba K (2007) Cdk5 is required for multipolar-to-bipolar transition during radial neuronal migration and proper dendrite development of pyramidal neurons in the cerebral cortex. Development 134:2273–2282.

Paglini G, Pigino G, Kunda P, Morfini G, Maccioni R, Quiroga S, Ferreira A, Caceres A (1998) Evidence for the participation of the neuron-specific CDK5 activator P35 during laminin-enhanced axonal growth. J Neurosci 18:9858–9869.

Pigino G, Paglini G, Ulloa L, Avila J, Caceres A (1997) Analysis of the expression, distribution and function of cyclin dependent kinase 5 (cdk5) in developing cerebellar macroneurons. J Cell Sci 110 (Pt 2):257–270.

Polleux F, Morrow T, Ghosh A (2000) Semaphorin 3A is a chemoattractant for cortical apical dendrites. Nature 404:567–573.

Qi RZ, Ching YP, Kung HF, Wang JH (2004) Alpha-chimaerin exists in a functional complex with the Cdk5 kinase in brain. FEBS Lett 561:177–180.

Rashid T, Banerjee M, Nikolic M (2001) Phosphorylation of Pak1 by the p35/Cdk5 kinase affects neuronal morphology. J Biol Chem 276:49043–49052.

Roselli F, Tirard M, Lu J, Hutzler P, Lamberti P, Livrea P, Morabito M, Almeida OF (2005) Soluble beta-amyloid1-40 induces NMDA-dependent degradation of postsynaptic density-95 at glutamatergic synapses. J Neurosci 25:11061–11070.

Sasaki Y, Cheng C, Uchida Y, Nakajima O, Ohshima T, Yagi T, Taniguchi M, Nakayama T, Kishida R, Kudo Y, Ohno S, Nakamura F, Goshima Y (2002) Fyn and Cdk5 mediate semaphorin-3A signaling, which is involved in regulation of dendrite orientation in cerebral cortex. Neuron 35:907–920.

Tada T, Sheng M (2006) Molecular mechanisms of dendritic spine morphogenesis. Curr Opin Neurobiol 16:95–101.

Tang X, Wang X, Gong X, Tong M, Park D, Xia Z, Mao Z (2005) Cyclin-dependent kinase 5 mediates neurotoxin-induced degradation of the transcription factor myocyte enhancer factor 2. J Neurosci 25:4823–4834.

Tokuoka H, Saito T, Yorifuji H, Wei F, Kishimoto T, Hisanaga S (2000) Brain-derived neurotrophic factor-induced phosphorylation of neurofilament-H subunit in primary cultures of embryo rat cortical neurons. J Cell Sci 113 (Pt 6):1059–1068.

Tseng HC, Ovaa H, Wei NJ, Ploegh H, Tsai LH (2005) Phosphoproteomic analysis with a solid-phase capture-release-tag approach. Chem Biol 12:769–777.

Whitford KL, Dijkhuizen P, Polleux F, Ghosh A (2002) Molecular control of cortical dendrite development. Annu Rev Neurosci 25:127–149.

Xie F, Raetzman LT, Siegel RE (2004) Neuregulin induces GABAA receptor beta2 subunit expression in cultured rat cerebellar granule neurons by activating multiple signaling pathways. J Neurochem 90:1521–1529.

Zukerberg LR, Patrick GN, Nikolic M, Humbert S, Wu CL, Lanier LM, Gertler FB, Vidal M, Van Etten RA, Tsai LH (2000) Cables links Cdk5 and c-Abl and facilitates Cdk5 tyrosine phosphorylation, kinase upregulation, and neurite outgrowth. Neuron 26:633–646.

Cyclin-Dependent Kinase 5 (Cdk5) Modulates Signal Transduction Pathways Regulating Neuronal Survival

Parvathi Rudrabhatla, Jyotshna Kanungo, Ya-Li Zheng, Niranjana D. Amin, Sashi Kesavapany, and Harish C. Pant

Abstract Cyclin-dependent kinase 5 (Cdk5) in the nervous system has evolved to become a "surveillance system" that, among its other functions, monitors and integrates fluctuations in the activities of signaling cascades involved in nervous system growth, differentiation, and survival. In this review, we have focused primarily on Cdk5 cross talk and neuronal survival conducted in our laboratory. Cdk5 activity is tightly regulated in the nervous system, but may be deregulated under neuronal stress such as oxidative injury, excitotoxic stimulations, and β-amyloid exposure resulting in apoptosis marked by aggregates of hyperphosphorylated Tau, neurofilament, and other cytoskeletal proteins. We have demonstrated that Cdk5 modulates mitogen-activated protein kinase (MAPK), c-Jun N-terminal kinase (JNK3), and phosphoinositol-3-kinase/ protein kinase B (PI3 K/Akt). In addition, it is shown that Cdk5 directly and indirectly mediates the Ras guanine nucleotide exchange factors-1/2 (RasGRF1/2) and glycogen synthase kinase 3β (GSK3β) signaling pathways, respectively.

Introduction

Phosphorylation of cytoskeletal proteins is topographically regulated in the nervous system under normal conditions. Most of this phosphorylation occurs on the proline-directed serine/threonine (S/T) residues of cytoskeletal proteins such as neurofilament high and medium molecular weight (NF-H and NF-M) and the microtubule-associated proteins (MAPs/Tau). Normally, these proteins are phosphorylated on S/T-P residues selectively in the axonal compartment of the neuron. Although all kinases, phosphatases, their substrates, and their regulators are synthesized in the cell body, no cytoskeletal protein phosphorylation on these proline-directed S/T-P residues is detected in the cell body.

H.C. Pant
Laboratory of Neurochemistry, National Institute of Neurological Disorders
and Stroke, National Institutes of Health, Bethesda, Maryland 20814, USA
e-mail: panth@ninds.nih.gov

N.Y. Ip, L.-H. Tsai (eds.), *Cyclin Dependent Kinase 5 (Cdk5)*,
DOI: 10.1007/978-0-387-78887-6_6, © Springer Science+Business Media, LLC 2008

This compartmentalization of phosphorylation is tightly regulated, and in a number of neuropathological conditions, such as amyotrophic lateral sclerosis (ALS), Alzheimer's disease (AD), Neiman Pick's Type C disease and others, it becomes deregulated. Proline-directed S/T-P residues on cytoskeletal proteins are aberrantly hyperphosphorylated within cell bodies, resulting in the accumulation of abnormal cellular aggregates and massive neuronal cell death. The mechanisms of the topographic phosphorylation are not understood. In this review we have focused on the role of Cdk5 cross talk in regulating the phosphorylation activity of proline-directed S/T kinases involved in neuronal survival.

While studying the kinases responsible for phosphorylation of neuronal cytoskeletal proteins specifically in their S/T-P residues, we discovered that Cdk5 is one of the principal kinases involved in their phosphorylation (Shetty et al., 1993). Although Cdk5 is ubiquitously expressed in all cells and shares a high degree of homology with other members of the cyclin-dependent kinase family (CDKs), its activity is found specifically in postmitotic neurons because its activators, p35 and p39, are expressed primarily in neurons (Ko et al., 2001). In addition, it is highly conserved among various species, particularly, in vertebrates (Fig. 1A). Studies from our own and other laboratories around the world have shown that Cdk5 is a multifunctional S/T protein kinase that is involved in a wide range of neuronal functions from neurite outgrowth and neuronal migration to synaptic activity and cell survival. Some of them are shown in Fig. 1B. In addition to cytoskeletal proteins, NF-M/H and MAPs/Tau, Cdk5 phosphorylates a large number of other molecules on their S/T-P residues as shown in Fig. 2A. To identify its consensus sequence, we designed a large number of S/T-P–containing peptides and found that it selectively phosphorylates XS/TPXK/R, where X is either a basic or neutral amino acid. A VKSPAKEEAKSPEK sequence of the rat NF-H repeated 10 times was synthesized to identify the specifically phosphorylated residues by Cdk5. It is important to note that only the first serine with a neutral amino acid followed by a basic residue, VKS(1)PAK, was found to be phosphorylated but not the second, KS(2)PEK, where it is acidic (Sharma et al., 1998). High-resolution and detailed NMR structural data analyses of this peptide suggested the presence of a transient loop or beta type II turn at the first S(1) residue within VKSPAK (X = A) repeat. In comparison, the second S(2) residue in KSPEK (X = E) did not present spectral data (Fig. 2B) indicative of a stable secondary structure (Sharma et al., 1998). We found that unlike mitogen-activated protein kinases (MAPKs; Erk1/2), Cdk5 is more selective for S/T-P K/R motifs as shown in Fig. 2C.

Cdk5 activity is tightly regulated in the developing nervous system (Fig. 2D). We have shown, for example, that Cdk5-null knockout (KO) mice are lethal, exhibiting abnormal corticogenesis and other neuronal abnormalities before dying between E16 and P0. Earlier, we have also demonstrated that experimental re-expression of Cdk5 in neurons of Cdk5 KO mice *in vivo* completely restores the wild-type phenotype, clearly demonstrating that neuronal, and not

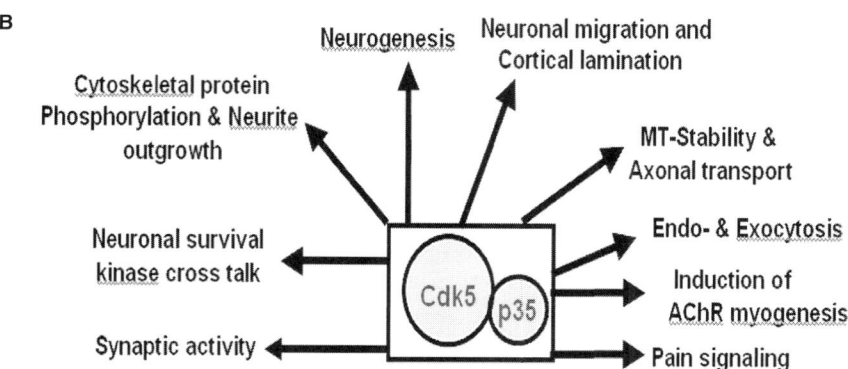

Fig. 1 (**A**) Comparison of Cdk5 amino acid sequence among various species. Amino acids in black represent conserved residues and that in red represent non-conserved residues. (**B**) Schematic presentation shows involvement of Cdk5 in a wide range of neuronal functions (*See* Color Insert)

glial, Cdk5 activity is necessary for normal development and survival (Takahashi et al., 2003). During neuronal insults, increase in intracellular calcium and activation of calpains result in the cleavages of p35 to p25, thereby inducing deregulation and hyperactivation of Cdk5. In outcome, aberrant hyperphosphorylation of cytoskeletal proteins (e.g., NFs, MAPs, Tau) occurs, forming aggregates of these proteins in the cell body and consequently inducing neuronal death. This process has been associated with a large number of neurodegenerative diseases (Ko et al., 2001). Cdk5 is involved not only in phosphorylating its substrate proteins but also directly or indirectly in modulating the other kinase activities, thus regulating the different signal transduction pathways leading to nervous system development and survival. Because of

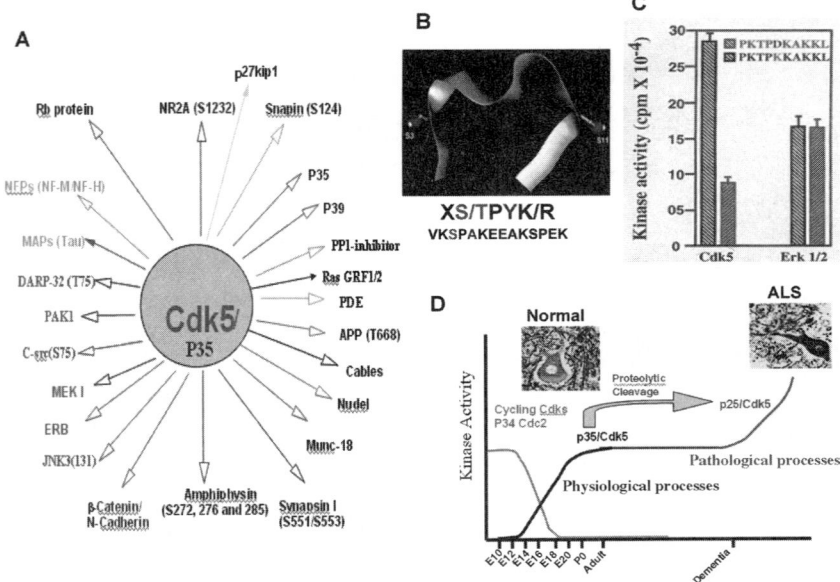

Fig. 2 (**A**) Some of the Cdk5 substrates (**B**) Ribbon diagram of the best-fit average structure of the peptide VKSPAKEEAKSPEK repeat in rat NF-H, in which the first serine is phosphorylated by Cdk5. (**C**) Comparison of Cdk5 and Erk1/2 activity using peptide substrates, PKTPKKAKKL and PKTPDKAKKL. (**D**) Cdk5 activity is tightly regulated during nervous system development; dividing cells lack Cdk5 activity; during nervous system development, increased expression of neuron-specific p35 protein induces Cdk5 activity by binding to Cdk5. Upon neuronal insult, proteolytic cleavage of p35 produces p25 that deregulates Cdk5 activity (*See* Color Insert)

its multifunctional role, and its positive and negative effects on neuronal function and survival, Cdk5 has been characterized as a "Jekyll and Hyde" kinase (Cruz and Tsai, 2004).

Cdk5-MAPK Cross Talk

To understand the molecular mechanisms that are involved in producing the phenotypes seen in Cdk5$^{-/-}$ mice, initially we measured the proline- and non-proline-directed kinase activities in E16 mice brain using three types of substrates. The endogeneous and casein phosphorylating activity was similar in wild-type and Cdk5-null neuronal tissues; however, to our surprise, in the case of myelin basic protein and histone H1), which are also substrates of Cdk5, the phosphorylation activity was elevated in Cdk5-null tissues compared to the wild type (Fig. 3A). Immunohistochemical staining of 18-day-old embryonic brain stem sections from Cdk5$^{-/-}$ and wild-type mice with SMI31 antibody, which

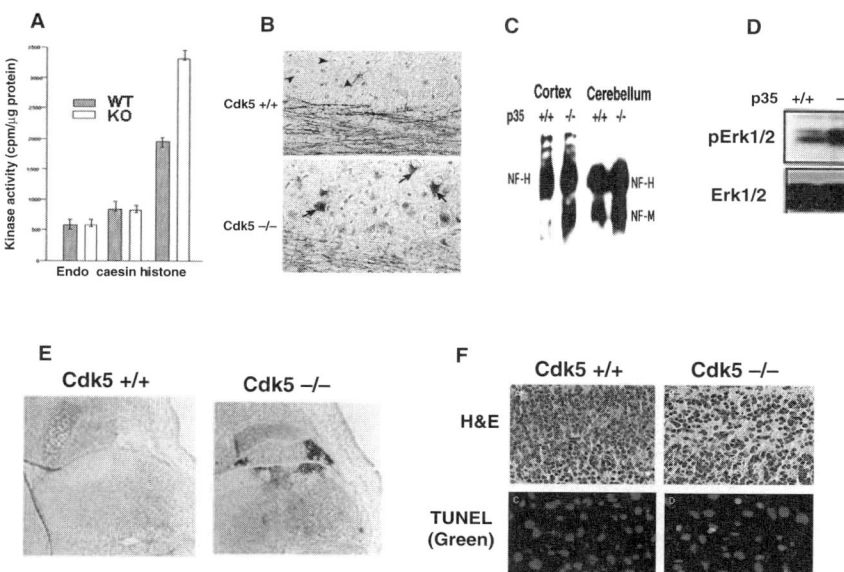

Fig. 3 (**A**) Comparison of endogenous Cdk5 activity in the brain extracts isolated from wild-type (WT or Cdk5$^{+/+}$) and Cdk5$^{-/-}$ mice, using casein, histone H1 (type III) as substrates. (**B**) Immunohistochemical analyses of E18 brain stem neurons from WT and Cdk5$^{-/-}$ mice using SMI31 antibody. (**C**) Western blot analysis of NF-M/H using SMI31 antibody in the WT and p35$^{-/-}$ brain lysates. (**D**) Phospho-Erk1/2 and total Erk1/2 expression in the WT and p35$^{-/-}$ brain homogenate. (**E**) Caspase-3 activity in E14 spinal cord neurons from WT and CDk5$^{-/-}$ mice. (**F**) Analysis of apoptosis in E16.5 cortical plate from WT and Cdk5$^{-/-}$ mice using TUNEL assay to identify apoptotic cells. *Upper panel* shows the H&E staining of sections (*See* Color Insert)

stains phosphorylated S/T-P residues in the neuronal cytoskeletal proteins (e.g., NF-M/H, Tau), showed intense immunostaining of phosphorylated cytoskeletal proteins in the neuronal cell bodies of Cdk5$^{-/-}$ mice. These neuronal cell bodies are not stained with these antibodies in the wild-type mice (Fig. 3B). The expression of the phosphorylated cytoskeletal proteins is restricted to the neuronal processes only. To further verify this observation, the cytoskeletal protein fractions from the cortex and cerebellum of 3–4-weeks-old p35$^{-/-}$ and wild-type mice were analyzed by immunoblotting with SMI31 antibody. Since phosphorylation of cytoskeletal proteins is higher in adult mice brain compared to the embryonic brain, and p35$^{-/-}$ mice survive to adulthood and have very low Cdk5 activity, we analyzed phosphorylation of NF-M/H proteins in p35$^{-/-}$ and wild-type mice. As shown in Fig. 3C, the immunoreactivity of NF-M to the SMI31 antibody in the cortex of p35$^{-/-}$ mice was higher than in the wild-type mice. Our previous studies have shown that the rodent NF-M is a preferred substrate for Erk1/2 phosphorylation (Veeranna et al., 1998). These observations indicated that in p35$^{-/-}$ mice, the reduction in Cdk5 activity might have upregulated Erk1/2 activity. Data presented in Fig. 3D show enhanced phosphorylation activity of Erk1/2 in p35$^{-/-}$ mice brains compared to the wild type, while the expression levels of Erk1/2 remained unchanged. Such sustained high activity of Erk1/2 in the neuronal cells has been implicated as the cause of cell death (Cheung and Slack, 2004). Thus, we measured neuronal survival in the embryonic E14 spinal cord neurons by immunostaining the sections with anti-caspase-3 antibody (Fig. 3E). In addition, we performed TUNEL (Terminal deoxynucleotidyl Transferase Biotin-dUTP Nick End Labeling) assay on sections of E16.5 cortical plate of Cdk5-null and wild-type mice (Fig. 3F). These results demonstrate that there is an extensive apoptosis in Cdk$^{-/-}$ brains compared to the wild-type brains. On the basis of these studies, we proposed that Cdk5 modulates MAPK (MEK/Erk1/2) pathway and is involved in neuronal survival.

Cdk5/p35 seemed to act as a "molecular switch" to modulate the duration of Erk1/2 activation in nerve growth factor (NGF)-stimulated PC12 cells (Harada et al., 2001; Sharma et al., 2002). Transient activation of Erk1/2 phosphorylation (1–2 h) is essential for neurite outgrowth and differentiation (Harada et al., 2001). Transient decline of Erk1/2 activity in NGF-treated PC12 cells after 1 h coincides with the observed increase in p35 level and Cdk5 activity, which "turns off" Erk1/2 kinase activity by phosphorylating and inactivating MEK1 as neurons differentiate (Sharma et al., 2002). If, as we have suggested, Cdk5 activity acts in neurons to temporally modulate the activated MAP kinase pathway, then inhibition of Cdk5 should deregulate Erk1/2 activity and affect neuronal survival. Although proliferating PC12 cells express Cdk5, its activity is low and only after NGF stimulation is p35 synthesized, activating Cdk5 which remains active during neurite outgrowth (Harada et al., 2001). NGF-stimulated transient upregulation of Erk1/2 activity begins to decline at 2 h as Cdk5 activity increases. We suggested that the decline in Erk activity reflected cross talk by activated Cdk5 (Sharma et al., 2002). To explore the role of Cdk5/p35 in neurons as a modulator of the MAP kinase cascade, we chose to use E18 rat cortical neurons cultured for 7 days. In contrast to PC12 cells, these cortical neurons express high endogenous levels of active Cdk5/p35. In order to study the kinetics of activation of the MAPK pathway, we exposed cortical neurons to NGF and sampled cells at intervals up to 12 h. Figure 4A summarizes these experimental results. Rapid activation of Erk1/2 phosphorylation reaching a peak at 15 min occurs, followed by a gradual decline to baseline levels at 1–12 h upon NGF stimulation. Does this kinetic pattern in cortical neurons reflect changes in Cdk5 activity similar to those seen in PC12 cells? To explore this possibility, these same lysates were used for Cdk5 immunoprecipitation and kinase assays. In contrast to PC12 cells, there was no significant change in the expression of Cdk5, nor p35, while Cdk5 kinase activity was constant and robust throughout the time period (Zheng et al., 2007). Does this mean that the initial rise and subsequent decline in Erk1/2 phosphorylation after 15 min is unrelated to Cdk5 activity? Can we assume that in these postmitotic neurons there is no cross talk regulation of the MAP kinase pathway by Cdk5? To answer these questions, we first treated cortical neurons over the same time

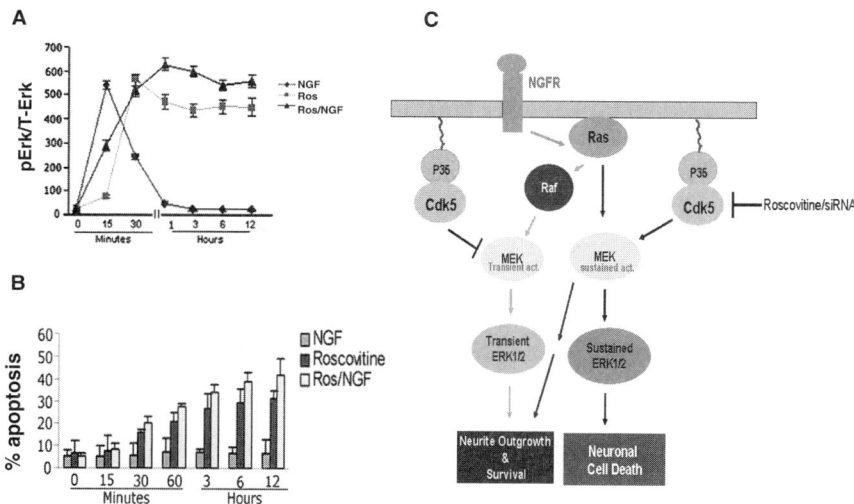

Fig. 4 (**A**) Line graph shows the corresponding quantification of phospho-Erk1/2 levels in the three different treatment groups. Data represent mean ±SEM of three experiments. Roscovitine (Ros) inhibition of Cdk5 activity induced sustained activation of Erk1/2. (**B**) Bar graph shows the time course of apoptosis as assayed by the TUNEL procedure. E18 rat embryonic cortical neurons were cultured for 7 days in B27/neurobasal medium and then treated with 50 ng/ml NGF, 20 μM roscovitine, or both for 12 h. Cells were fixed at different times and stained for TUNEL assay. The percentage of TUNEL-positive cells gradually increased from 1 to 12 h after roscovitine or roscovitine plus NGF treatment. Data represent mean ±SEM of three experiments. (**C**) A model of Cdk5/p35 cross talk modulation of the MAPK cascade to ensure neuronal survival. In normal-functioning neurons, the left side of the figure shows how Cdk5/p35 inhibits MEK1 activity to terminate the MAPK transient activation response after a short duration, thereby ensuring survival. On the right side is shown the induction of sustained activation of Erk1/2 and apoptosis by roscovitine inhibition of Cdk5 activity that supports the modulating role of Cdk5/p35 shown on the left

period with roscovitine, an inhibitor of Cdk5, to study its effect on Erk1/2 phosphorylation. At the dose of roscovitine used (20 μM), Cdk5 activity was completely inhibited after 30 min. In non-treated cortical neurons, baseline levels of Erk1/2 activity were low, but within 15 min of treatment, Erk1/2 phosphorylation increased more slowly than after NGF treatment, then rising rapidly to a peak at 30 min, and sustaining this level up to 12 h. In the absence of Cdk5 activity, the activation of Erk1/2 is delayed, as if released from an inhibition, and then sustained at high levels (Fig. 4A). It appears that in this system, inhibition of Cdk5 activity not only delays the Erk1/2 activation but also sustains it. We suggest that in these cortical neurons, there is a balanced cross talk interaction between Cdk5 and the endogenous MAP kinase activity. Abolishing Cdk5 activity induces sustained activation of Erk1/2, suggesting a Cdk5 feedback regulation of MAPK signaling cascade in neurons. The kinetics of Erk1/2 phosphorylation were similar to that observed after roscovitine treatment alone, sustained at a slightly higher level. It appears that the time course of Erk1/2 phosphorylation, when Cdk5 is inactive, is unaffected by NGF stimulation. A gradual decline does not occur and, instead, Erk1/2 phosphorylation is sustained at high levels. The data from all three experimental situations are quantified in Fig. 4A (Zheng et al., 2007). These results suggest that endogenous Cdk5 in cortical neurons acts to modulate the activity of the MAP kinase pathway.

Inhibition of Cdk5 Activity Induces Neuronal Apoptosis by Increased and Sustained Erk1/2 Activation

The above studies show that the inhibition of Cdk5 activity results in increased and sustained Erk1/2 activation in cortical neurons. If Erk1/2 activation is sustained over a much longer period (6–24 h), as in stressed or injured neurons, this chronic activation is responsible for cell death (Cheung and Slack, 2007). Therefore, we asked whether sustained Erk1/2 activation by inhibition of Cdk5 causes cortical neuron apoptosis. To answer this question, we chose to study the kinetics of apoptosis after NGF and/or roscovitine induction of Erk activation using the same experimental paradigm as before. TUNEL assays were performed on samples of cells fixed at different times after treatment. A total of 500 cells were counted at each time point, and the percentages of TUNEL-positive cells were determined (Fig. 4B). Baseline levels of apoptosis (5%) were recorded in cells treated only with NGF. Roscovitine-treated cells, with or without NGF, however, exhibited a detectable increment of apoptosis beginning at 30 min after the peak of Erk1/2 kinase activity had been reached and sustained. The level of apoptosis gradually increased to 30–40% after 12 h of sustained activity and continued through 24 h reaching almost 50% apoptosis. To further confirm the long-term apoptotic effect of sustained Erk1/2 activity, we repeated the experiment with roscovitine and followed Erk1/2 activity and cleaved caspase-3 expression (an apoptotic marker) for 48 h. Cortical neurons deficient in Cdk5 activity began to show signs of apoptosis early on, after 1 h, as soon as peak levels of sustained Erk activity were attained. We proposed that in the absence of Cdk5 activity, apoptosis is induced by sustained Erk1/2 activity. Apoptosis induced by Cdk5 inhibition was rescued by inhibition of MEK, the upstream activator of Erk (Zheng et al., 2007). Similar results were obtained by Cdk5 knockdown achieved by infecting neurons with Cdk5 siRNA. In addition, inhibition of Cdk5 deregulated and aberrantly hyperphosphorylated Tau and NF-M/H in the neuronal cell body, thereby inducing apoptosis (Zheng et al., 2007). Figure 4C illustrates and describes our model of the role of Cdk5 in modulating the expression of the MAP kinase pathway in cortical neurons.

Cdk5-JNK3 Cross Talk

The c-Jun NH2-terminal kinase (JNK) family of protein kinases, also known as stress-activated protein kinases (SAPK), are pro-apoptotic signaling molecules that phosphorylate serine residues 63 and 73 of the transcription factor c-Jun in the activation domain, leading to increased AP-1 transcriptional activity and apoptosis (Derijard et al., 1994; Kyriakis et al., 1994). In addition to c-Jun, JNK also phosphorylates ATF2 and other Jun family proteins that function as components of the AP-1 transcription factor complex (Gupta et al., 1995, 1996). Although JNK1 and JNK2 are widely expressed in murine tissues, including the nervous system, JNK3 is selectively expressed in the brain (Martin et al., 1996). Mice lacking JNK3 exhibit increased resistance to kainic acid–induced apoptosis in the hippocampus (Yang et al., 1997), indicating a preferential role of JNK3 in stress-induced neuronal apoptosis. Analysis of JNK1 and JNK2 KO mice has shown that JNK1 and JNK2 regulate region-specific apoptosis during early brain development (Kuan et al., 1999). Transfection studies using constitutively active and dominant-negative components of the JNK signaling pathway established that JNK activity and c-Jun phosphorylation were involved in apoptosis of NGF-deprived sympathetic neurons (Eilers et al., 1998). As shown earlier, apoptotic cell death is prevalent in brains of Cdk5-deficient mice and that Cdk5$^{-/-}$ neurons display higher sensitivity to UV irradiation (Li et al., 2002). These observations suggested that there might be a link between JNK3 and Cdk5 in regulating neuronal apoptosis, thus allowing us to assess the role played by Cdk5 in neuronal apoptosis. We propose that Cdk5 phosphorylates JNK3, which may downregulate its activity, c-Jun phosphorylation, and neuronal apoptosis (Fig. 5A).

Cdk5 Phosphorylates JNK3

One putative Cdk5 phosphorylation motif (T/SPXK) is present in JNK1, JNK2, and JNK3 (Thr93 in rat JNK1 and JNK2; Thr131 in rat JNK3) (Kyriakis et al., 1994). Analysis of the phosphorylated residues using various peptides and proteins revealed that XS/TPXK/R motifs (X = neutral/basic residues) are the most suitable substrates for Cdk5 (Shetty et al., 1993). These studies suggest that FTPQK, the conserved motif in JNK1, JNK2 and JNK3, is an ideal substrate for Cdk5/p35 (Fig. 5B). Indeed, it was found that Cdk5 could phosphorylate JNK3 on Thr131.

   To establish whether JNK3 protein is a potential substrate for Cdk5 phosphorylation *in vitro*, HEK293T cells co-transfected with the wild-type or the kinase-inactive mutant Cdk5 (K33T) were assessed for the status of phosphorylation of the recombinant JNK3. Cdk5 immunoprecipitates (IP) from cells transfected with active Cdk5/p35 phosphorylated histone H1 and recombinant JNK3, whereas the IP from cells transfected with vector or

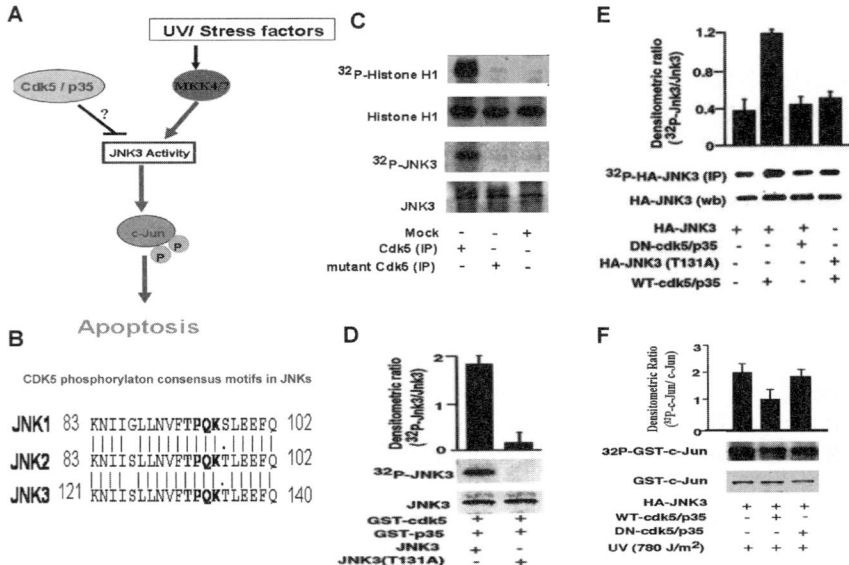

Fig. 5 (**A**) A model for the neuronal survival by Cdk5 cross talk. (**B**) The Cdk5 phosphorylation motif TPQK is indicated in JNKs. (**C**) HEK293T cells were transfected with Cdk/p35,
mutant Cdk5 (K33T)/p35, and lysates were immunoprecipitated using the anti-Cdk5 antibody (C-8) and placed in an *in vitro* kinase reaction with histone H1 or recombinant JNK3 as
substrates. *Upper panel* shows incorporation of ^{32}P into the substrates; *bottom panel* shows the
amount of substrate by Coomassie Blue staining of the gel. (**D**) Recombinant wild-type JNK3
or mutant JNK3 (T131A) were purified and equal amounts placed in an *in vitro* kinase
reaction with GST–Cdk5 and GST–p35. *Upper panel* shows incorporation of ^{32}P into
JNK3; *bottom panel* shows the amount of substrate by Coomassie Blue staining of the gel.
The histogram ($n = 3$) reflects the relative amount of labeled JNK3 to the amount of JNK3
(Coomassie Blue) in the *in vitro* kinase reaction. (**E**) ^{32}P-labeled wild-type or threonine mutant
(T131A) JNK3 were affinity purified from transfected HEK293T cells and subjected to
autoradiography (*top*) or Western blotting (*bottom*). The histogram shows the relative
amount of labeled protein to the amount of immunoreactive JNK3 in the gel. (**F**) Cdk5
inhibits UV-irradiation-induced JNK3 activation. HEK293T cells were transfected with
plasmids harboring HA-JNK3, wild-type or mutant Cdk5 (K33T), and p35, with or without
UV irradiation. JNK3 was immunoprecipitated using anti-HA antibody and subjected to an
in vitro kinase reaction with GST–c-Jun as the substrate; *top*, incorporation of ^{32}P into
GST–c-Jun; *bottom*, amount of GST–c-Jun by Coomassie Blue staining of the gel. The histogram ($n = 3$) reflects the amount of labeled GST–c-Jun relative to the mass of GST–c-Jun
(Coomassie Blue) in the *in vitro* kinase reaction

dominant-negative mutant Cdk5 did not (Fig. 5C). To investigate the putative
Cdk5 phosphorylation site in JNK3, Thr131 in JNK3 was mutated to Ala
(T131A). The ability of Cdk5 to phosphorylate wild-type and mutant HA-
JNK3 (T131A) was examined by an *in vitro* kinase assay with GST-Cdk5/p35
(Fig. 5D). Wild-type JNK3 was phosphorylated but not the mutant. More-
over, these data were confirmed in intact cells. HEK293T cells were co-transfected

with HA-JNK3 and wild-type Cdk5/p35 or dominant-negative Cdk5/p35 and then metabolically labeled with [^{32}P]-orthophosphate. The level of phosphorylated HA-JNK3 was measured by immunoprecipitation using anti-HA-tag antibody, and its phosphorylation state was determined. Co-expression of wild-type Cdk5/p35 and HA-JNK3 resulted in a 3-fold enhancement of HA-JNK3 phosphorylation compared with co-expression of mutant Cdk5/p35 and HA-JNK3 (Fig. 5E). In addition, mutation of Thr131 to Ala in JNK3 significantly reduced Cdk5-dependent phosphorylation of JNK3 compared with the wild-type protein.

Cdk5 Phosphorylation of JNK3 Inhibits c-Jun Phosphorylation

To examine the functional significance of the Cdk5 phosphorylation site in JNK3, HEK293T cells were co-transfected with wild-type Cdk5/p35 or inactive mutant Cdk5/p35 and HA-JNK3 and exposed to UV irradiation (780 J/m2). Phosphorylation of GST–c-Jun was significantly inhibited by co-expression of wild-type Cdk5/p35, but not by the inactive mutant Cdk5/p35 (Fig. 5F). To further demonstrate the *in vivo* effect of Cdk5 on JNK3 activity and JNK3 kinase activity on c-Jun phosphorylation, their activities were measured in the wild-type and Cdk5$^{-/-}$ mice brains. First, wild-type and Cdk5$^{-/-}$ mice exhibited a similar level of *c-fos* and *c-Jun* mRNA expression (Fig. 6A). JNK3 protein expression and c-Jun protein phosphorylation levels by Western blot analysis with anti-JNK3 and anti-phospho-c-Jun (Ser-63) antibodies in the cortex of wild-type and Cdk5$^{-/-}$ mice are shown in Fig. 6B. Additionally, phosphorylated c-Jun was confirmed by immunocytochemical analysis with anti-phospho-c-Jun (Ser63) antibody (Fig. 6D,E). Expression levels of JNK3 and c-Jun proteins were no different in either wild-type or Cdk5$^{-/-}$ mice (Fig. 6B). However, there was a more than 3-fold increase in JNK3 activity (Fig. 6C) and a robust c-Jun phosphorylation in Cdk5$^{-/-}$ brain extracts compared with that of the wild-type mice (Fig. 6B). In addition, cortical neuronal cultures derived from Cdk5$^{-/-}$ mice showed a 2-fold higher activity of JNK3 compared with the wild-type mice. We further tested whether Cdk5-deficient cells exhibited increased sensitivity to apoptotic stimuli. The rate of cell death upon UV irradiation in Cdk5$^{-/-}$ and wild-type neurons showed higher apoptosis (>70%) compared to the wild-type cells following 24 h post-UV irradiation (780 J/m^2) (Fig. 6F). The increased sensitivity of Cdk5$^{-/-}$ cells to apoptotic stimuli suggests that Cdk5 activity plays an important role in neuronal cell survival by regulating the JNK3 activity (Li et al., 2003). The evidence presented here identifies a crucial regulator of cell apoptosis, JNK3, as a target of Cdk5 activity. Cdk5 with its activator p35 regulates phosphorylation and activity of JNK3 on Thr131 in neurons, and regulates c-Jun activity and survival (Fig. 6G).

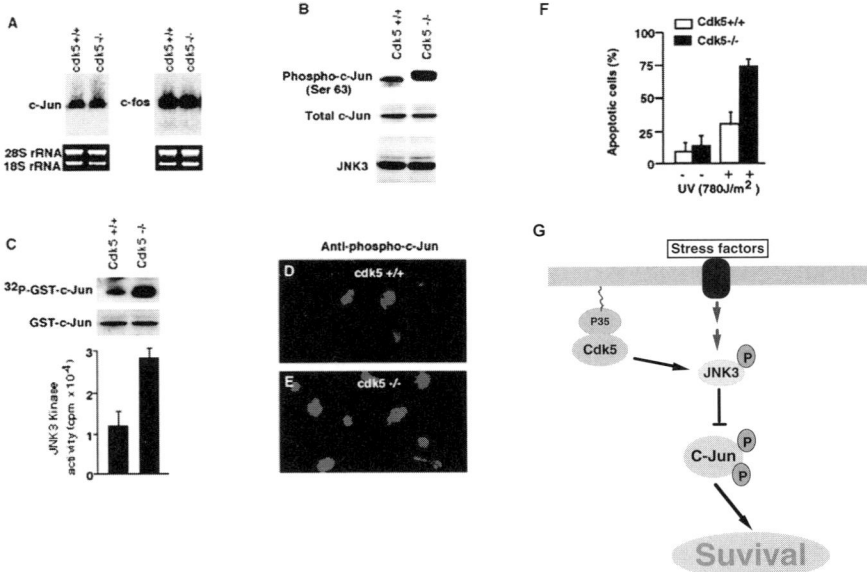

Fig. 6 (A) Total RNA was isolated from wild-type and Cdk5$^{-/-}$ mouse embryo brains (E16) and probed with murine c-Fos and c-Jun cDNA probes. Ethidium bromide staining of 18S and 28S ribosomal RNA monitors RNA loading and transfer efficiency. (B) Western blot analyses of total JNK3, c-Jun protein expression, and phospho-c-Jun in Cdk5$^{-/-}$ and wild-type mouse brain extracts. (C) Comparison of JNK3 kinase activity in wild-type and Cdk5$^{-/-}$ mice brain lysates. JNK3 was immunoprecipitated using anti-JNK3 antibody from wild-type and Cdk5$^{-/-}$ mice (E16) brain and subjected to an *in vitro* kinase assay with GST–c-Jun as a substrate. *Top*, incorporation of ^{32}P into GST–c-Jun; *bottom*, amount of GST–c-Jun by Coomassie Blue staining of the gel. The histogram ($n = 3$) reflects the relative amount of labeled GST–c-Jun to the mass of GST–c-Jun (Coomassie Blue). (D–E) Immunocytochemical analysis of phospho-c-Jun in Cdk5$^{-/-}$ and wild-type mice, using the anti-phospho-c-Jun antibody. Expression of phosphorylated c-Jun at Ser63 is greatly increased in the cortex of Cdk5$^{-/-}$ mice (E) compared with the wild-type mice (D). (F) Cdk5-deficient neurons are more sensitive to apoptotic stimuli. Apoptotic cells are normalized. Apoptosis was assessed by TUNEL staining. *Open bars* correspond to wild type; *solid bars* correspond to Cdk5$^{-/-}$ cells. The data are presented from three independent experiments. Error bars represent the standard error of the mean expressed as a percentage. (G) A model for the neuronal survival by Cdk5-JNK3 and c-Jun cross talk

Cdk5-Receptor Tyrosine Kinase (Erb2/3) Cross Talk

A clue to other possible sites of Cdk5 activity in regulating neuronal survival is suggested by the observation that Cdk5 activity, mediated by the neuregulin/ErbB pathway, upregulates acetylcholine receptor expression at the neuromuscular junction in embryonic myotubes and adult muscle (Fu et al., 2001). Neuregulin proteins mediate their action through the ErbB family of receptor tyrosine kinases, including ErbB2, ErbB3, and ErbB4 (Burden and Yarden,

1997; Gassmann and Lemke, 1997). NRG receptors (NRGRs) differ in kinase activity and substrate selectivity. Each ErbB protein has an extracellular ligand-binding domain, a single transmembrane domain, a short intracellular juxta-membrane region, a tyrosine kinase domain, and a proline-rich carboxylterminal tail (Burden and Yarden, 1997). NRG-1 binds to ErbB3 and induces the formation of heterodimers between ErbB2 and ErbB3 and thus activates the receptor (Burden and Yarden, 1997). Neuregulins have been implicated in a number of events in cell survival, mitosis, migration, and differentiation (Shah et al., 1994; Dong et al., 1995; Anton et al., 1997; Rio et al., 1997). The survival of Schwann cells, for example, is mediated by neuregulin signaling through the phosphoino-sitide-3-kinase (PI3 K)/Akt pathway, a critical survival pathway in neurons and in most cell types (Datta et al., 1997). Accordingly, it seems possible that an alternative site for Cdk5 regulation of neuronal cell survival might be the neuregulin-mediated PI3 K/Akt signaling pathway. Phospholipid kinases or the phosphoinositide kinases are responsible for the phospholipid phosphoryla-tion. Among three general families, PI3 K selectively phosphorylates phosphati-dylinositol 4,5-diphosphate [PtdIns(4,5)P2] and produces phosphatidylinositol 3,4,5-trisphosphate [PtdIns(3,4,5)P3] upon stimulation by a variety of ligands *in vivo* as well as *in vitro* (Stephens et al., 1993). PtdIns(3,4,5)P3 is a key molecule involved in cell growth and survival signaling (Stambolic et al., 1998). To explore this possibility, we took advantage of the Cdk5-null mice, which exhibit no Cdk5 activity (Ohshima et al., 1996) (Fig. 7A) and could potentially affect PI3 K/Akt activities. The analysis of endogenously phosphorylated proteins and lipids from Cdk5$^{-/-}$ brain showed a marked reduction in phosphorylation of low molecular weight components compared to the wild-type mice (Fig. 7B,C). This observa-tion led to the analysis of kinases involved in lipid metabolism. The PI3 K activity was 3-fold reduced in the brains of Cdk5$^{-/-}$ mice compared to the wild-type mice (Fig. 7D), suggesting that Cdk 5 regulates the PI3 K pathway. Further, Akt expression and PI3 K activity were analyzed, which demonstrated that expression of Akt was unaffected while PI3 K catalytic activity was significantly reduced (Fig. 7E,F). The above observations suggest that Cdk5 activity is involved in the PI3 K signaling pathway. These results indicated that Cdk5 might play a role in neuronal survival through NRG-mediated PI3 K/Akt path-way. To determine the target substrate for Cdk5, we based our analysis on the observation that the Cdk5/p35 complex is implicated in neuregulin-induced acetylcholine receptor expression at the neuromuscular junction (Fu et al., 2001). This suggested that ErbB receptors might be a substrate for Cdk5 phos-phorylation. Indeed, we found that the putative Cdk5 phosphorylation consen-sus sequence motifs are present in the ErbB2 and ErbB3 receptor molecules. Ser1176 in the sequence SPGK of ErbB2, Thr871 in the TPIK sequence, and Ser1204 in the SPPR sequence of ErbB3 are putative phosphorylation sites in Cdk5 consensus motifs. Using various proline-directed S/T wild-type and mutant peptides derived from Erb2/3 in phosphorylation assays, we provided evidence that among several S/T-P peptides, the Ser1176 in the sequence SPGK of Erb2, Thr871 in the TPIK sequence, and Ser1204 in the SPPR of Erb3 are

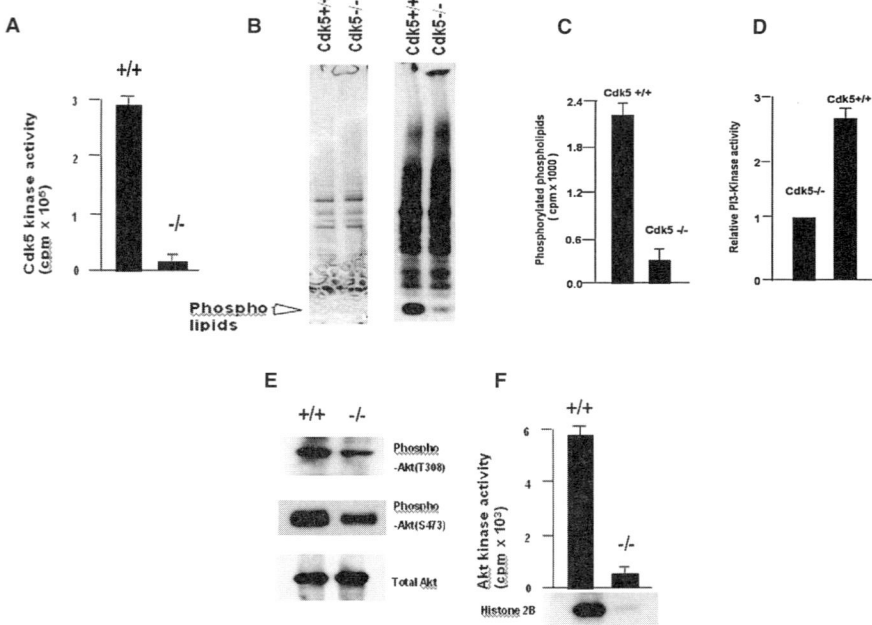

Fig. 7 (A) Cdk5$^{-/-}$ mice brain lysates show reduced Cdk5 activity. (B) Phosphorylation of
proteins and lipids extracted from wild-type and Cdk5$^{-/-}$ mice brain homogenates were
analyzed by *in vitro* phosphorylation in the presence of [γ-^{32}P]-ATP. Lipid phosphorylation
was reduced in Cdk5$^{-/-}$ mouse brain. Phospholipids were extracted from brain extracts of
wild-type and Cdk5$^{-/-}$ mice after phosphorylating the brain extracts. (C) The level of lipid
phosphorylation was greatly reduced in the Cdk5$^{-/-}$ mice brain. (D) In order to determine the
PI3 K activity, cortical tissue lysates from Cdk5$^{-/-}$ mice and wild-type mice (200 µg of total
protein) were immunoprecipitated with the anti-p85α (subunit of PI3 K) antibody. PI3 K
activity was assayed using the immunoprecipitates, which showed reduced PI3 K activity in
the Cdk5$^{-/-}$ brain lysates. (E) Western blot analysis of Akt phosphorylation using anti-
phospho-Thr^{308} and anti-phospho-Ser^{473} antibodies in the wild-type and Cdk5$^{-/-}$ mice
brain extracts; and (F) Akt activity determined by immunoprecipitating wild-type and
Cdk5$^{-/-}$ mice brain extracts with anti-Akt antibody and using histone H2B as a substrate
shows reduced Akt kinase activity in Cdk5$^{-/-}$ mice brain lysates

phosphorylated (Li et al., 2003). The expression and protein phosphorylation of
these sites were confirmed *in vivo* using the ErbB3 IP from wild-type and Cdk5$^{-/-}$
brain extracts and lysates of cultured cortical neurons (7 days) that were subjected to Western blot analyses using anti-phospho-tyrosine, anti-phospho-
threonine, and anti-phospho-serine antibodies (Fig. 8A,B). One of the key
signaling pathways involved in the regulation of neuronal survival is the
PI3 K/Akt kinase intracellular signaling transduction pathway (Kennedy et al.,
1997; Datta et al., 1999). Neuregulin-promoted neuronal survival in cortical
neurons was reduced significantly by inhibiting Cdk5 activity by its specific
inhibitor, roscovitine. Roscovitine, in the presence of neuregulin, induced

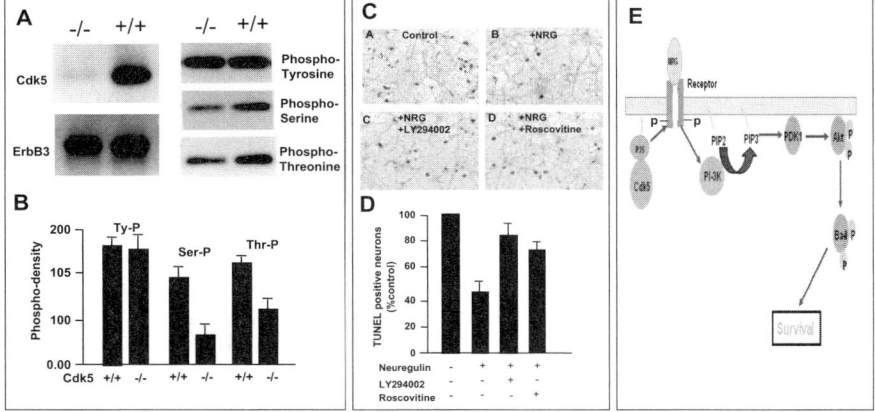

Fig. 8 Cdk5 is involved in promoting neuronal survival by neuregulin. (**A, B**) Cdk5 phosphorylation of ErbB3 on specific residues is reduced in Cdk5$^{-/-}$ brain extracts. (**C, D**) Rat cortical neurons were cultured for 5 days and switched to Dulbecco's modified Eagle's medium without any growth factors and serum-deprived for 20 h to induce apoptosis. Control (**A**); with 100 ng/ml neuregulin β_1 (**B**); with a 30-min pretreatment with the PI3 K inhibitor, LY294002 (50 μM) (**C**); with a 30-min pretreatment with the Cdk5 inhibitor, roscovitine (25 μM) (**D**). Apoptotic cells were detected by TUNEL staining. In each experiment, the number of TUNEL-positive cells was determined for each condition. The number of TUNEL-positive cells in control culture was defined as 100%. The number of TUNEL-positive cortical neurons in each experimental condition was then expressed as a percentage of control values. (**E**) A model for Cdk5 kinase in neuregulin-dependent activation of PI3 K and Akt activity mediating neuronal survival

apoptosis that was comparable to the effect of neuregulin added along with the PI3 K inhibitor (Fig. 8C,D). The above results suggest that phosphorylation of Akt induced by neuregulin is inhibited by PI3 K and Cdk5 inhibitors, implying that Cdk5 may also play an important role in neuronal survival through the regulation of Akt activity. Thus, it became apparent that Cdk5 activity is involved in neuron protection induced by neuregulin. The model presented here shows that Cdk5 phosphorylates neuregulin receptor Erb2/3 and regulates PI3 K/Akt phosphorylation and neuronal survival pathways (Fig. 8E). Our data on cortical neurons are in agreement with these observations but introduce Cdk5 as an important modulator of the pathway by virtue of its regulation of ErbB2/3 receptor phosphorylation (Fu et al., 2001).

Other Signaling Pathways Regulated by Cdk5

Cdk5 and GSK3 Cross Talk Affects Fast Axonal Transport

Extruded axoplasm derived from the giant axon of the squid has been ideal for the visualization and study of both fast and slow axonal transport (Brady et al.,

1985; Sheetz et al., 1986; Brady et al., 1990; Bearer et al., 2000), making it possible to study the molecular motors involved and their mode of action in transporting vesicles from cell body to terminals. One of these motors is kinesin, which is responsible for fast anterograde transport of vesicles and is also involved in transport of cytoskeletal proteins. In an early study on the effect of tumor suppressor proteins that bind to microtubules and interact with membrane-associated proteins, it was found that an adenomatous polyposis coli (APC) protein inhibited anterograde and retrograde movement in the squid axoplasm system (Ratner et al., 1998). It was observed that low concentrations of olomucine, which inhibits Cdk5 activity, also inhibited fast axonal transport. Subsequent studies by this same laboratory demonstrated that GSK3 phosphorylation of kinesin light chains significantly inhibited anterograde, but not retrograde, fast transport (Morfini et al., 2002). In collaboration with this laboratory, we agreed to study the role of Cdk5 in the regulation of fast transport by kinesin. What emerged was a complex interaction between Cdk5, PP1 phosphatase, and GSK3 kinase (Morfini et al., 2004). Inhibiting Cdk5 with roscovitine or olomucine activated PP1 phosphatase, which, in turn, dephosphorylated and activated GSK3. Active GSK3 then phosphorylated the kinesin light chain, releasing it from its cargo, thereby decreasing vesicle transport. Here, we see a more elaborate cross talk interaction involving Cdk5, a phosphatase and GSK3 kinase, integrated into a network regulating transport of membrane-bound vesicles and targeting them to terminals (Fig. 9A).

Cdk5-RasGRF Cross Talk, Upstream of the MAPK Pathway

RasGRF1/2 are members of the Ras guanine nucleotide exchange factor (RasGRF) family of proteins, which are directly responsible for the activation

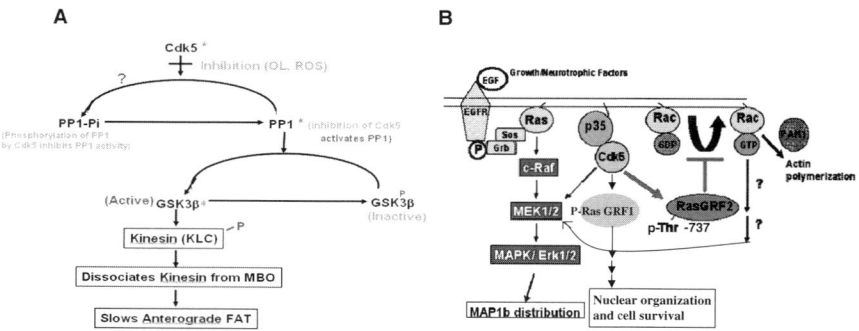

Fig. 9 (**A**) A model shows Cdk5, protein phosphatase (PP1), and GSK3 kinase (GSK3β), integrated into a network regulating fast transport of vesicles and targeting them to terminals. (**B**) A model depicts Cdk5/RasGRF cross talk mediating organization of cytoplasmic and nuclear architecture

of Ras and Rac GTPases. Although RasGRF1 is phosphorylated by protein kinase A and by the non-receptor tyrosine kinases, we have shown that RasGRF1 interacts with and is phosphorylated by Cdk5 on Ser731 to regulate its steady-state levels in neurons. Phosphorylation on this site by Cdk5 leads to RasGRF1 degradation through a calpain-dependent mechanism. Significantly, cortical neurons from Cdk5 KO mice have higher levels of RasGRF1 that were reduced after Cdk5 transfection. Overexpression of RasGRF1 in mitotic cells produces disorganized nuclei, which can be rescued when RasGRF1 is co-expressed with active Cdk5. Moreover, overexpression of wild-type Ras-GRF1, or a phosphorylation mutant of RasGRF1 together with the kinase-dead Cdk5 mutant (K33T), results in fragmented nuclei and cell death. On the other hand, a reduction in RasGRF1 levels through Cdk5/p35 overexpression also leads to nuclear condensation in neurons. These data show that tight regulation of the levels of RasGRF1 by Cdk5 phosphorylation is essential to maintain nuclear organization and cell survival (Fig. 9B). Using p35 as bait in a yeast two-hybrid system, we found p35 associates with RasGRF2. Coimmuno-precipitation and colocalization studies using transfected cell lines as well as primary cortical neurons showed close association of p35 with RasGRF2 (Kesavapany et al., 2004). Additionally, Cdk5 phosphorylates RasGRF2 both *in vitro* and *in vivo*, leading to a decrease in RacGRF activity and a subsequent reduction in Erk1/2 activity. We have shown that Cdk5/p35 phosphorylates RasGRF2 on Ser737, which leads to an accumulation of RasGRF2 in the neuronal cell bodies coinciding with an accumulation of MAP 1b. Therefore, membrane association of p35 and subsequent localization of Cdk5 upstream toward RasGRF2 and Rac can have important downstream effects on signal transduction cascades regulating neuronal differentiation and survival (Fig. 9B).

Recently, O'Hare et al. (2005) provided evidence that p35 can regulate a prosurvival activity of Cdk5 within the cytoplasm, whereas p25 regulates Cdk5 pro-death activity within the nucleus. They further show that although nuclear p25/cdk5 complexes accumulate in both excitotoxic and apoptotic death, they are functionally relevant only in the former, in which they are produced early in the death process. In apoptotic death, p25 complexes are only produced as a late consequence of the mitochondrial death pathway.

Concluding Remarks

The mechanisms underlying the positive and negative roles of Cdk5 in neuronal survival are still not clearly understood. Contrary to its reported pathological role, recent studies now implicate p25 as a "normal" player in modulating synaptic function, LTD, learning, and memory in specific brain regions in young animals (Fischer et al., 2003, 2005; Angelo et al., 2006). Transgenic mice expressing either low level of p25, or with expression restricted spatiotemporally to specific brain regions, show a Cdk5/p25-positive transient effect on

LTD in the hippocampus and water maze learning. Prolonged expression of p25, in older animals, however, does exert a predictable effect on β-amyloid and Tau phosphorylation and neurodegeneration (Cruz et al., 2003; Cruz and Tsai, 2004). Evidently, the role of Cdk5/p25 is far more complex than previously assumed.

Because of its "dual personality," the role of Cdk5 in neuronal apoptosis and survival can only be explained in a context-specific manner. It has been implicated in apoptosis in neuronal and other cell systems, even during normal development (Ahuja et al., 1997; Zhang et al., 1997; Zhu et al., 2002; Guo, 2003; Weishaupt et al., 2003; Penaloza et al., 2006). On the other hand, in a series of experiments, we have shown that Cdk5/p35 sustains neuronal survival by virtue of its cross talk interactions with survival and apoptotic signal transduction pathways. Initially, we demonstrated that Cdk5 prevents neuronal apoptosis by phosphorylation and inactivation of JNK3, a major apoptotic pathway (Li et al., 2002). At the same time we showed that Cdk 5 phosphorylation of MEK1 at a specific site, downregulates the MAP kinase pathway in PC12 cells and can account for the kinetics of NGF-induced differentiation (Sharma et al., 2002). Recently, an analysis of cortical neurons with endogenous high levels of active Cdk5 showed that Cdk5 prevented a sustained Erk1/2 activation and induced apoptosis as a result of modulating the activity of MEK1 (Zheng et al., 2007). We have also demonstrated a specific role of Cdk5 in the PI3 K/Akt survival pathway. Phosphorylation of specific sites on the ErbB2, or ErbB3 neuregulin receptors, activates both PI3 K and Akt kinases and sustains neuronal survival (Li et al., 2003). Cdk5 cross talk and their role in neuronal survival are summarized in Fig. 10. Finally, in two separate investigations, we have demonstrated that Cdk5 cross talk upstream, at the level of Ras and Rac regulation, in the MAP kinase pathway is essential for neuronal survival (Kesavapany et al., 2004, 2006). These represent only some of the upstream sites of MAP kinase regulation that modulate neuronal function, of which there are many (Grewal et al., 1999). Cdk5 cross talk also

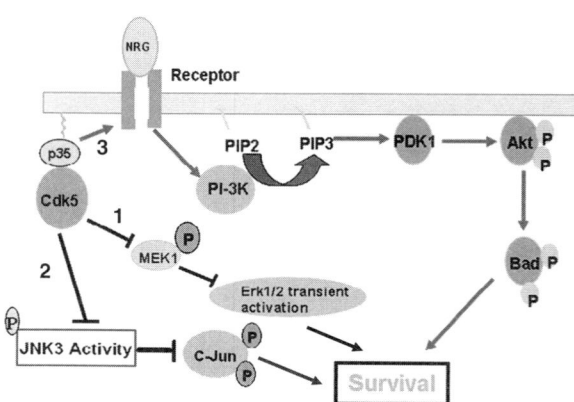

Fig. 10 A model for the neuronal survival by Cdk5 cross talk with MAPK (1), JNK3 (2), and PI3 K (3) pathways

affects axonal transport. We have also observed that Cdk5 cross talk with the MAP kinase pathway inhibits Erk1/2 activity, and anterograde transport of NFs (Moran et al., 2005). Finally, the activity of another kinase, GSK3, seems to be modulated by Cdk5 phosphorylation as it affects kinesin function during fast vesicle transport in axoplasm extruded from the squid giant axon (Morfini et al., 2004). To summarize, we have proposed that Cdk5, very much enriched and active in neuronal cells, plays a key "surveillance" role in modulating neuronal survival. Consistent with this proposed function, in a study of Cdk5 activity in zebrafish development, Cdk5 knockdown after injection of Cdk5 siRNA into a blastomere at the two-cell stage resulted in a significant reduction in Rohon-Beard and trigeminal ganglion cells and the appearance of apoptotic cells in the brain (Kanungo et al., 2006, 2007). Yet, not only are we perplexed by the lingering question of how Cdk5 deregulation modulates compartment-specific patterns of cytoskeletal protein phosphorylation and induces abnormal phosphorylated aggregates in cell bodies, but also a principal gap in our knowledge persists in explaining how deregulation of Cdk5 affects its cross talk interactions with survival and apoptotic signal transduction systems.

References

Ahuja HS, Zhu Y, Zakeri Z (1997) Association of cyclin-dependent kinase 5 and its activator p35 with apoptotic cell death. Dev Genet 21:258–267.

Angelo M, Plattner F, Giese KP (2006) Cyclin-dependent kinase 5 in synaptic plasticity, learning and memory. J Neurochem 99:353–370.

Anton ES, Marchionni MA, Lee KF, Rakic P (1997) Role of GGF/neuregulin signaling in interactions between migrating neurons and radial glia in the developing cerebral cortex. Development 124:3501–3510.

Bearer EL, Breakefield XO, Schuback D, Reese TS, LaVail JH (2000) Proc Natl Acad Sci USA 97:8146–8150.

Brady ST, Lasek RJ, Allen RD (1985) Videomicroscopy of fast axonal transport in extruded axoplasm: A new model for study of molecular mechanisms. Cell Motil 5:81–101.

Brady ST, Pfister KK, Bloom GS (1990) A monoclonal antibody against kinesin inhibits both anterograde and retrograde fast axonal transport in squid axoplasm. Proc Natl Acad Sci USA 87:1061–1065.

Burden S, Yarden Y (1997) Neuregulins and their receptors: a versatile signaling module in organogenesis and oncogenesis. Neuron 18:847–855.

Cheung EC, Slack RS (2004) Emerging role for ERK as a key regulator of neuronal apoptosis. Sci STKE 2004:PE45.

Cruz JC, Tsai LH (2004) A Jekyll and Hyde kinase: roles for Cdk5 in brain development and disease. Curr Opin Neurobiol 14:390–394.

Cruz JC, Tseng HC, Goldman JA, Shih H, Tsai LH (2003) Aberrant Cdk5 activation by p25 triggers pathological events leading to neurodegeneration and neurofibrillary tangles. Neuron 40:471–483.

Datta SR, Brunet A, Greenberg ME (1999) Cellular survival: a play in three Akts. Genes Dev 13:2905–2927.

Datta SR, Dudek H, Tao X, Masters S, Fu H, Gotoh Y, Greenberg ME (1997) Akt phosphorylation of BAD couples survival signals to the cell-intrinsic death machinery. Cell 91:231–241.

Derijard B, Hibi M, Wu IH, Barrett T, Su B, Deng T, Karin M, Davis RJ (1994) JNK1: a protein kinase stimulated by UV light and Ha-Ras that binds and phosphorylates the c-Jun activation domain. Cell 76:1025–1037.

Dong Z, Brennan A, Liu N, Yarden Y, Lefkowitz G, Mirsky R, Jessen KR (1995) Neu differentiation factor is a neuron-glia signal and regulates survival, proliferation, and maturation of rat Schwann cell precursors. Neuron 15:585–596.

Eilers A, Whitfield J, Babij C, Rubin LL, Ham J (1998) Role of the Jun kinase pathway in the regulation of c-Jun expression and apoptosis in sympathetic neurons. J Neurosci 18:1713–1724.

Fischer A, Sananbenesi F, Spiess J, Radulovic J (2003) Cdk5: a novel role in learning and memory. Neurosignals 12:200–208.

Fischer A, Sananbenesi F, Pang PT, Lu B, Tsai LH (2005) Opposing roles of transient and prolonged expression of p25 in synaptic plasticity and hippocampus-dependent memory. Neuron 48:825–838.

Fu AK, Fu WY, Cheung J, Tsim KW, Ip FC, Wang JH, Ip NY (2001) Cdk5 is involved in neuregulin-induced AChR expression at the neuromuscular junction. Nat Neurosci 4:374–381.

Gassmann M, Lemke G (1997) Neuregulins and neuregulin receptors in neural development. Curr Opin Neurobiol 7:87–92.

Grewal SS, York RD, Stork PJ (1999) Extracellular-signal-regulated kinase signalling in neurons. Curr Opin Neurobiol 9:544–553.

Guo Q (2003) Cyclin-dependent kinase 5–a neuronal killer? Sci Aging Knowledge Environ 2003:pe36.

Gupta S, Campbell D, Derijard B, Davis RJ (1995) Transcription factor ATF2 regulation by the JNK signal transduction pathway. Science 267:389–393.

Gupta S, Barrett T, Whitmarsh AJ, Cavanagh J, Sluss HK, Derijard B, Davis RJ (1996) Selective interaction of JNK protein kinase isoforms with transcription factors. Embo J 15:2760–2770.

Harada T, Morooka T, Ogawa S, Nishida E (2001) ERK induces p35, a neuron-specific activator of Cdk5, through induction of Egr1. Nat Cell Biol 3:453–459.

Kanungo J, Li BS, Zheng Y, Pant HC (2006) Cyclin-dependent kinase 5 influences Rohon-Beard neuron survival in zebrafish. J Neurochem 99:251–259.

Kanungo J, Li BS, Goswami M, Zheng YL, Ramchandran R, Pant HC (2007) Cloning and characterization of zebrafish (Danio rerio) cyclin-dependent kinase 5. Neurosci Lett 412:233–238.

Kennedy SG, Wagner AJ, Conzen SD, Jordan J, Bellacosa A, Tsichlis PN, Hay N (1997) The PI 3-kinase/Akt signaling pathway delivers an anti-apoptotic signal. Genes Dev 11:701–713.

Kesavapany S, Pareek TK, Zheng YL, Amin N, Gutkind JS, Ma W, Kulkarni AB, Grant P, Pant HC (2006) Neuronal nuclear organization is controlled by cyclin-dependent kinase 5 phosphorylation of ras guanine nucleotide releasing factor-1. Neurosignals 15:157–173.

Kesavapany S, Amin N, Zheng YL, Nijhara R, Jaffe H, Sihag R, Gutkind JS, Takahashi S, Kulkarni A, Grant P, Pant HC (2004) p35/cyclin-dependent kinase 5 phosphorylation of ras guanine nucleotide releasing factor 2 (RasGRF2) mediates Rac-dependent extracellular signal-regulated kinase 1/2 activity, altering RasGRF2 and microtubule-associated protein 1b distribution in neurons. J Neurosci 24:4421–4431.

Ko J, Humbert S, Bronson RT, Takahashi S, Kulkarni AB, Li E, Tsai LH (2001) p35 and p39 are essential for cyclin-dependent kinase 5 function during neurodevelopment. J Neurosci 21:6758–6771.

Kuan CY, Yang DD, Samanta Roy DR, Davis RJ, Rakic P, Flavell RA (1999) The Jnk1 and Jnk2 protein kinases are required for regional specific apoptosis during early brain development. Neuron 22:667–676.

Kyriakis JM, Banerjee P, Nikolakaki E, Dai T, Rubie EA, Ahmad MF, Avruch J, Woodgett JR (1994) The stress-activated protein kinase subfamily of c-Jun kinases. Nature 369:156–160.

Li BS, Zhang L, Takahashi S, Ma W, Jaffe H, Kulkarni AB, Pant HC (2002) Cyclin-dependent kinase 5 prevents neuronal apoptosis by negative regulation of c-Jun N-terminal kinase 3. Embo J 21:324–333.

Li BS, Ma W, Jaffe H, Zheng Y, Takahashi S, Zhang L, Kulkarni AB, Pant HC (2003) Cyclin-dependent kinase-5 is involved in neuregulin-dependent activation of phosphatidylinositol 3-kinase and Akt activity mediating neuronal survival. J Biol Chem 278:35702–35709.

Martin JH, Mohit AA, Miller CA (1996) Developmental expression in the mouse nervous system of the p493F12 SAP kinase. Brain Res Mol Brain Res 35:47–57.

Moran CM, Donnelly M, Ortiz D, Pant HC, Mandelkow EM, Shea TB (2005) Cdk5 inhibits anterograde axonal transport of neurofilaments but not that of tau by inhibition of mitogen-activated protein kinase activity. Brain Res Mol Brain Res 134:338–344.

Morfini G, Szebenyi G, Elluru R, Ratner N, Brady ST (2002) Glycogen synthase kinase 3 phosphorylates kinesin light chains and negatively regulates kinesin-based motility. Embo J 21:281–293.

Morfini G, Szebenyi G, Brown H, Pant HC, Pigino G, DeBoer S, Beffert U, Brady ST (2004) A novel CDK5-dependent pathway for regulating GSK3 activity and kinesin-driven motility in neurons. Embo J 23:2235–2245.

Ohshima T, Ward JM, Huh CG, Longenecker G, Veeranna, Pant HC, Brady RO, Martin LJ, Kulkarni AB (1996) Targeted disruption of the cyclin-dependent kinase 5 gene results in abnormal corticogenesis, neuronal pathology and perinatal death. Proc Natl Acad Sci USA 93:11173–11178.

O'Hare MJ, Kushwaha N, Zhang Y, Aleyasin H, Callaghan MS, Slack RS, Albert PR, Vincent I, Park DS (2005) Differential roles of nuclear and cytoplasmic cyclin-dependent kinase 5 in apoptotoc and excitotoxic neuronal death. J Neurosci 25:8954–8966.

Penaloza C, Lin L, Lockshin RA, Zakeri Z (2006) Cell death in development: shaping the embryo. Histochem Cell Biol 126:149–158.

Ratner N, Bloom GS, Brady ST (1998) A role for Cdk5 kinase in fast anterograde axonal transport: novel effects of olomoucine and the APC tumor suppressor protein. J Neurosci 18:7717–7726.

Rio C, Rieff HI, Qi P, Khurana TS, Corfas G (1997) Neuregulin and erbB receptors play a critical role in neuronal migration. Neuron 19:39–50.

Shah NM, Marchionni MA, Isaacs I, Stroobant P, Anderson DJ (1994) Glial growth factor restricts mammalian neural crest stem cells to a glial fate. Cell 77:349–360.

Sharma P, Barchi Jr. JJ, Huang X, Amin ND, Jaffe H, Pant HC (1998) Site-specific phosphorylation of Lys-Ser-Pro repeat peptides from neurofilament H by cyclin-dependent kinase 5: structural basis for substrate recognition. Biochemistry 37:4759–4766.

Sharma P, Veeranna, Sharma M, Amin ND, Sihag RK, Grant P, Ahn N, Kulkarni AB, Pant HC (2002) Phosphorylation of MEK1 by cdk5/p35 down-regulates the mitogen-activated protein kinase pathway. J Biol Chem 277:528–534.

Sheetz MP, Vale R, Schnapp B, Schroer T, Reese TS (1986) Vesicle movements and microtubule-based motors. J Cell Sci 5:181–188

Shetty KT, Link WT, Pant HC (1993) cdc2-like kinase from rat spinal cord specifically phosphorylates KSPXK motifs in neurofilament proteins: isolation and characterization. Proc Natl Acad Sci USA 90:6844–6848.

Stambolic V, Suzuki A, de la Pompa JL, Brothers GM, Mirtsos C, Sasaki T, Ruland J, Penninger JM, Siderovski DP, Mak TW (1998) Negative regulation of PKB/Akt-dependent cell survival by the tumor suppressor PTEN. Cell 95:29–39.

Stephens L, Eguinoa A, Corey S, Jackson T, Hawkins PT (1993) Receptor stimulated accumulation of phosphatidylinositol (3,4,5)-trisphosphate by G-protein mediated pathways in human myeloid derived cells. Embo J 12:2265–2273.

Takahashi S, Saito T, Hisanaga S, Pant HC, Kulkarni AB (2003) Tau phosphorylation by cyclin-dependent kinase 5/p39 during brain development reduces its affinity for microtubules. J Biol Chem 278:10506–10515.

Veeranna, Amin ND, Ahn NG, Jaffe H, Winters CA, Grant P, Pant HC (1998) Mitogen-activated protein kinases (Erk1,2) phosphorylate Lys-Ser-Pro (KSP) repeats in neurofilament proteins NF-H and NF-M. J Neurosci 18:4008–4021.

Weishaupt JH, Neusch C, Bahr M (2003) Cyclin-dependent kinase 5 (CDK5) and neuronal cell death. Cell Tissue Res 312:1–8.

Yang DD, Kuan CY, Whitmarsh AJ, Rincon M, Zheng TS, Davis RJ, Rakic P, Flavell RA (1997) Absence of excitotoxicity-induced apoptosis in the hippocampus of mice lacking the Jnk3 gene. Nature 389:865–870.

Zhang Q, Ahuja HS, Zakeri ZF, Wolgemuth DJ (1997) Cyclin-dependent kinase 5 is associated with apoptotic cell death during development and tissue remodeling. Dev Biol 183:222–233.

Zheng YL, Li BS, Kanungo J, Kesavapany S, Amin N, Grant P, Pant HC (2007) Cdk5 Modulation of mitogen-activated protein kinase signaling regulates neuronal survival. Mol Biol Cell 18:404–413.

Zhu Y, Lin L, Kim S, Quaglino D, Lockshin RA, Zakeri Z (2002) Cyclin dependent kinase 5 and its interacting proteins in cell death induced in vivo by cyclophosphamide in developing mouse embryos. Cell Death Differ 9:421–430.

CDK5 and Mitochondrial Cell Death Pathways

Katrin Meuer, Mathias Bähr, and Jochen H. Weishaupt

Abstract Programmed cell death (PCD) or apoptosis is regarded as an evolutionary conserved program that physiologically controls cell number and morphology of multicellular organisms during development. The same apoptotic machinery can be reactivated in the adult organism under disease conditions. Mitochondria play a central role in many apoptotic cascades, both as a storage place for pro-apoptotic molecules and for the production of ATP that is required for apoptosis. Moreover, mitochondria can acquire a broad range of morphologies, and mitochondrial fission has been shown to be a necessary step in early apoptosis. In recent years, abundant evidence accumulated that cyclin-dependent kinase 5 (CDK5) is an important death-promoting kinase in neurodegeneration. However, it remained unclear for a long time where to integrate CDK5 into the established apoptosis pathways. In this book chapter, we review the involvement of CDK5 in mitochondrial cell death pathways, with special emphasis on the aspect of mitochondrial morphology during neuronal apoptosis.

Programmed Cell Death

Programmed cell death (PCD) is regarded as an evolutionary conserved program that physiologically controls cell number and morphology of multicellular organisms during development. PCD is a general term that comprises several distinct types of cell death, which are distinguished by morphological criteria, different cell death triggers, and the signal transduction pathways involved [23]. Much of our knowledge about apoptosis, that is best-characterized type I PCD also referred to as nuclear PCD [4], comes from studies of the immune system, where apoptosis plays a crucial role in the regulation of immune cell maturation and immune homoeostasis. Moreover, homologues of the molecular players of apoptosis in vertebrates can already be found in simple single-cell life like yeast.

J.H. Weishaupt
University of Göttingen, Department of Neurology, Robert-Koch-Str. 40,
37075 Göttingen, Germany
e-mail: jweisha@gwdg.de

N.Y. Ip, L.-H. Tsai (eds.), *Cyclin Dependent Kinase 5 (Cdk5)*,
DOI: 10.1007/978-0-387-78887-6_7, © Springer Science+Business Media, LLC 2008

Studies on *C. elegans* provided insights into the basic molecular mechanisms of apoptotic cell death also relevant for vertebrates, including the family of bcl-2 proteins or a group of cysteinyl aspartate–specific proteinases called caspases [55].

However, it became evident that the same apoptotic machinery, which has crucial functions during development, can be reactivated in the adult organism under disease conditions, depending on several possible upstream events that trigger the death cascade. Many of the main proteins driving apoptotic cascades are even constitutively stored in healthy, mature cells, and ready to become active or subcellularly redistributed when the balance between pro- and anti-survival signaling mechanisms is disturbed. Generally, apoptosis is regarded as an active, energy-requiring process. In tissue areas where energy supply is abrogated (e.g., in cerebral ischemia), cells die by another type of cell death, by necrosis [36]. Therefore, mitochondria play a central role in many apoptotic cascades, both as a storage place for pro-apoptotic molecules and for the production of ATP that is required for apoptosis.

Apoptosis in Neuronal Cells

Although synaptic degeneration, Wallerian degeneration, and aspects of neuronal or glial dysfunction have emerged as important contributors to pathology and functional impairment observed in neurodegenerative diseases, the definite apoptotic loss of neurons remains an important aspect and a potential therapeutic target.

Apoptosis and necrosis most likely represent two ends of a wide-ranged continuum of possible ways a cell can die. Consistent with this view, detection of classical signs of apoptosis remained controversial in several rodent *in vivo* models for neurodegenerative diseases, for example amyotrophic lateral sclerosis (ALS), Parkinson's disease, or Huntington's disease, and also in respective human post-mortem brain samples. However, the time window necessary for the completion of apoptosis has been estimated to comprise hours to days, once the cell death cascade has been fully activated. Thus, at a given time point, only a small subset of cells can be expected to be in the apoptotic execution phase. Moreover, a lot of experimental evidence strongly supports the contribution of apoptotic cell death to the pathology of most neurodegenerative diseases. For example, caspase inhibition or overexpression of the anti-apoptotic bcl-2 protein effectively prevented neuronal cell death and disease symptoms in respective cell culture models or in *in vivo* paradigms [22,27].

Despite largely converging downstream events, distinct upstream apoptosis pathways are described according to the different ways of apoptosis initiation and cell types. One classical apoptosis pathway is referred to as the so-called extrinsic pathway, which requires death receptor activation (e.g. of Fas or the tumor necrosis factor-α (TNF-α) receptor) for instigating the death cascade. This pathway classically circumvents the mitochondrial step in apoptosis, that is cytochrome c release and formation of the "apoptosome" (consisting of cyto-chrome c, caspase-9, and the Apaf-1 protein). Subsequent to binding of FasL or TNF-α to their receptor, recruitment and autocatalytic cleavage of caspase-8 is a crucial step in the extrinsic cascade, followed by activation of downstream effector caspases 3 and 7 [34]. However, the extrinsic apoptosis pathway has not been established as a major contributor to neuronal cell death [11,43]. In contrast to the extrinsic apoptosis pathway, features of the so-called intrinsic PCD cascade are more often found in neurons. This pathway is not triggered by death receptor activation, but starts at the level of mitochondria. Both death-promoting and anti-apoptotic bcl-2 family members act upon these organelles, and mitochondria are the integration points sensing the balance between survival and death signaling. Thus, mitochondrial function and dysfunction define the "apoptotic state" of neurons at risk to undergo apoptosis. As outlined below, CDK5 turned out to be a key player acting on this cell death decision centers.

Mitochondrial Apoptosis Pathways and CDK5 Toxicity

In recent years, abundant evidence accumulated that CDK5 is an important death-promoting kinase in neurodegeneration. Increased CDK5 activity has been measured in various models for neurodegenerative diseases. The best estab-lished and classical way of CDK5 deregulation is the formation of p25/p29 by calpain-mediated cleavage of the CDK5 activators p35/p39. [40]. Similarly, p39, in analogy to p35 and p25, is cleaved to p29 [41]. As a consequence, CDK5 bound to p25 has a different subcellular distribution than CDK5/p35, with higher amounts of CDK5 found in peri- and intranuclear regions. As the subcellular distribution is one determinant of substrate specificity, misallocation of CDK5 due to proteolysis of p35 and p39 may lead to phosphorylation of "pathological" substrates by CDK5. Also the balance between cytoplasmic and nuclear CDK5 activity could be crucial for neuronal maintenance and death [37].

Despite overwhelming evidence supporting that CDK5 is an important player in neurodegeneration [10,53], it remained unclear for a long time where to integrate CDK5 into established apoptosis pathways. Initially, CDK5 was con-sidered to be the enzymatic activity responsible for abnormal phosphorylation of Tau or neurofilaments in neurodegenerative diseases [10]. Hyperphosphorylated Tau binds to microtubules less well, resulting in microtubule destabilization [29]. Neurofilaments have also been suggested to influence dynamics of microtubules and actin filaments. Therefore, it is tempting to speculate that p25/CDK5 leads to cytoskeletal disruption followed by the induction of PCD. However, whether these substrates are really the death-mediating downstream targets of p25/CDK5 has not been proven yet. The first substrate that was shown to be relevant for CDK5 toxicity *in vivo* was the transcription factor myocyte-specific enhancer factor 2 (MEF2). Gong et al. [17] first described nuclear translocation of CDK5 and a nuclear pathway leading to inactivation and degradation of this protective transcription factor upon phosphorylation by nuclear CDK5. However, nuclear CDK5 toxicity is probably only one part of the story. p35, which can also be neurotoxic in some paradigms, is primarily localized in the cytoplasm, and also p25 translocates only partially to the nucleus, albeit to a greater extend than p35. Moreover, a number of CDK5 substrates and thus candidates for mediators of toxicity, for example cytoskeletal elements, have been identified that are CDK5 substrates in disease conditions.

As outlined in the general description of neuronal apoptosis pathways, mitochondria are critically involved in the initiation of the intrinsic apoptosis pathway [42]. Thus, mitochondria are a candidate for one of the direct or indirect cytoplasmic targets of CDK5. In fact, CDK5 inhibition was shown to prevent mitochondrial dysfunction and cytochrome c release. Application of the highly potent CDK5 inhibitor indolinone A, which displays an IC_{50} for CDK5 in the low nanomolar range, abolished staurosporine-induced cytochrome c release in primary cortical cultures [53]. In a further set of experiment, another indolinone derivative, indolinone D, was used. This compound has a higher IC_{50} for CDK5, but, to our knowledge, is the only small-molecule kinase inhibitor with selectivity for CDK5, lacking cytostatic effects and significant inhibitory activity against other members of the CDK family. In agreement with a protective action on mitochondria, application of indolinone D prevented the staurosporine-induced decline of the mitochondrial transmembrane potential [53]. Nevertheless, there is only scarce knowledge about the links between CDK5 deregulation and the specific mitochondrial parameters that are modified by deregulated CDK5. CDK5 could influence or instigate the apoptotic process at several stages of the apoptotic mitochondrial decline, ranging from an early impact on upstream signaling pathways that converge on mitochondria to later events, for example cytochrome c release.

In a recent study it was shown that CDK5 inhibition attenuated the apoptotic rise in mitochondrial calcium levels during ceramid-induced apoptosis. In this cell death paradigm, calcium is transferred from the ER to mitochondria, by a mechanism that requires t-Bid. One requirement for ER–mitochondria calcium transfer is a close proximity of the two organelles. It was found that CDK5 deregulation induced the clustering of both mitochondria and ER around the centrosomes, thereby reducing ER–mitochondria distance and facilitating calcium transfer [7]. The CDK5-dependent subcellular redistribution of organelles was dependent on Tau phosphorylation, as phosphorylation-site mutant of Tau that could not be phosphorylated by CDK5 any more prevented ER and mitochondria clustering, respective calcium transfer, as well as neuronal cell death [7].

Similarly, in other situations of apoptotic cell death also, subcellular mitochondrial redistribution, specifically perinuclear clustering, has been observed, which is possibly induced by CDK5 overactivation, too, and may represent a more general mechanism how CDK5 contributes to cell death.

Similar to the mitochondrial clustering described by Darios et al. [7], mitochondrial division represents a very early change in the mitochondrial network, which is again tightly connected to cytoskeletal functions. In the following section, we will describe the molecular machinery that regulates mitochondrial morphology and, according to most recent insights, also apoptosis, in an at least partially CDK5-dependent manner in neurons.

Dynamics of Mitochondrial Morphology

Although the internal structure of mitochondria is highly conserved, the outer shape of these organelles can vary substantially. Mitochondria acquire a broad range of morphologies, from a punctuate shape to a tubular appearance or even extended networks [1].

In most cell types, mitochondrial morphology is complex and dynamic. It is defined by an actively regulated and continuous balance between fission and fusion [1]. Moreover, several proteins belonging to the family of large GTPases have been identified that energy-dependently regulate mitochondrial morphology and subcellular distribution by promoting either fission (Fis1, Drp1, MTP18) or fusion of mitochondria (Mfn1/2 and Opa1) [2,20,23,50] (Table 1). Consequently,

Table 1 Molecular components of the mitochondrial fission and fusion machinery

Mammals	Yeast	Loss-of-function phenotype	Localization	Function	References
Mfn1/2	Fzo	fragmented mitochondria	mitochondrial outer membrane	integral membrane GTPase, involved in tethering of mitochondria for fusion	[14]
OPA1	Mgm1	fragmented and aggregated mitochondria	intermembrane space/inner membrane	dynamin-like GTPase, inner membrane fusion	[42]
Drp1	Dnm1	network-like mitochondria	cytosol and mitochondrial outer membrane	dynamin-like GTPase	[28,33]
Fis1	Fis1	network-like mitochondria	mitochondrial outer membrane	integral membrane protein	[21,37]
–	Ugo1	fragmented mitochondria	mitochondrial outer membrane	integral membrane protein	[46,50]
–	Mdv1	network-like mitochondria	cytosol	adapter protein	[35,39,45]
–	Caf4p	network-like mitochondria	cytosol	adapter protein	[49]

Mfn1/2 and OPA1 are GTPases which are involved in fusion of mitochondrial membranes. Drp1 and Fis1 regulate mitochondrial fission. Ugo1, Mdv1, and Caf4p have only been described in yeast. Mdv1 and Caf4p act as scaffolds for Mgm1 and Fzo.

loss of proteins that participate in mitochondrial fission results in long, extended, reticulated organelles due to the continuing fusion events.

Dynamics of mitochondrial morphology crucially influence a range of physiological processes. Some cells contain morphologically and functionally distinct types of mitochondria [6], and both mitochondrial morphology and subcellular distribution greatly vary between different cell types. Because mitochondria cannot emerge de novo, they have to divide and be inherited to the respective daughter cells during cell division. Moreover, fusion of mitochondria results in mixture and recombination of mitochondrial DNA, which has been proposed to delay cellular aging. Furthermore, changes in the subcellular mitochondrial network might be an important way to match local energy demands in the cell.

The Mitochondrial Fission Machinery

Mitochondria consist of a double membrane, and two principally independent fission machineries seem to exist for both membranes. Studies of *C. elegans* showed that fission of the outer membrane can be blocked while fission of the inner membrane proceeds [25]. The only protein that has so far been discovered to be involved in inner membrane fission is the yeast protein Mdm33. Loss of Mdm33 leads to enhanced mitochondrial septum formation and fragmentation of the inner mitochondrial space [31]. No mammalian Mdm33 orthologue has been identified so far.

The main fission proteins of the outer membrane are Drp1 (dynamin-related protein 1; the orthologue of yeast Dnm1p) and Fis1 (corresponding to yeast Fis1p; Table 1). Similar to dynamin 1, Drp1 belongs to the family of large GTPases, which are participating in signal transduction, cell proliferation regulation of vesicular transport, and membrane translocation of proteins. It is rapidly degraded, unless protected by covalent attachment to small ubiquitin-like modifier 1 (SUMO1) [18]. Furthermore, Drp1 is recruited to the mitochondrial outer membrane in a cytoskeleton-dependent manner, where it complexes with Fis1. Fis1 itself is constitutively localized in the outer mitochondrial membrane via its C-terminal hydrophobic-basic domain. It seems to serve as an anchor for Drp1, which forms spiral homo-oligomers at specific sites defined by Fis1 during mitochondrial fission. Overexpression of Fis1 induces fission in a Drp1-dependent manner, suggesting that it acts as a limiting factor in mitochondrial fission by recruiting Drp1 [54]. Besides recruiting Drp1 to the outer mitochondrial membrane, Fis1 also seems to cooperate with Drp1 during the subsequent fission event itself. During this process, the GTPase activity of Drp1 may lead to contraction to the ring-like Drp1 homo-oligomers. Another model suggests that the distance between Drp1 spirals and helices is increased by its GTPase activity, thereby tethering the mitochondria apart.

Drp1 is necessary for mitochondrial fission, as knockdown of Drp1 or blocking its function with a GTPase-domain mutant form that exerts dominant-negative effects blocks mitochondrial fission, resulting in elongated reticular appearance of mitochondria. However, protein levels of Drp1 do not seem to be a key limiting factor in the fission process, as mild-to-moderate overexpression of wild-type Drp1 does not lead to increased fission.

Although the molecular machinery executing the fission process at the mitochondrial membrane is currently being elucidated, the relevant upstream signals regulating Drp1 activity and mitochondrial fission remains unclear. Drp1 levels can be regulated by SUMOylation [18], and also actin or dynein-dependent mitochondrial translocation of Drp1 may be a modulatory factor [9,52]. Interestingly, the pro-apoptotic bcl-2-like protein Bax translocates to mitochondria and was found to co-localize with Drp1 at mitochondrial fission sites. Knockdown of Drp1 or the use of dominant-negative mutants of Drp1 showed that Bax translocation to mitochondria is indeed Drp1 dependent, although this finding was not reproduced in other experimental settings [21,39].

The Mitochondrial Fusion Machinery

The proteins that are involved in mitochondrial fusion are distinct from the mitochondrial fission machinery. Again, fusion of the inner and the outer membrane have to be coordinated. Principally, outer and inner membranes can even fuse in separate steps.

Three essential fusion factors have been identified in yeast. The GTPases Fzo1 and Ugo1 are transmembrane proteins of the outer membrane, while Mgm1 is an intermembrane space protein associated with the inner membrane (Table 1).

In mammals, mitofusin 1 and 2 (Mfn1/2), mammalian orthologues of the yeast Fzo1, reside within the outer membrane. They contain two transmembrane domains close to their C-terminus, anchoring Mfn1/2 in the outer membrane. The protein sequence between the two transmembrane domains interacts with a yet unknown inner membrane protein, which mediates association of outer and inner membranes.

Mitofusins have been shown to tether mitochondria together via homotypic coiled-coil domain interaction. While no mammalian orthologue of Ugo1 has been found in mammals, Mgm1 has its mammalian counterpart in the protein OPA1. Active OPA1 is freely localized in the intermembrane space after being released by proteolytic cleavage of its intermembrane-anchored domain. OPA1 is involved in mitochondrial cristae remodeling and probably fusion of the inner mitochondrial membrane [5]. How, if at all, both processes are directly interconnected is still not completely clear.

The coordinated fusion of mitochondria could be initiated by a close contact between the two outer membranes. The first step could resemble SNARE-mediated membrane fusion [30], with Mfn1/2 making the first connection that initiates the fusion process. This may be following by conformational changes during assembly of the fusion complex that further facilitates contact and fusion of outer membrane lipids. After completion of the outer membrane fusion, association of outer and inner membranes at fusion sites could be responsible for fusion of the inner membranes, finally resulting in a mixture of mitochondrial contents [14].

Mitochondrial Morphology During Neuronal Apoptosis

Mitochondrial shape changes during apoptosis have emerged as a very recent focus and new aspect of cell death research. While ultrastructural morphological changes and a subcellular redistribution of mitochondria in the context of cell death have been already observed earlier [8], the attention that this topic has attracted recently may be at least partially due to the availability of fluorescent proteins that allow highly specific staining of mitochondria in living cells.

A substantial reduction in mean mitochondrial length and an increase in the number of mitochondria can often be observed after induction of apoptosis. This could principally be due to increased fission or decreased fusion. Frank et al. used a dominant-negative mutant of Drp1, which prevented not only staurosporine-induced apoptotic mitochondrial fragmentation, but also cytochrome c release and apoptotic DNA fragmentation, and thus provided first evidence for a causal role of mitochondrial fission for apoptosis in the immortalized cell line that was used (COS-7 cells) [12].

Subsequently, further evidence accumulated that mitochondrial fission might be closely linked to apoptotic cell death [26], also in neuronal cells [32,44]. However, it turned out that not only increased fission activity but also impaired fusion machinery can contribute to cellular Demise. RNAi knockdown of components of the mitochondrial fusion apparatus, for example downregulation of OPA1, sensitized HeLa cells to apoptosis [26]. Interestingly, in at least two instances, mutations in mitochondrial fusion proteins are the cause for inherited neurodegenerative diseases in men. Mutations in the GTPase domain of OPA1 induces a form of optic atrophy due to retinal ganglion cell death, and is the most frequent cause for dominantly inherited vision loss in children. Concomitantly, alterations in the mitochondrial inner membrane cristae are observed, suggesting that OPA1 is required for maintaining their shape [13,38]. In addition, inactivating mutations in the fusion protein Mfn2 have been identified as the cause for a Charcot-Marie-Tooth type IIa axonal neuropathy, further implicating mitochondrial fusion, and more generally mitochondrial shape changes, in neuronal apoptosis [3,57].

In general, knockdown or inhibition of proteins promoting mitochondrial fission, or overexpression of mitochondrial fusion proteins, inhibits apoptosis. Thus, at the first glance there seems to be a clear linear correlation between mitochondrial fission, or decreased fusion, and apoptosis. However, the picture turned out not to be as simple. Whether the morphologically observed fragmentation itself is directly responsible for apoptosis induction and finally cytochrome c release remained a matter of debate. For example, in many cases, mitochondrial fission is necessary for apoptosis, but fission itself is not sufficient to induce apoptosis. Bcl-xL blocks cell death, but not mitochondrial fission, and over-expression of the mitochondrial fission protein Fis1 always induces fission, but does not necessarily result in efficient cell death induction [20,26,32].

Several models are thus possible that explain the concomitant fission occurring before or together with the release of pro-apoptotic factors. Principally, fission might be an event directly required for the release of pro-apoptotic factor, or theoretically be only coincidentally related to the permeabilization of the mitochondrial membrane via two independent functions of the mitochondrial fission machinery. In this case, pro-fission proteins such as Drp1 would exert two principally separate functions: The promotion of mitochondrial fission and alteration of the mitochondrial membrane that facilitates the complete release of cytochrome c and other pro-apoptotic factors. However, it seems unlikely that both fission proteins as Fis1 or Drp1 and fusion-promoting factors as Mfn1/2 or OPA1 regulate mitochondrial shape and apoptosis by completely independent functions. A unifying model proposes that cytochrome c release requires fission-related remodeling of the inner membrane cristae that occurs during mitochondrial fission, as cytochrome c and other components of the respiratory chain are concentrated in the mitochondrial cristae. Cristae remodeling has been observed in conditions of increased fission and decreased fusion. Recent data suggest that Drp1, primarily involved in outer membrane fission, could signal to the inner membrane by unknown mechanisms and initiate cristae remodeling. Thereby, fragmented mitochondria would be prompted for maximal release of cytochrome c. This would provide a model how mitochondria shaping proteins could influence the progression of PCD [15].

Interestingly, time-lapse microscopy revealed that mitochondrial fission proceeds with a considerably constant time response. Largely independent of the cell type and pro-apoptotic stimulus employed, mitochondrial fission starts in most instances within minutes, and is completed within 20–30 min [32]. The rigid time window is reminiscent of data obtained from life video microscopy of GFP–cytochrome c–transfected cells [16]. Both mitochondrial fission and release of mitochondrial cytochrome c are basic, and possibly directly interconnected, observations with conserved, invariant time dynamics.

The principal phenomenon of mitochondrial fission in the context of apoptosis was first studied using non-neuronal immortalized cell lines. However, despite the important contribution of mitochondrial dysfunction specifically to the demise of neurons, there is only limited knowledge about mitochondrial fission during neuronal apoptosis so far. Few studies demonstrated mitochondrial fission during PCD of neurons. Rintoul et al. described mitochondrial fission, and decreased motility, under excitotoxic conditions in primary forebrain neurons [44]. Similarly, rapid and early mitochondrial shortening was described in dopaminergic neuronal cell lines (Fig. 1) and primary neuronal midbrain cultures [32].

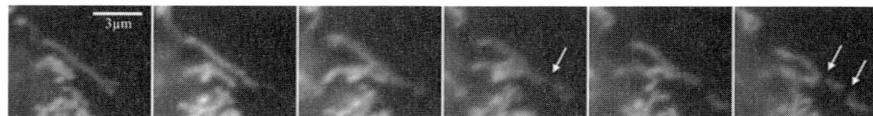

Fig. 1 Mitochondrial fragmentation during neuronal apoptosis. The neuronal dopaminergic cell line CSM14.1 was transfected with the fluorescent protein DsRed2 fused to a mitochondrial targeting signal, resulting in a highly specific mitochondrial staining. Cells were treated with staurosporine (1 μM) and pictures were taken every minute. Note the mitochondrial fission process taking place within few minutes (*arrows*)

Regulation of Mitochondrial Fission by CDK5

To date, several conditions are known which are accompanied by mitochondrial fission, for example cell division or cell death. However, upstream regulators, that is molecules transmitting the pro-fission signal to mitochondria, have not been described. One of the most upstream cell death regulators with regulatory function on mitochondrial fission has been described in *C. elegans*. Jagasia et al. found that Egl-1, which is the worm orthologue of a pro-apoptotic bcl-2-homology-domain protein, promotes mitochondrial fission [19]. It interacts and inhibits transcription of Ced-9, the *C. elegans* orthologue of bcl-2. Interestingly, a gain-of-function mutation in the anti-apoptotic Ced-9 protein prevented Egl-1-induced mitochondrial fission, showing that the upstream death-promoting transcription factor Egl-1 indeed induces mitochondrial fission. However, the knowledge about upstream regulators of mitochondrial fission remained restricted to regulation of members of the bcl-2 protein family. In mammalian cells, Bcl-xL, despite its well-established protective effects at the mitochondrial level, did not prevent mitochondrial fission in at least two studies [20,32]. In contrast to Bcl-xL, bcl-2 was shown to block mitochondrial fission [24]. Here, a direct interaction with the fission machinery was shown to be responsible for the reduction in mitochondrial fission [24], as bcl-2 and Drp1 competed for interaction with Fis1.

CDK5 is so far the only signal transduction kinase with proven influence on mitochondrial fission. Before CDK5 came into play, a role of cytoskeletal functions, specifically an intact actin and cytoskeleton, for mitochondrial translocation of Drp1 was described [9,52]. In addition to actin, tubulin was also required for apoptotic mitochondrial fission in neuronal cells (Meuer et al.; unpublished observation). This suggested that CDK5 is already a potential modulator for this process, given the well-established importance of CDK5 as a regulator of cytoskeletal functions and cellular aspects that require reorganization of the cytoskeleton, for example cell morphology, neurite outgrowth, or cell migration.

In fact, it turned out that overexpression of CDK5 together with p25 induced mitochondrial fission in a neuronal cell line. A similar, albeit not as pronounced, effect was achieved by expression of CDK5 with its physiological activator p35, indicating that indeed cytoplasmic CDK5 activity was responsible for the observed effect on mitochondria [32]. However, as expected, the deregulated CDK5 activity did also induce apoptotic cell death, and thus mitochondrial fission could have been attributed to a merely secondary effect resulting form CDK5-induced apoptosis and a general decline of mitochondrial function in this situation. Therefore, the influence of CDK5 on mitochondrial fission was studied in the context of staurosporine-induced apoptosis. Interestingly, the timecourse of CDK5 overactivation was compatible with the early mitochondrial fission, both occurring within the first 30 min. CDK5 overactivation and a decline in mitochondrial length were observed already within the first 10 min after addition of staurosporine [32]. Three different means were then employed in order to inhibit CDK5 activity. Pharmacological CDK5 inhibition with indolinone A, expression of a dominant-negative CDK5 mutant, as well as RNAi knockdown of CDK5 reduced apoptosis-related mitochondrial fission in a dopaminergic neuronal cell line. Moreover, this effect could be overcome by overexpression of wild-type Drp1, suggesting CDK5 activity was related to the function of Drp1 and, more generally, to the mitochondrial fission machinery [32]. Further linking mitochondrial fission and neuronal cell death, overexpression of Drp1 did also abolish the anti-apoptotic effect of CDK5 inhibition, a finding that provides direct evidence that the mitochondrial fission machinery is indeed one mediator of CDK5 toxicity (Fig. 2). As already mentioned above, without CDK5 inhibition, Drp1 protein

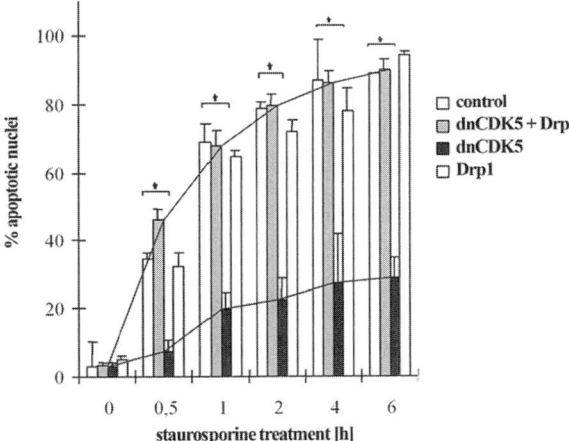

Fig. 2 Neuroprotection by CDK5 inhibition is abolished by wild-type Drp1 expression. CSM141.1 cells were transfected with a dominant-negative mutant of CDK5 (CDK5$_{N144}$) and wild-type Drp1 or a control vector. After 18 h, cultures were treated with staurosporine (1 μM), and the proportion of cells with apoptotic nuclei (pyknotic or fragmented) was determined by DAPI staining at different time points after application of staurosporine. The dominant-negative CDK5 mutant protected against staurosporine-induced apoptosis. This effect was abolished by parallel overexpression of Drp1. Note that Drp1 itself did not induce apoptosis

levels are not a limiting factor for mitochondrial fission, because mere Drp1 overexpression does not promote fission in mammalian cells.

Mitochondrial CDK5 Substrates

Taken together, accumulating evidence suggests a direct influence of CDK5 on early parameters of mitochondrial dysfunction, for example altered mitochondrial morphology, subcellular redistribution, and a decrease in mitochondrial transmembrane potential, finally resulting in cytochrome c release, caspase activation, and cell death. Nevertheless, the answer to the question which substrates are the CDK5 targets that convey those mitochondrial effects and CDK5 toxicity requires further studies.

As far as the CDK5 effect on mitochondrial fission is concerned, it is tempting to speculate that Drp1 is itself a CDK5 target. Intriguing parallels can be found between mitochondrial fission and endocytosis. CDK5 has previously been demonstrated to play a critical role in synaptic vesicle endocytosis by phosphorylation of dynamin 1 [48]. Similarly, CDK5 may directly modulate other members of the dynamin-related family of GTPases, to which Drp1 belongs. Drp1 indeed contains a CDK5 consensus site, but direct evidence that Drp1 is a CDK5 substrate is currently lacking and needs further investigation.

Moreover, CDK5 may need to phosphorylate several different targets to orchestrate mitochondrial fission, including, as already suggested above, components of the cytoskeleton. In addition to cytoskeletal elements and a possible direct modulation of the mitochondrial fission machinery, CDKs have been shown to phosphorylate and modulate the function of bcl-2 [51], although bcl-2 family members have not been shown to be a substrate specifically for CDK5.

Although several pieces of evidence support a direct action of CDK5 on mitochondria via cytoplasmic targets, additional mechanisms other than mitochondrial fission are likely to contribute to toxic CDK5 effects on mitochondria, to a varying degree depending on the paradigm. For example, CDK5-dependent phosphorylation of p53 activates Bax transcription that, as delineated above, is an important contributor to mitochondrial apoptosis pathways and more specifically mitochondrial fission [56]. Similarly, nuclear CDK5 activity inactivates the anti-apoptotic transcription factor MEF2, which participates in transcription of a variety of pro-survival genes that could also influence the mitochondrial status [17,47,49].

Summary

PCD has emerged as an important type of cellular demise in neurodegenerative diseases. Among different apoptotic pathways, the mitochondrial apoptosis cascade is the most prominent in neurons. While nuclear CDK5 activity contributes to CDK5 neurotoxicity at least in some experimental paradigms, other evidence suggests cytoplasmic or mitochondrial targets, too. Recently, several pathways connecting CDK5 deregulation and mitochondrial apoptosis could be delineated.

The enzymatic substrate that is responsible for the modulatory effect of CDK5 on mitochondrial morphology has not been determined yet. Nevertheless, it became clear that CDK5 is an upstream regulator of this new aspect of mitochondrial alterations occurring during apoptosis. Some uncertainty still exists about the precise relationship between mitochondrial fission and its contribution to apoptotic cytochrome c release. Nevertheless, it turned out that CDK5 toxicity it at least partially conveyed by enhanced mitochondrial fission, while the direct CDK5 targets mediating this effect on mitochondrial morphology still need to be identified.

References

1. Bereiter-Hahn J, Voth M (1994) Dynamics of mitochondria in living cells: shape changes, dislocations, fusion, and fission of mitochondria. Microsc Res Tech 27: 198–219.
2. Bossy-Wetzel E, Barsoum MJ, Godzik A, Schwarzenbacher R, Lipton SA (2003) Mitochondrial fission in apoptosis, neurodegeneration and aging. Curr Opin Cell Biol 15: 706–16.
3. Bradbury J (2004) Mitochondrial fusion protein mutated in CMT2A. Lancet Neurol 3: 326.
4. Bredesen DE, Rao RV, Mehlen P (2006) Cell death in the nervous system. Nature 443: 796–802.
5. Cipolat S, Martins de Brito O, Dal Zilio B, Scorrano L (2004) OPA1 requires mitofusin 1 to promote mitochondrial fusion. Proc Natl Acad Sci USA 101: 15927–32.
6. Collins TJ, Berridge MJ, Lipp P, Bootman MD (2002) Mitochondria are morphologically and functionally heterogeneous within cells. Embo J 21: 1616–27.
7. Darios F, Muriel MP, Khondiker ME, Brice A, Ruberg M (2005) Neurotoxic calcium transfer from endoplasmic reticulum to mitochondria is regulated by cyclin-dependent kinase 5-dependent phosphorylation of tau. J Neurosci 25: 4159–68.
8. De Vos K, Goossens V, Boone E, Vercammen D, Vancompernolle K, Vandenabeele P, Haegeman G, Fiers W, Grooten J (1998) The 55-kDa tumor necrosis factor receptor induces clustering of mitochondria through its membrane-proximal region. J Biol Chem 273: 9673–80.
9. De Vos KJ, Allan VJ, Grierson AJ, Sheetz MP (2005) Mitochondrial function and actin regulate dynamin-related protein 1-dependent mitochondrial fission. Curr Biol 15: 678–83.
10. Dhavan R, Tsai LH (2001) A decade of CDK5. Nat Rev Mol Cell Biol 2: 749–59.
11. Diem R, Meyer R, Weishaupt JH, Bahr M (2001) Reduction of potassium currents and phosphatidylinositol 3-kinase-dependent AKT phosphorylation by tumor necrosis factor-(alpha) rescues axotomized retinal ganglion cells from retrograde cell death in vivo. J Neurosci 21: 2058–66.
12. Frank S, Gaume B, Bergmann-Leitner ES, Leitner WW, Robert EG, Catez F, Smith CL, Youle RJ (2001) The role of dynamin-related protein 1, a mediator of mitochondrial fission, in apoptosis. Dev Cell 1: 515–25.

13. Frezza C, Cipolat S, Martins de Brito O, Micaroni M, Beznoussenko GV, Rudka T, Bartoli D, Polishuck RS, Danial NN, De Strooper B, Scorrano L (2006) OPA1 controls apoptotic cristae remodeling independently from mitochondrial fusion. Cell 126: 177–89.

14. Fritz S, Rapaport D, Klanner E, Neupert W, Westermann B (2001) Connection of the mitochondrial outer and inner membranes by Fzo1 is critical for organellar fusion. J Cell Biol 152: 683–92.

15. Germain M, Mathai JP, McBride HM, Shore GC (2005) Endoplasmic reticulum BIK initiates DRP1-regulated remodelling of mitochondrial cristae during apoptosis. Embo J 24: 1546–56.

16. Goldstein JC, Waterhouse NJ, Juin P, Evan GI, Green DR (2000) The coordinate release of cytochrome c during apoptosis is rapid, complete and kinetically invariant. Nat Cell Biol 2: 156–62.

17. Gong X, Tang X, Wiedmann M, Wang X, Peng J, Zheng D, Blair LA, Marshall J, Mao Z (2003) Cdk5-mediated inhibition of the protective effects of transcription factor MEF2 in neurotoxicity-induced apoptosis. Neuron 38: 33–46.

18. Harder Z, Zunino R, McBride H (2004) Sumo1 conjugates mitochondrial substrates and participates in mitochondrial fission. Curr Biol 14: 340–5.

19. Jagasia R, Grote P, Westermann B, Conradt B (2005) DRP-1-mediated mitochondrial fragmentation during EGL-1-induced cell death in C. elegans. Nature 433: 754–60.

20. James DI, Parone PA, Mattenberger Y, Martinou JC (2003) hFis1, a novel component of the mammalian mitochondrial fission machinery. J Biol Chem 278: 36373–9.

21. Karbowski M, Lee YJ, Gaume B, Jeong SY, Frank S, Nechushtan A, Santel A, Fuller M, Smith CL, Youle RJ (2002) Spatial and temporal association of Bax with mitochondrial fission sites, Drp1, and Mfn2 during apoptosis. J Cell Biol 159: 931–8.

22. Kermer P, Klocker N, Labes M, Bahr M (1998) Inhibition of CPP32-like proteases rescues axotomized retinal ganglion cells from secondary cell death in vivo. J Neurosci 18: 4656–62.

23. Kerr JF, Wyllie AH, Currie AR (1972) Apoptosis: a basic biological phenomenon with wide-ranging implications in tissue kinetics. Br J Cancer 26: 239–57.

24. Kong D, Xu L, Yu Y, Zhu W, Andrews DW, Yoon Y, Kuo TH (2005) Regulation of Ca2+-induced permeability transition by Bcl-2 is antagonized by Drpl and hFis1. Mol Cell Biochem 272: 187–99.

25. Labrousse AM, Zappaterra MD, Rube DA, van der Bliek AM (1999) C. elegans dyna-min-related protein DRP-1 controls severing of the mitochondrial outer membrane. Mol Cell 4: 815–26.

26. Lee YJ, Jeong SY, Karbowski M, Smith CL, Youle RJ (2004) Roles of the mammalian mitochondrial fission and fusion mediators Fis1, Drp1, and Opa1 in apoptosis. Mol Biol Cell 15: 5001–11.

27. Li M, Ona VO, Guegan C, Chen M, Jackson-Lewis V, Andrews LJ, Olszewski AJ, Stieg PE, Lee JP, Przedborski S, Friedlander RM (2000) Functional role of caspase-1 and caspase-3 in an ALS transgenic mouse model. Science 288: 335–9.

28. Li Z, Okamoto K, Hayashi Y, Sheng M (2004) The importance of dendritic mitochondria in the morphogenesis and plasticity of spines and synapses. Cell 119: 873–87.

29. Lund ET, McKenna R, Evans DB, Sharma SK, Mathews WR (2001) Characterization of the in vitro phosphorylation of human tau by tau protein kinase II (cdk5/p20) using mass spectrometry. J Neurochem 76: 1221–32.

30. Meeusen S, McCaffery JM, Nunnari J (2004) Mitochondrial fusion intermediates revealed in vitro. Science 305: 1747–52.

31. Messerschmitt M, Jakobs S, Vogel F, Fritz S, Dimmer KS, Neupert W, Westermann B (2003) The inner membrane protein Mdm33 controls mitochondrial morphology in yeast. J Cell Biol 160: 553–64.

32. Meuer K, Suppanz IE, Lingor P, Planchamp V, Goricke B, Fichtner L, Braus GH, Dietz GP, Jakobs S, Bahr M, Weishaupt JH (2007) Cyclin-dependent kinase 5 is an

upstream regulator of mitochondrial fission during neuronal apoptosis. Cell Death Differ 14: 651–61.

33. Mozdy AD, McCaffery JM, Shaw JM (2000) Dnm1p GTPase-mediated mitochondrial fission is a multi-step process requiring the novel integral membrane component Fis1p. J Cell Biol 151: 367–80.

34. Muzio M, Chinnaiyan AM, Kischkel FC, O'Rourke K, Shevchenko A, Ni J, Scaffidi C, Bretz JD, Zhang M, Gentz R, Mann M, Krammer PH, Peter ME, Dixit VM (1996) FLICE, a novel FADD-homologous ICE/CED-3-like protease, is recruited to the CD95 (Fas/APO-1) death – inducing signaling complex. Cell 85: 817–27.

35. Naylor K, Ingerman E, Okreglak V, Marino M, Hinshaw JE, Nunnari J (2006) Mdv1 interacts with assembled dnm1 to promote mitochondrial division. J Biol Chem 281: 2177–83.

36. Nicotera P, Leist M, Ferrando-May E (1999) Apoptosis and necrosis: different execution of the same death. Biochem Soc Symp 66: 69–73.

37. O'Hare MJ, Kushwaha N, Zhang Y, Aleyasin H, Callaghan SM, Slack RS, Albert PR, Vincent I, Park DS (2005) Differential roles of nuclear and cytoplasmic cyclin-dependent kinase 5 in apoptotic and excitotoxic neuronal death. J Neurosci 25: 8954–66.

38. Olichon A, Emorine LJ, Descoins E, Pelloquin L, Brichese L, Gas N, Guillou E, Delettre C, Valette A, Hamel CP, Ducommun B, Lenaers G, Belenguer P (2002) The human dynamin-related protein OPA1 is anchored to the mitochondrial inner membrane facing the inter-membrane space. FEBS Lett 523: 171–6.

39. Parone PA, James DI, Da Cruz S, Mattenberger Y, Donze O, Barja F, Martinou JC (2006) Inhibiting the mitochondrial fission machinery does not prevent Bax/Bak-dependent apoptosis. Mol Cell Biol 26: 7397–408.

40. Patrick GN, Zukerberg L, Nikolic M, de la Monte S, Dikkes P, Tsai LH (1999) Conversion of p35 to p25 deregulates Cdk5 activity and promotes neurodegeneration. Nature 402: 615–22.

41. Patzke H, Tsai LH (2002) Calpain-mediated cleavage of the cyclin-dependent kinase-5 activator p39 to p29. J Biol Chem 277: 8054–60.

42. Putcha GV, Deshmukh M, Johnson EM, Jr. (1999) BAX translocation is a critical event in neuronal apoptosis: regulation by neuroprotectants, BCL-2, and caspases. J Neurosci 19: 7476–85.

43. Raoul C, Estevez AG, Nishimune H, Cleveland DW, deLapeyriere O, Henderson CE, Haase G, Pettmann B (2002) Motoneuron death triggered by a specific pathway downstream of Fas. potentiation by ALS-linked SOD1 mutations. Neuron 35: 1067–83.

44. Rintoul GL, Filiano AJ, Brocard JB, Kress GJ, Reynolds IJ (2003) Glutamate decreases mitochondrial size and movement in primary forebrain neurons. J Neurosci 23: 7881–8.

45. Schauss AC, Bewersdorf J, Jakobs S (2006) Fis1p and Caf4p, but not Mdv1p, determine the polar localization of Dnm1p clusters on the mitochondrial surface. J Cell Sci 119: 3098–106.

46. Sesaki H, Jensen RE (2001) UGO1 encodes an outer membrane protein required for mitochondrial fusion. J Cell Biol 152: 1123–34.

47. Smith PD, Mount MP, Shree R, Callaghan S, Slack RS, Anisman H, Vincent I, Wang X, Mao Z, Park DS (2006) Calpain-regulated p35/cdk5 plays a central role in dopaminergic neuron death through modulation of the transcription factor myocyte enhancer factor 2. J Neurosci 26: 440–7.

48. Tan TC, Valova VA, Malladi CS, Graham ME, Berven LA, Jupp OJ, Hansra G, McClure SJ, Sarcevic B, Boadle RA, Larsen MR, Cousin MA, Robinson PJ (2003) Cdk5 is essential for synaptic vesicle endocytosis. Nat Cell Biol 5: 701–10.

49. Tang X, Wang X, Gong X, Tong M, Park D, Xia Z, Mao Z (2005) Cyclin-dependent kinase 5 mediates neurotoxin-induced degradation of the transcription factor myocyte enhancer factor 2. J Neurosci 25: 4823–34.

50. Tondera D, Czauderna F, Paulick K, Schwarzer R, Kaufmann J, Santel A (2005) The mitochondrial protein MTP18 contributes to mitochondrial fission in mammalian cells. J Cell Sci 118: 3049–59.

51. Vantieghem A, Xu Y, Assefa Z, Piette J, Vandenheede JR, Merlevede W, De Witte PA, Agostinis P (2002) Phosphorylation of Bcl-2 in G2/M phase-arrested cells following photodynamic therapy with hypericin involves a CDK1-mediated signal and delays the onset of apoptosis. J Biol Chem 277: 37718–31.

52. Varadi A, Johnson-Cadwell LI, Cirulli V, Yoon Y, Allan VJ, Rutter GA (2004) Cytoplasmic dynein regulates the subcellular distribution of mitochondria by controlling the recruitment of the fission factor dynamin-related protein-1. J Cell Sci 117: 4389–400.

53. Weishaupt JH, Kussmaul L, Grotsch P, Heckel A, Rohde G, Romig H, Bahr M, Gillardon F (2003) Inhibition of CDK5 is protective in necrotic and apoptotic paradigms of neuronal cell death and prevents mitochondrial dysfunction. Mol Cell Neurosci 24: 489–502.

54. Yoon Y, Krueger EW, Oswald BJ, McNiven MA (2003) The mitochondrial protein hFis1 regulates mitochondrial fission in mammalian cells through an interaction with the dynamin-like protein DLP1. Mol Cell Biol 23: 5409–20.

55. Yuan J, Shaham S, Ledoux S, Ellis HM, Horvitz HR (1993) The C. elegans cell death gene ced-3 encodes a protein similar to mammalian interleukin-1 beta-converting enzyme. Cell 75: 641–52.

56. Zhang J, Krishnamurthy PK, Johnson GV (2002) Cdk5 phosphorylates p53 and regulates its activity. J Neurochem 81: 307–13.

57. Zuchner S, Mersiyanova IV, Muglia M, Bissar-Tadmouri N, Rochelle J, Dadali EL, Zappia M, Nelis E, Patitucci A, Senderek J, Parman Y, Evgrafov O, Jonghe PD, Takahashi Y, Tsuji S, Pericak-Vance MA, Quattrone A, Battaloglu E, Polyakov AV, Timmerman V, Schroder JM, Vance JM (2004) Mutations in the mitochondrial GTPase mitofusin 2 cause Charcot-Marie-Tooth neuropathy type 2A. Nat Genet 36: 449–51.

Regulation and Function of Cdk5 in the Nucleus

Qian Yang and Zixu Mao

Abstract Early studies have shown that Cyclin-dependent kinase 5 (Cdk5) and its regulators function in the neuronal cytoplasm to regulate a range of cellular processes. This seems to distinct Cdk5 from other more traditional Cdks. However, recent discoveries have now extended the role of Cdk5 to the nucleus. This chapter summarizes these novel findings. These include the detection of Cdk5 and its regulators in the nucleus, identification of nuclear targets of Cdk5, and regulation of nuclear Cdk5 and its activators.

Introduction

Cyclin-dependent kinase 5 (Cdk5) is a proline-directed Ser/Thr kinase involved in many aspects of neuronal functions both during development and in diseases (Lew et al., 1992; Meyerson et al., 1992) (Dhavan and Tsai, 2001). The kinase activity of Cdk5 is stimulated via its association with protein activators, p35 or p39, or their truncated forms p25 or 29, respectively. Despite its ubiquitous expression, Cdk5 activity seems to be restricted primarily to postmitotic neurons. This is likely due to the neuronal-specific expression of p35 and p39. Earlier characterization studies also indicated that Cdk5 and p35 are present abundantly in the cytoplasm of postmitotic neurons. Because of their apparent absence from the nucleus, Cdk5 and p35 quickly distinguished themselves from other traditional Cdks and cyclins that reside in the nucleus to regulate cell cycle progression. Therefore, it is not surprising that the prevailing view in the field has been that Cdk5 functions only in the cytoplasm to phosphorylate substrates and affect neuronal functions. Although the basic concept that the level and activity of Cdk5 are probably the highest in the cytoplasm of postmitotic neurons is correct, an increasing number of studies have now provided multiple lines of correlative and functional evidence to expand the role of Cdk5 beyond

Z. Mao
Departments of Pharmacology and Neurology, Center for Neurodegenerative Disease,
Emory University School of Medicine, 615 Michael St, Atlanta, GA 30322, USA
e-mail: zmao@pharm.emory.edu

N.Y. Ip, L.-H. Tsai (eds.), *Cyclin Dependent Kinase 5 (Cdk5)*,   107
DOI: 10.1007/978-0-387-78887-6_8, © Springer Science+Business Media, LLC 2008

the cytoplasmic domain to the nucleus and to extra-neuronal cells. This chapter summarizes some of the findings which support the presence and functions of Cdk5 in the nucleus.

Detection of Cdk5 and Its Activators in the Nucleus

Immunohistochemical studies by Tsai et al. first discovered that Cdk5, in a striking difference from Cdk2 and Cdc2, is expressed in postmitotic neurons but not in mitotic cells in the embryonic mouse nervous system (Tsai et al., 1993). More detailed analysis of mouse forebrain revealed that the expression of Cdk5 is concentrated in fasciculated axons of postmitotic neurons, suggesting a cytoplasmic localization of Cdk5. Using two distinct antibodies that recognize the N- and C-termini of Cdk5, respectively, Ino and Chiba further characterized Cdk5 subcellular localization. These authors observed no immunostaining at reasonable antibody concentrations in frozen or paraffin-embedded sections, but found that Cdk5 is highly expressed in neuronal axons using heat-treated sections. Interestingly, they also detected intensive Cdk5 staining in the nucleus of Purkinje cells, cerebellar nuclear neurons, hippocampus, dentate gyrus, and hypothalamus in the CNS and sensory neurons in peripheral nervous system (Ino and Chiba, 1996). The nuclear staining seemed to be specific because it largely matches mRNA pattern observed by in situ hybridization and can be blocked by excess peptide corresponding to the epitope. The revelation of the possible nuclear presence of Cdk5 led the authors to speculate that Cdk5 may have a "physiological function" in the nucleus.

This initial observation by Ino and Chiba was followed by a series of studies that examined the expression of Cdk5 and p35 in the nucleus under various conditions. Zhang et al. first studied the localization of Cdk5 in several tissues during development and remodeling. They observed a striking expression of Cdk5 in the developing CNS, the developing eye, and the developing limb. In those tissues, Cdk5 appeared to be highly expressed in the apoptotic cells with a pattern of staining consistent with either a strong nuclear expression or a nuclear and cytoplasmic expression (Zhang et al., 1997). These findings suggested a possible role of Cdk5 in apoptotic cell death. Shortly after Zhang's study, Ahuja et al. reported the detection of Cdk5 and p35 but not other Cdks specifically in dying cells in the embryonic mouse limbs (Ahuja et al., 1997). Cdk5 and p35 staining revealed apparent nuclear localization in the interdigital cells undergoing apoptosis, further strengthening a link between Cdk5/p35 and cell death. The association of nuclear Cdk5 with cellular development was also observed in the developing rat lens. Lens fiber cells undergo several discrete steps of differentiation during development that includes denucleation. Gao et al. found that developing rat lens fiber cells are positive for Cdk5 in the nucleus prior to nuclear degeneration (Gao et al., 1997), underscoring the potential role of Cdk5 in the nucleus.

In addition to being detected in the nucleus of apoptotic cells, nuclear Cdk5 has also been found in developing cells that do not undergo overt apoptosis. For example, Musa et al. reported the detection of Cdk5 in the developing testis. Immunohistochemical staining revealed abundant presence of Cdk5 in Leydig cells and Sertoli cells in both nucleus and cytoplasm (Musa et al., 1998). Similarly, Cdk5 was also found in the cytoplasm and nucleus of oocyte (Lee et al., 2004). Interestingly, the same study revealed that p35 seems to be present only in the cytoplasm. It is not clear what this apparent dissociation of their localization in the nucleus reflects.

Besides development, Cdk5 has been documented in the nucleus of matured post-developmental cells. Neystat et al. examined the expression of Cdk5 in the substantia nigra in models of neurodegenerative Parkinson's disease. They reported the exclusive expression of Cdk5 in the nucleus of substantia nigra neurons undergoing late phase of apoptosis induced by 6-hydroxydopamine and axotomy (Neystat et al., 2001). Similarly, using androgen withdrawal-induced regression of the prostate gland in adult male mice as a model, Zhang et al. observed an increase in nuclear immune staining and kinase activity of Cdk5 in mouse ventral prostate gland following orchidectomy (Zhang et al., 1997).

In addition to the immunohistochemical characterization, the nuclear localization of Cdk5 and its activators was also confirmed by biochemical and cellular studies. While studying the protease calpain-mediated cleavage of p35 to its C-terminal fragment p25, Patrick et al. noted that p25, when overexpressed in COS-7 cells, shows a pattern of subcellular distribution distinct from that of p35. Unlike the predominantly peripheral plasma membrane association exhibited by p35, p25, which lacks the myristoylation signal motif present at the N-terminus of p35, is localized in great amount in the perinuclear region and in the nucleus (Patrick et al., 1999). This finding raised the possibility that p25 and p35 may have overlapping but distinct subcellular distributions, which could potentially serve to target Cdk5 activity to different regions of the cell. Gong et al. sought to confirm the nuclear presence of Cdk5 and its activators using traditional biochemical means. Gong's study showed that Cdk5 and p35 are clearly detectable in the nuclear fractions prepared from cultured primary cortical neurons (Gong et al., 2003), a finding re-confirmed in cultured primary cerebellar granule neurons (Tang et al., 2005). Compared to the cytoplasmic fractions, the amount of Cdk5 and p35 in the nuclear fractions is relatively small. However, the small amount of nuclear Cdk5 and p35 clearly represents a genuine signal and not a cross-contamination from the cytoplasm, given the clear segregations of other nuclear and cytoplasmic markers. Together, these studies helped pave the way for functional investigation of the roles of Cdk5 and its activators in the nucleus.

Nuclear Function of Cdk5 and p35

The functions of Cdk5 and p35 and their regulation in the nucleus have just begun to be illustrated. From the limited number of studies published, it is clear that one of the major roles of Cdk5/p35 is to regulate nuclear transcription factors. In addition, Cdk5/p35 is also involved in the regulation of other factors residing in the nucleus that do not have a direct role in transcriptional regulation. This section is therefore organized in two parts, regulation of nuclear transcription factors and other non-transcription factor targets.

Nuclear Transcription Factors Targeted by Cdk5

The very first clue that Cdk5 may regulate a transcription factor came from the study led by Lee et al. in Dr. Jerry Wang's group. By a series of biochemical assays, Lee et al. showed that the tumor suppressor retinoblastoma gene product Rb binds p25 *in vitro* and co-precipitates with p25 from embryonic mouse brain (Lee et al., 1997). Moreover, bacterially expressed Cdk5/p25 phosphorylates GST-Rb. A subsequent study by Hamdane et al. confirmed the phosphorylation of Rb by Cdk5 *in vitro* (Hamdane et al., 2005). The authors speculated that phosphorylation of Rb by Cdk5 may play a role in regulation of Rb function, perhaps in neuronal apoptosis. However, as these studies did not present any functional and cellular data to establish Cdk5-dependent regulation of Rb in cells, the data remained correlative. Direct evidence to demonstrate phosphorylation and regulation of Rb activity and its cellular function by Cdk5 in cells is still lacking.

Nevertheless, they raise the possibility that such a functional relationship between Cdk5 and Rb might exist.

The first piece of evidence that clearly establishes a nuclear role of Cdk5 came from Gong's study of nuclear transcription factor myocyte enhancer factor 2 (MEF2). Prior to Gong's work, earlier studies have shown that MEF2s are expressed in the nucleus of different types of neurons and are required for neuronal survival in several experimental paradigms (Mao et al., 1999). Gong et al. showed in a series of experiments that Cdk5 directly phosphorylates MEF2D at Ser444 *in vitro* and in neuronal cells (Gong et al., 2003). Phosphorylation of MEF2 by Cdk5 led to inhibition of MEF2 transcriptional activity and induced neuronal death. Further studies by Tang et al. demonstrated that phosphorylation of MEF2 by Cdk5 makes MEF2 a better substrate for caspases, leading to more efficient caspase-dependent degradation of MEF2 (Tang et al., 2005). In addition, phosphorylation of MEF2D by Cdk5 was also shown to stimulate MEF2 SUMOylation (Gregoire et al., 2006), which inhibits MEF2 function during myogenesis. Cdk5-mediated phosphorylation and inhibition of MEF2 has been shown to play a role in both oxidative stress–induced and glutamate excitotoxicity–induced neuronal death in cultured primary neurons (Gong et al., 2003). More importantly, phosphorylation of MEF2 by Cdk5 in dopamine neurons in the substantial nigra region was shown to modulate neuronal death in response to neurotoxin MPTP in a mouse model of Parkinson's disease (Smith et al., 2006). Together, this series of studies establish a critical role for Cdk5-mediated phosphorylation and inhibition of nuclear survival factor MEF2 in regulation of neuronal death during neurodegeneration *in vivo*.

The above studies prompted O'Hare et al. to take a closer examination of the roles of nuclear and cytoplasmic Cdk5 in apoptotic and excitotoxic neuronal death. Using camptothecin-induced DNA damage and glutamate-induced toxicity as models, O'Hare et al. showed that nuclear p25 is induced by both stimuli but with different kinetics (O'Hare et al., 2005). By targeting a dominant-negative Cdk5 to either the cytoplasm or the nucleus, O'Hare demonstrated that nuclear Cdk5 plays an important role in excitotoxin-induced neuronal death, while cytoplasmic Cdk5 mediates survival signal. Interestingly, nuclear Cdk5 activity was not required for DNA damage–induced neuronal death, revealing the complexity of nuclear role of Cdk5 in mediating neuronal death.

In addition to MEF2, Cdk5 has also been postulated to regulate transcription factor p53 to modulate neuronal death. Zhang et al. first demonstrated by *in vitro* kinase assay that Cdk5/p25 phosphorylates recombinant p53 (Zhang et al., 2002). Using PC12 cells as a model, Zhang et al. further showed that increase in Cdk5 activity enhances the level of p53 protein, its transcriptional activity, and the expression of p53-responsive gene p21 and Bax. In a subsequent study, Lee and Kim demonstrated that Cdk5 regulates p53 in response to mitomycin C–induced DNA damage in mouse cortical neurons and human neuroblastoma SH-SY5Y cells (Lee and Kim, 2007). They mapped Ser33 and Ser46 in p53 as the sites of phosphorylation by Cdk5. Phosphorylation at these sites increased the stability of p53, leading to neuronal death.

Study by Fu et al. extended the regulation by Cdk5 to yet another transcription factor signal transducer and activator of transcription (STAT), a family of proteins known to play important roles in embryogenesis, development, and immune response. Fu et al. reported that Cdk5 phosphorylates Ser727 of STAT3, a site shown to be phosphorylated by MAPK, in myotubes, which mediates the effects following NGR stimulation (Fu et al., 2004). Phosphorylation of Ser727 increased the DNA binding by STAT3 and specific target gene expression. A later study by Lin et al. confirmed the regulation of STAT3 by Cdk5 in medullary thyroid carcinoma cells. They demonstrated that Cdk5-dependent phosphorylation of Ser727 was involved in the proliferation of medullary thyroid carcinoma cells (Lin et al., 2007). Inhibition of Cdk5 seemed to reduce nuclear distributions of Cdk5/p35 complex and phospho-STAT3 in these cells, suggesting that Cdk5 may regulate Ser727 phosphorylation in the nucleus.

Ubeda et al. studied Cdk5 in the pancreatic beta cells. Following their initial report that Cdk5 and p35 are expressed in pancreatic beta cells (Ubeda et al., 2004), Ubeda et al. recently examined the role of Cdk5 in these cells. They found that glucose toxicity increases the levels of p35 protein and Cdk5 kinase activity, which parallels a reduction in the levels of transcription factor PDX-1 in the nucleus (Ubeda et al., 2006). PDX-1 regulates the expression of insulin, which is reduced in response to glucose toxicity. The authors showed that inhibition of Cdk5 selectively restores nuclear PDX-1 level without affecting its cytoplasmic concentration. These findings suggest that nuclear Cdk5 may be particularly relevant to glucose toxicity–induced loss of pancreatic beta cell function and to the homeostasis of the endocrine system.

The role of Cdk5 in endocrine system was further strengthened by a more recent report describing the regulation of glucocorticoid receptor (GR) by Cdk5. GR is a nuclear receptor superfamily protein that functions as a hormone-dependent transcription factor. Kino et al. showed that both p35 and p25 interact with the ligand-binding domain of the GR (Kino et al., 2007). Cdk5 directly phosphorylated GR at multiple sites including Ser203 and Ser211. These phosphorylation events suppressed the transcriptional activity of GR by disrupting its interaction with DNA. Microarray analysis in cortical neurons suggest that roscovitine differentially regulates the expression of more than 90% of the endogenous glucocorticoid-responsive genes, indicating that Cdk5 has gene-specific effects on GR activity.

In addition to regulation of transcription factors, Cdk5 has also been shown to interact with chromatin-remodeling factors such as SET and histone deacetylase. SET oncogene is associated with acute undifferentiated leukemia. SET oncoprotein is present in the nucleus within a complex that binds to chromatin and blocks histone from acetylation by p300/CBP and PCAF. Using an affinity isolation approach, Qu et al. demonstrated that nuclear SET protein specifically binds to the N-terminal region of p35 or p39 (Qu et al., 2002). This interaction stimulated Cdk5/p35 kinase activity. Curiously, such a model of interaction between SET and Cdk5 activators would exclude the interaction between SET and p25 or p29. It is not clear what role the selective regulation of p35 or p39 by SET may play in cells and whether Cdk5 directly regulates SET in a reciprocal manner.

mSds3 is an essential component of the mSin3-histone deacetylase (HDAC) complex. mSds3 represses transcription by recruiting HDAC. Genetic inactivation of mSds3 is lethal to cells, which is characterized by severe proliferation defects and increased apoptosis. Li et al. identified mSds3 as a protein that interacts with p35 by yeast two-hybrid screen (Li et al., 2004). Overexpression of p35 can augment mSds3-mediated transcriptional repression *in vitro*. Furthermore, Cdk5 directly phosphorylated mSds3 at Ser228, and this phosphorylation seemed to play a role in the ability of exogenous mSds3 to rescue cell growth and viability in mSds3-null cells. Together with the findings of SET–p35 interaction, these studies provide examples of how Cdk5/p35 may modulate or be modulated by chromatin-remodeling machinery to affect gene transcription. The link between Cdk5 complex and chromatin structure is further supported by a more recent study which showed that in a mouse model of p25-induced temporally and spatially restricted loss of neurons, recovery of learning and memory is associated with histone acetylation and chromatin remodelling (Fischer et al., 2007).

Non-Transcription Factor Targets of Cdk5 in the Nucleus

Protein p27[kip1] is a member of the Cip-Kip family of Cdk inhibitors. They bind cyclin–Cdk complex in the nucleus and inhibit their catalytic activity to control exit of cell cycle and cell migration. Lacy et al. first revealed that p27 directly interacts with Cdk5 but not with p25 in the Cdk5/p25 complex (Lacy et al., 2005). Kawauchi et al. confirmed the interaction between p27[Kip1] and Cdk5 (Kawauchi et al., 2006) and showed that Cdk5 phosphorylates p27[Kip1] at Ser10 to stabilize the protein. Cdk5-p27 signaled to activate an actin-binding protein cofilin to regulate neuronal migration. The authors propose a model in which Cdk5-p27 regulates neuronal migration primarily via their cytoplasmic control of cytoskeletal reorganization. However, since both p27 and Cdk5 are present in the nucleus, this raises the possibility that Cdk5 may also modulate other neuronal functions by interacting with p27 in the nucleus.

Members of ERM family proteins (ezrin, radixin, and moesin) are known to play important roles in microvilli formation, cell–cell junctions, membrane ruffles, substrate adhesion, migration, proliferation, and senescence. ERM accomplishes this by forming a connection between membrane and cytoskeleton, enabling proper signal transduction. Yang and Hinds showed that Cdk5 plays a role in pRb-induced senescence by phosphorylating ezrin at Thr235 (Yang et al., 2003). This phosphorylation prevented the intermolecular association of ezrin and promoted the release of Rho GDP dissociation inhibitor from ezrin. Released Rho GDP dissociation inhibitor interacted with and inhibited Rac 1 GTPase, allowing proper conduction of cytoskeleton-related signaling. More importantly, in addition to their cytoplasmic localization, both endogenous and over-expressed ERM proteins including ezrin have also been shown to be present in the nucleus, suggesting a possible role of Cdk5-ezrin signaling in the nucleus.

Ras guanine nucleotide releasing factor 1 (RasGRF1) is a member of exchange factor (RasGEF) family that is involved in the activation of Ras and Rac GTPases. RasGRF1 is expressed exclusively in the brain, and phosphorylation of RasGRF1 at Ser916 by PKA leads to activation of Ras. Kesavapany et al. showed that Cdk5 phosphorylates RasGRF1 at Ser731 (Kesavapany et al., 2006). Phosphorylation at Ser731 leads to the degradation of RasGRF1 via a calpain-dependent mechanism, which controls the steady-state levels of RasGRF1 in mammalian cells including neurons. Kesavapany et al. observed that in mitotic cells, overexpression of RasGRF1 leads to disorganization of nucleus, which can be rescued by active Cdk5 (Kesavapany et al., 2006). Furthermore, the phosphorylation mutant of RasGRF1 also caused nuclear condensation and fragmentation. The authors concluded that Cdk5-RasGRF1 controls the nuclear organization. Interestingly, the observed morphological features of the neurons undergoing nuclear disorganization are very similar to those of apoptotic cells. At present, it is not clear whether the nuclear disorganization is due to a direct role of Cdk5-RasGRF1 in regulation of nuclear function or an indirect consequence of neurons undergoing apoptosis under the experimental conditions. Besides RasGRF1, Cdk5 also regulates the signaling of other small GTPases by direct phosphorylation of Rac, Rac effector PAK1, and RasGRF2 (Kesavapany et al., 2004; Nikolic et al., 1998). Many of these small GTPases are present in the nucleus, raising the possibility that Cdk5 may play a similar role in regulating the signal of nuclear Ras family members.

Regulation of Nuclear Cdk5 and Its Activators

Not much is known about how Cdk5 and its activators themselves are regulated in the nucleus. Current evidence indicates that the level of Cdk5 and its activators as well as the activity of Cdk5 may be subjected to regulation.

Regulation of the Level and Nuclear Localization of p35 and p25

p25, which lacks the myristoylation site present in p35, adopts a different subcellular localization from p35 and becomes enriched in the peri-nuclear region and in the nucleus (Patrick et al., 1999). Since protease calpain cleaves p35 to p25, pathological activation of calpain has been proposed as a primary mode of generating p25, which in turn controls the nuclear levels of Cdk5 activator. However, subsequent studies have indicated that nuclear localization of Cdk5 activator is not restricted to p25. Gong et al. demonstrated that primary cortical neurons express p35 in the nucleus under normal culture conditions and increasing amount of nuclear p35 and/or p25 can be detected as neurons become more mature in culture (Gong et al., 2003) (Gong and Mao, personal communication), suggesting a developmental- or maturation-stage-dependent regulation of nuclear p35 and/or p25. A subsequent study by Saito et al. examined this issue in greater detail. Saito et al. showed that conversion of p35 to p25 is regulated by the phosphorylation status of p35 in a development-dependent manner. Specifically, a phosphorylated form of p35, which is resistant to the cleavage by calpain and targets p35 for proteasomal degradation, is more prevalent in the fetal brain (Saito et al., 2003). This reduced the probability of generating p25 in the fetal brain. In contrast, the un-phosphorylated form of p35, which is sensitive to calpain cleavage, appears to be dominant in the adult brain. These findings indicate that developmentally regulated phosphorylation status of p35 may serve as a mechanism modulating the generation of p25 in the brain. However, the precise nature of this phosphorylation status that dictates the susceptibility of p35 to calpain remains to be illustrated.

The study by Saito provides an explanation for how developing brain may control the generation of p25. A recent study by Fu et al. investigated the mechanisms that regulate the levels of p35 in the nucleus. It is known that proteins are transported between the cytoplasm and the nucleus by the nuclear pore complex. Proteins importins mediate active nuclear imports by association with protein cargos in a manner regulated by a small GTPase Ran. Fu et al. reported that importins-5, -7, and -β all interact with an N-terminal region of p35, which serves as a nuclear localization signal for p35, to transport p35 to the nucleus (Fu et al., 2006). Since p25 lacks this nuclear localization signal and fails to bind to importins, its nuclear localization would necessitate a distinct mechanism. While something is known about p35 nuclear import, the equivalent mechanism for Cdk5 is not understood. It is interesting to note that importins and Cdk5 bind to p35 in a mutually exclusive manner, suggesting that Cdk5 may employ a different pathway from that used by p35 for its nuclear localization.

Association of Cdk5 with Cyclins

It is widely accepted that Cdk5 is regulated by non-cyclin activators p35 and p39, and does not play a role in cell cycle progression. However, there are reports on the association of Cdk5 with cyclins. Xiong et al. first showed that cyclin D1 can form a complex with Cdk5 although such a complex does not exhibit any kinase activity (Xiong et al., 1992). Using a full-length human Cdk5 as bait, Guidato et al. detected interaction between Cdk5 and cyclin D2 in two independent yeast two-hybrid screens (Guidato et al., 1998). This interaction was confirmed by immunoprecipitation assays in mammalian cells. Similar to Xiong's finding, association with cyclin D2 did not appear to activate Cdk5. Curiously, overexpression of cyclin D2 seemed to reduce the Cdk5/p35 kinase activity. Yin et al. observed a co-increase in the expression of cyclin D1 and Cdk5 in human head and neck squamous cell carcinoma cells transplanted in nude mice. Yin et al. noted that Cdk5 kinase activity and the activity associated with cyclin D1 appear to follow the same kinetics. Based on these, the authors proposed that cyclin D1 and Cdk5 might function as a pair in regulation of carcinoma apoptosis. However, it is not clear whether cyclin D1 and Cdk5 actually associate with each other to function as an active complex in these cancer cells.

In addition to cyclin D, cyclin E has also been shown to associate with Cdk5. Matsunaga showed by immunohistochemistry approach that cyclin E is detected in the differentiating regions during neuronal development and in adult brain where Cdk5 is also abundant (Matsunaga, 2000). Cyclin E formed a complex with Cdk5 but cyclinE/Cdk5 complex did not exhibit any activity. Given that the author noted a change in subcellular localization of cyclin E from nucleus to cytoplasm during postnatal development of the brain, it is not certain whether the association of cyclin E with Cdk5 occurs in the nucleus, cytoplasm, or both compartments.

Finally, Rosales et al. has recently purified a bovine brain enzyme that phosphorylates Cdk5 at Ser159 (Rosales et al., 2003). The enzyme contains Cdk7 and cyclin H. The nuclear roles of Cdk7 include control of cell cycle, gene transcription, and DNA repair (Lolli and Johnson, 2005). The enzymatic complex immunoprecipitated from mouse brain with anti-Cdk7 or cyclin H antibodies phosphorylated Cdk5, leading to an enhancement in Cdk5/p25 activity. On the other hand, blocking Cdk7 reduced Cdk5 phosphorylation. These findings, if confirmed, would add a new layer of complexity to the regulation of Cdk5 and point to other nuclear processes that Cdk5 may be involved in.

Final Remarks

It is clear that we are only at the beginning of understanding the roles and regulation of Cdk5 and its activators in the nucleus. There are some contradictory findings waiting to be reconciled, key observations to be expanded, and exciting questions to be asked. Continuing this line of exploration should yield important insights and help us fully understand the roles of Cdk5 in cells.

Acknowledgements The authors apologize for not being able to cite all the relevant studies due to space limitation. This work is supported by NIH AG023695, NS048254, and ES0153170, and by the Robert W. Woodruff Health Sciences Center Fund (Z.M).

References

Ahuja, H.S., Zhu, Y., and Zakeri, Z. (1997). Association of cyclin-dependent kinase 5 and its activator p35 with apoptotic cell death. Dev Genet *21*, 258–267.

Dhavan, R., and Tsai, L.H. (2001). A decade of CDK5. Nat Rev Mol Cell Biol *2*, 749–759.

Fischer, A., Sananbenesi, F., Wang, X., Dobbin, M., and Tsai, L.H. (2007). Recovery of learning and memory is associated with chromatin remodelling. Nature *447*, 178–182.

Fu, A.K., Fu, W.Y., Ng, A.K., Chien, W.W., Ng, Y.P., Wang, J.H., and Ip, N.Y. (2004). Cyclin-dependent kinase 5 phosphorylates signal transducer and activator of transcription 3 and regulates its transcriptional activity. Proc Natl Acad Sci USA *101*, 6728–6733.

Fu, X., Choi, Y.K., Qu, D., Yu, Y., Cheung, N.S., and Qi, R.Z. (2006). Identification of nuclear import mechanisms for the neuronal Cdk5 activator. J Biol Chem *281*, 39014–39021.

Gao, C.Y., Zakeri, Z., Zhu, Y., He, H., and Zelenka, P.S. (1997). Expression of Cdk5, p35, and Cdk5-associated kinase activity in the developing rat lens. Dev Genet *20*, 267–275.

Gong, X., Tang, X., Wiedmann, M., Wang, X., Peng, J., Zheng, D., Blair, L.A., Marshall, J., and Mao, Z. (2003). Cdk5-mediated inhibition of the protective effects of transcription factor MEF2 in neurotoxicity-induced apoptosis. Neuron *38*, 33–46.

Gregoire, S., Tremblay, A.M., Xiao, L., Yang, Q., Ma, K., Nie, J., Mao, Z., Wu, Z., Giguere, V., and Yang, X.J. (2006). Control of MEF2 transcriptional ac-tivity by coordinated phosphory-lation and sumoylation. J Biol Chem *281*, 4423–4433.

Guidato, S., McLoughlin, D.M., Grierson, A.J., and Miller, C.C. (1998). Cyclin D2 interacts with cdk-5 and modulates cellular cdk-5/p35 activity. J Neurochem *70*, 335–340.

Hamdane, M., Bretteville, A., Sambo, A.V., Schindowski, K., Begard, S., Delacourte, A., Bertrand, P., and Buee, L. (2005). p25/Cdk5-mediated retinoblastoma phosphorylation is an early event in neuronal cell death. J Cell Sci *118*, 1291–1298.

Ino, H., and Chiba, T. (1996). Intracellular localization of cyclin-dependent kinase 5 (CDK5) in mouse neuron: CDK5 is located in both nucleus and cytoplasm. Brain Res *732*, 179–185.

Kawauchi, T., Chihama, K., Nabeshima, Y., and Hoshino, M. (2006). Cdk5 phosphorylates and stabilizes p27kip1 contributing to actin organization and cortical neuronal migration. Nat Cell Biol *8*, 17–26.

Kesavapany, S., Amin, N., Zheng, Y.L., Nijhara, R., Jaffe, H., Sihag, R., Gutkind, J.S., Takahashi, S., Kulkarni, A., Grant, P., and Pant, H.C. (2004). p35/cyclin-dependent kinase 5 phosphorylation of ras guanine nucleotide releasing factor 2 (RasGRF2) mediates Rac-dependent Extracellular Signal-regulated kinase 1/2 activity, altering RasGRF2 and microtubule-associated protein 1b distribution in neurons. J Neurosci *24*, 4421–4431.

Kesavapany, S., Pareek, T.K., Zheng, Y.L., Amin, N., Gutkind, J.S., Ma, W., Kulkarni, A.B., Grant, P., and Pant, H.C. (2006). Neuronal nuclear organization is controlled by cyclin-dependent kinase 5 phosphorylation of ras guanine nucleotide releasing factor-1. Neuro-signals *15*, 157–173.

Kino, T., Ichijo, T., Amin, N.D., Kesavapany, S., Wang, Y., Kim, N., Rao, S., Player, A., Zheng, Y.L., Garabedian, M.J., et al. (2007). Cyclin-dependent kinase 5 differentially regulates the transcriptional activity of the glucocorticoid receptor through phosphorylation: clinical implications for the nervous system response to glucocorticoids and stress. Mol Endocrinol *21*, 1552–1568.

Lacy, E.R., Wang, Y., Post, J., Nourse, A., Webb, W., Mapelli, M., Musacchio, A., Siuzdak, G., and Kriwacki, R.W. (2005). Molecular basis for the specificity of p27 toward cyclin-dependent kinases that regulate cell division. J Mol Biol *349*, 764–773.

Lee, J.H., and Kim, K.T. (2007). Regulation of cyclin-dependent kinase 5 and p53 by ERK1/2 pathway in the DNA damage-induced neuronal death. J Cell Physiol *210*, 784–797.

Lee, K.Y., Helbing, C.C., Choi, K.S., Johnston, R.N., and Wang, J.H. (1997). Neuronal Cdc2-like kinase (Nclk) binds and phosphorylates the retinoblastoma protein. J Biol Chem *272*, 5622–5626.

Lee, K.Y., Rosales, J.L., Lee, B.C., Chung, S.H., Fukui, Y., Lee, N.S., Lee, K.Y., and Jeong, Y.G. (2004). Cdk5/p35 expression in the mouse ovary. Mol Cells *17*, 17–22.

Lew, J., Winkfein, R.J., Paudel, H.K., and Wang, J.H. (1992). Brain proline-directed protein kinase is a neurofilament kinase which displays high sequence homology to p34cdc2. J Biol Chem *267*, 25922–25926.

Li, Z., David, G., Hung, K.W., DePinho, R.A., Fu, A.K., and Ip, N.Y. (2004). Cdk5/p35 phosphorylates mSds3 and regulates mSds3-mediated repres-sion of transcription. J Biol Chem *279*, 54438–54444.

Lin, H., Chen, M.C., Chiu, C.Y., Song, Y.M., and Lin, S.Y. (2007). Cdk5 regulates STAT3 activation and cell proliferation in medullary thyroid carcinoma cells. J Biol Chem *282*, 2776–2784.

Lolli, G., and Johnson, L.N. (2005). CAK-Cyclin-dependent activating kinase: a key kinase in cell cycle control and a target for drugs? Cell Cycle *4*, 572–577.

Mao, Z., Bonni, A., Xia, F., Nadal-Vicens, M., and Greenberg, M.E. (1999). Neuronal activity-dependent cell survival mediated by transcription factor MEF2. Science *286*, 785–790.

Matsunaga, Y. (2000). [Expression of cyclin E in postmitotic cells in the central nervous system]. Kokubo Gakkai Zasshi *67*, 169–181.

Meyerson, M., Enders, G.H., Wu, C.L., Su, L.K., Gorka, C., Nelson, C., Harlow, E., and Tsai, L.H. (1992). A family of human cdc2-related protein kinases. Embo J *11*, 2909–2917.

Musa, F.R., Tokuda, M., Kuwata, Y., Ogawa, T., Tomizawa, K., Konishi, R., Takenaka, I., and Hatase, O. (1998). Expression of cyclin-dependent kinase 5 and associated cyclins in Leydig and Sertoli cells of the testis. J Androl *19*, 657–666.

Neystat, M., Rzhetskaya, M., Oo, T.F., Kholodilov, N., Yarygina, O., Wilson, A., El-Khodor, B.F., and Burke, R.E. (2001). Expression of cyclin-dependent kinase 5 and its activator p35 in models of induced apoptotic death in neurons of the substantia nigra in vivo. J Neurochem 77, 1611–1625.

Nikolic, M., Chou, M.M., Lu, W., Mayer, B.J., and Tsai, L.H. (1998). The p35/Cdk5 kinase is a neuron-specific Rac effector that inhibits Pak1 activity. Nature 395, 194–198.

O'Hare, M.J., Kushwaha, N., Zhang, Y., Aleyasin, H., Callaghan, S.M., Slack, R.S., Albert, P.R., Vincent, I., and Park, D.S. (2005). Differential roles of nuclear and cytoplasmic cyclin-dependent kinase 5 in apoptotic and exci-totoxic neuronal death. J Neurosci 25, 8954–8966.

Patrick, G.N., Zukerberg, L., Nikolic, M., de la Monte, S., Dikkes, P., and Tsai, L.H. (1999). Conversion of p35 to p25 deregulates Cdk5 activity and promotes neurodegeneration. Nature 402, 615–622.

Qu, D., Li, Q., Lim, H.Y., Cheung, N.S., Li, R., Wang, J.H., and Qi, R.Z. (2002). The protein SET binds the neuronal Cdk5 activator p35nck5a and modulates Cdk5/p35nck5a activity. J Biol Chem 277, 7324–7332.

Rosales, J., Han, B., and Lee, K.Y. (2003). Cdk7 functions as a cdk5 activating kinase in brain. Cell Physiol Biochem 13, 285–296.

Saito, T., Onuki, R., Fujita, Y., Kusakawa, G., Ishiguro, K., Bibb, J.A., Kishimoto, T., and Hisanaga, S. (2003). Developmental regulation of the proteolysis of the p35 cyclin-dependent kinase 5 activator by phosphorylation. J Neurosci 23, 1189–1197.

Smith, P.D., Mount, M.P., Shree, R., Callaghan, S., Slack, R.S., Anisman, H., Vincent, I., Wang, X., Mao, Z., and Park, D.S. (2006). Calpain-regulated p35/cdk5 plays a central role in dopaminergic neuron death through modulation of the transcription factor myocyte enhancer factor 2. J Neurosci 26, 440–447.

Tang, X., Wang, X., Gong, X., Tong, M., Park, D., Xia, Z., and Mao, Z. (2005). Cyclin-dependent kinase 5 mediates neurotoxin-induced degradation of the transcription factor myocyte enhancer factor 2. J Neurosci 25, 4823–4834.

Tsai, L.H., Takahashi, T., Caviness, Jr., V.S., and Harlow, E. (1993). Activity and expression pattern of cyclin-dependent kinase 5 in the embryonic mouse nervous system. Development 119, 1029–1040.

Ubeda, M., Kemp, D.M., and Habener, J.F. (2004). Glucose-induced expression of the cyclin-dependent protein kinase 5 activator p35 involved in Alzheimer's disease regulates insulin gene transcription in pancreatic beta-cells. Endocrinology 145, 3023–3031.

Ubeda, M., Rukstalis, J.M., and Habener, J.F. (2006). Inhibition of cyclin-dependent kinase 5 activity protects pancreatic beta cells from glucotoxicity. J Biol Chem 281, 28858–28864.

Xiong, Y., Zhang, H., and Beach, D. (1992). D type cyclins associate with multiple protein kinases and the DNA replication and repair factor PCNA. Cell 71, 505–514.

Yang, H.S., Alexander, K., Santiago, P., and Hinds, P.W. (2003). ERM proteins and Cdk5 in cellular senescence. Cell Cycle 2, 517–520.

Zhang, J., Krishnamurthy, P.K., and Johnson, G.V. (2002). Cdk5 phosphorylates p53 and regulates its activity. J Neurochem 81, 307–313.

Zhang, Q., Ahuja, H.S., Zakeri, Z.F., and Wolgemuth, D.J. (1997). Cyclin-dependent kinase 5 is associated with apoptotic cell death during development and tissue remodeling. Dev Biol 183, 222–233.

Cdk5 May Be an Atypical Kinase, but Not in the Way You Think

Li Wang, Jie Zhang and Karl Herrup

Abstract Cyclin-dependent kinase 5 (Cdk5) is a non-traditional CDK. It relies on two specific activators—p35 and p39—that are structurally similar to cyclins but genetically distinct. Analysis of the Cdk5 knockout (or the double p35/p39 knockout) has led to the view that the primary function of Cdk5 is in the migration and maturation of embryonic post-mitotic neurons. The literature has no reference to a role of Cdk5 in normal cell cycle regulation. Recent data from our lab, however, suggest that while it may not function as a traditional CDK and facilitate cell cycle progression, it does play a crucial role as a cell cycle suppressor in normal post-mitotic neurons. In this chapter, we review the evidence that this unique function is important for neuronal cell survival and differentiation. The action of Cdk5 in neurons appears to have sub-cellular specificity as well. We present early evidence that it is the nuclear form of Cdk5 that is crucial for holding the cell cycle in check. Cdk5 is found to exit the nucleus in stressed neurons at risk for death. The shift in sub-cellular location is accompanied by cell cycle re-entry and neuronal death. This "new" function of Cdk5 raises cautions in the design of Cdk5-directed drugs for the therapy of neurodegenerative diseases

Introduction

The Cyclin-Dependent Kinase Family

Cyclin-dependent kinases (CDKs) are a family of serine/threonine kinases that are important for the progression of normal cell cycle events. Eleven members of the CDK family have been identified to date. They are well conserved at the level of primary amino acid sequence (sharing 40–75% identity), and six of them, Cdk1–4, 6, and 7 are considered important for the cell cycle regulation

K. Herrup
Department of Cell Biology and Neuroscience, Nelson Biological Laboratories,
Rutgers University, 604 Allison Road, Piscataway, NJ 08854, USA
e-mail: herrup@biology.rutgers.edu

N.Y. Ip, L.-H. Tsai (eds.), *Cyclin Dependent Kinase 5 (Cdk5)*,
DOI: 10.1007/978-0-387-78887-6_9, © Springer Science+Business Media, LLC 2008

(Morgan, 1997). Of the others, Cdk5 is known to be important for nervous system development and function, while Cdk8 and Cdk9 are appreciated as regulators of transcription (Akoulitchev et al., 2000; Sano and Schneider, 2003). The function of Cdk10 is not known, but it has been shown to interact with the transcription factor Ets2 (Kasten and Giordano, 2001).

Cyclin-Dependent Kinases and Normal Cell Cycle Regulation

The cell cycle consists of four sequential and tightly regulated phases (Fig. 1). G0 is the resting state of the cells. In a normal cell cycle, there are four recognized phases—G1 phase (Gap1), S phase (DNA synthesis), G2 phase (Gap 2), and M phase (mitosis). This progression is governed by different CDK activities at different stages. For full activity, a CDK must partner with

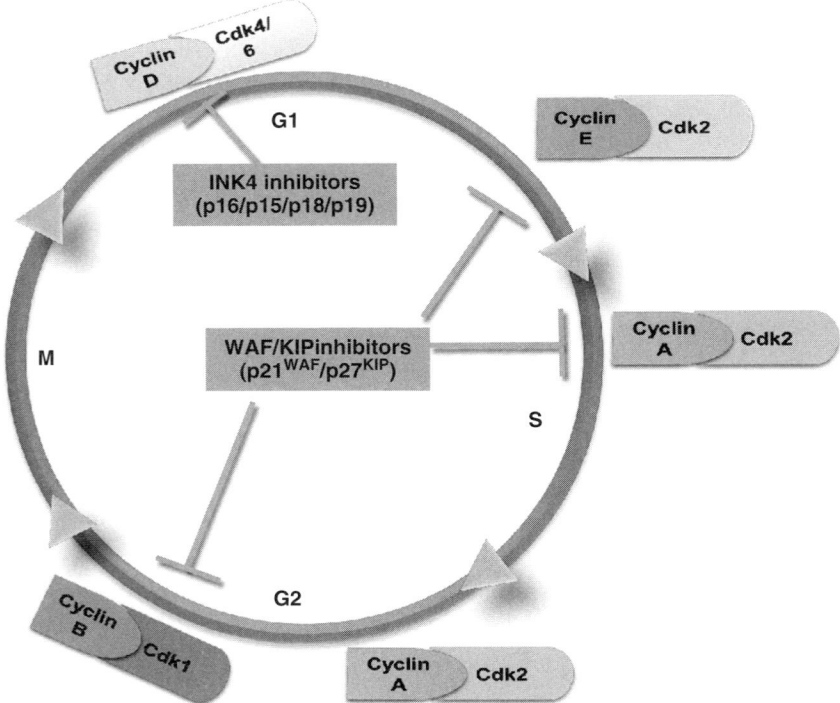

Fig. 1 A highly simplified diagram of the proteins that drive a vertebrate cell cycle through its four phases, G1, S, G2, and M. We have emphasized the sequential action of the cyclin-dependent kinases, their activating cyclins, and the cyclin-dependent kinase inhibitors. (*See* Color Insert)

a regulatory subunit known as a cyclin. While the levels of CDK proteins tend to remain constant during the cell cycle, each cyclin has a unique pattern of increases and decreases. The cyclin pattern thus determines the activity of its partner CDK at each stage of the cycle.

For example, the Cdk3–cyclin C complex helps cells exit from G0 and enter G1 phase by stimulating the tumor suppressor retinoblastoma protein (pRb) phosphorylation (Ren and Rollins, 2004). The D-type cyclins are important integrators of mitogenic signaling, as they are the key transcriptional products stimulated by the RAS/RAF/MAPK pathway. Cyclin D activates Cdk4 and Cdk6, which then phosphorylate pRb. This hyperphosphorylated form of pRb will release the transcription factor, E2F, from the Rb/E2F repressor complex. The free E2F can now activate the expression of many genes that are needed for the transition from G1 to S phase such as cyclin E and cyclin A as well as other genes that are needed at later phases. The expression of cyclin E activates Cdk2 to effect further phosphorylation of pRb, thereby enabling the cell to proceed into late G1 phase. During S phase, cyclin E is replaced by cyclin A on Cdk2. The Cdk2–cyclin A complex controls the progression through S phase by phosphorylation of a series of proteins which are necessary for DNA synthesis such as DP-1 (an inhibitor of DNA binding) and CDC6 (an initiator of DNA replication). After S-phase DNA replication is completed, the cell enters G2 phase and prepares for mitosis. At the end of S phase, cyclin A also associates and activates Cdk1. Both the Cdk2–cyclin A and the Cdk1–cyclin A complex are important for the S–G2 transition. During G2 phase, cyclin A is degraded whereas the level of cyclin B increases. Cyclin B binds and activates Cdk1, which is thought to be important for triggering mitosis. Cdk1–cyclin B controls many proteins that are required for M-phase progression and cell cycle exit. At the end of M phase, one cell separates into two daughter cells and replication is complete (Malumbres and Barbacid, 2005).

Regulation of Cyclin-Dependent Kinases

The foregoing description makes it plain that a well-regulated cell cycle requires the activity of the mitotic CDKs to be strictly regulated. The regulation occurs through both activation and inhibition. Monomeric, unmodified CDKs have no kinase activity. As their name implies, they must bind to their regulatory cyclins as an initial step in their activation. The cyclins are themselves a diverse group of proteins. There are eight different types of cyclins, distinguished by the letters A–H. They bind and activate distinct CDKs at distinct cell cycle stages (Malumbres and Barbacid, 2005; Bloom and Cross, 2007). Even after cyclin binding, however, CDKs are only partially activated. To be fully activated, they must be phosphorylated at a conserved threonine residue located in a region known as the catalytic T loop. Phosphorylation is accomplished by the CDK activating kinase (CAK)—Cdk7–cyclin H—a distantly related

CDK–cyclin pair (Shuttleworth, 1995). This modification flattens the T –loop, resulting in critical changes in the substrate-binding site. Phosphorylation acts primarily to enhance the binding of protein substrates (Russo et al., 1996), but it also enhances the affinity of the CDK–cyclin interactions (Desai et al., 1995).

CDKs are also regulated by inhibition. The activity of the CDK–cyclin complex can be reduced by the phosphorylation of the CDK subunit on inhibitory sites by dual-specificity kinases (such as WEE1 and MYT1). This inhibition is relieved when the CDC25 phosphatases dephosphorylate these residues. Second, the binding of CDK inhibitors known as CKIs to the CDK–cyclin complex also can inhibit the kinase activity. There are two types of CKIs—the INK family and the WAF/KIP family. There are four members in INK family (p16^{INK4a}, p15^{INK4b}, p18^{INK4c}, and p19^{INK4d}). They serve as competitive inhibitors of the Cdk4/6–cyclin D complex, from which Cdk4 dissociates when they bind. They can also bind Cdk4 directly and prevent its binding to cyclin D (Lees, 1995). There are three members in the WAF/KIP family—p21^{WAF1}, p27^{KIP1}, and p57^{KIP2}. They share a conserved region at their N-terminus, which is important for binding and inhibition of CDK–cyclin complexes. Unlike the INK4 family, WAF/KIP family members bind to the entire CDK–cyclin complex, inhibiting its activity. They are able to inhibit all of the G1 CDK–cyclin complexes as well as the Cdk1–cyclin B complex (Harper et al., 1993, 1995; Xiong et al., 1993; Matsuoka et al., 1995).

Cell Cycle Regulation in Post-Mitotic Neurons

Neurons of the central nervous system are generated during development in a mitogenic region known as the ventricular zone (and the associated subventricular zone). The mechanisms that ensure the orderly progression of the many rounds of cell cycling in these regions and block further cycling once differentiation begins are virtually unknown. What is known is that once the young neuroblasts emigrate from this zone, they will never divide again. While ensuring this block to future division might seem to be primarily a problem of differentiation rather than cell cycle regulation, accumulating evidence suggests that the neuron must constantly suppress the cell cycle process in order to remain in the post-mitotic state throughout life (reviewed in Herrup and Yang, 2007)

Development and In Vitro Systems

In the past few years, more and more studies have shown that neurons that have left the ventricular zone, but are still immature, can go back to the cell cycle. Rather than leading to cell division, however, this event leads to a rapid cell death. For example, over-expression of the SV40 T-antigen oncogene can force a maturing nerve cell into a cell cycle. When this occurs in Purkinje cells, they are found to incorporate bromodeoxyuridine (BrdU), but instead of dividing, they die (Feddersen et al., 1992, 1995). A similar phenomenon is observed in retinal photoreceptors (al-Ubaidi et al., 1992, 1997). Hair cells in the ears of mice with a targeted deletion of the p19IND4d (one of the CKIs) undergo a similar degeneration (Chen et al., 2003). Further, mouse embryos that are homozygous for null alleles of the retinoblastoma gene contain neurons at all levels of the neuraxis that re-enter a cell cycle and die (Clarke et al., 1992; Jacks et al., 1992; Lee et al., 1992).

The death of neurons following target deprivation *in vivo* can also proceed by a process involving an unscheduled cell cycle. In *staggerer* ($Rora^{sg/sg}$) mice, cerebellar Purkinje cells never fully develop because of the deletion of the orphan nuclear receptor, RORα. This leads to the loss of 100% of the cerebellar granule cells, and before their death, the granule cells engage in DNA synthesis and express at least two cell cycle proteins—cyclin D and PCNA (Herrup and Busser, 1995). The dying granule cells in *lurcher* mice show a similar engagement of cell cycle processes (Herrup and Busser, 1995).

The linkage between cell death and cell cycle re-entry has also been shown in cell culture systems. Similar to their *in vivo* counterparts, cerebellar granule neurons *in vitro* incorporate BrdU and express cell cycle–related proteins prior to death induced by different stimuli (Giovanni et al., 1999). These culture experiments have particular relevance, as the death of the stressed neurons can be effectively blocked or delayed by blocking the cell cycle events (Farinelli et al., 1996; Park et al., 1996, 1997a, b; Appert-Collin et al., 2006). These *in vitro* results provide direct evidence that ectopic cell cycle events play a causal role in neuronal death.

Aberrant Cell Cycle Events in Neurodegenerative Diseases

Neuronal cell cycle initiation is also found in several different neurodegenerative diseases and their mouse models. The best studied of these is Alzheimer's disease (AD). Several groups have shown that there is increased expression of multiple cell cycle–related protein such as cyclin E, PCNA, Ki67, cyclin B1, cyclin D, and Cdk4 in affected regions of the AD brain—including hippocampus, locus coeruleus, and dorsal raphe (Nagy et al., 1997; Vincent et al., 1997; Busser et al., 1998). The neuronal loss is severe in these regions in AD patients. In unaffected regions such as the cerebellar cortex, however, the expression of cell cycle–related proteins is not found. Thus, as in development, ectopic cell cycle events are associated with neuronal death. In addition to the presence of these proteins, there is increased activity of CDC2–cyclin B complex and its activating phosphatase, CDC25, in the AD brain (Vincent et al., 1997, 2001; Ding et al., 2000).

Further, through the use of fluorescent in situ hybridization (FISH), Yang et al. (2001) offer strong support for the presumption that this ectopic cell cycle activity is productive. This work shows that actual DNA replication has taken place in affected regions of the AD brain. Yet while there is clear evidence of DNA replication and G2/M-phase marker expression, no cell division (M phase) is found. Instead, this ectopic cell cycle progression appears to proceed to neuronal cell death, and it does so at all stages of AD (Yang et al., 2001, 2003). Recent studies show that forcing a neuron into a cell cycle can induce neurodegenerative changes which are similar to those seen in AD. These data provide support for the notion that ectopic cell cycle events are involved in the generation of the characteristic pathological hallmarks of AD (McShea et al., 2007; Park et al., 2007).

Ectopic cell cycle protein expression is not only found in AD, but has also been reported in the brain of patients with Parkinson's disease (PD), Ataxia-telangiectasia (AT), amyotrophic lateral sclerosis (ALS), Down's syndrome, Pick's Disease, Niemann–Pick syndrome type C, and brain injury induced by focal ischemia (Nagy et al., 1997; Husseman et al., 2000; Love, 2003; Nguyen et al., 2003; Ranganathan and Bowser, 2003; Wen et al., 2005; Yang and Herrup, 2005; Höglinger et al., 2007). Recent results have shown that blocking CDK activity can block neuronal death induced by focal ischemia. This strongly suggests that, as in the *in vitro* studies, cell cycle events are not merely a correlate of cell death, but play an active role in bringing them about (Rashidian et al., 2005; Wen et al., 2005).

Cell Cycle Events in Mouse Disease Models

Cell cycle activity is found in neurons of the mouse models of AD and AT diseases. Yang et al. report abnormal cell cycle events in the three transgenic models of AD—R1.40, Tg2576, and APP23 (Yang et al., 2006)—as well as in a mouse genocopy of AT (Yang and Herrup, 2005). There is ectopic cell cycle protein expression and DNA replication in these transgenic lines. In the AD mouse, ectopic cell cycle events first appear at 6 months of age, which is much earlier than the β–amyloid plaque deposition; plaques are not found until the animals are over 1 year old. Curiously, however, there is no neuronal cell loss in the mouse models, even in regions that show heavy cell cycle involvement for many months. Thus, the appearance of cell cycle events in these neurons is not directly linked to neuronal death. The appearance of cell cycle proteins coupled with DNA replication strongly indicates that those neurons are under some stress and likely to be unhealthy. This leads to the suggestion that cell cycle activity is a sensitive detector of neuronal distress and that this distress can be uncoupled from the final death of the cell.

Cdk5: An Atypical CDK

As their name suggests, most CDKs are involved in cell cycle progression. From this viewpoint, Cdk5 is an atypical member of the CDK family. It was originally identified by its close sequence homology (60%) to CDC2 (Lew et al., 1992; Meyerson et al., 1992). Yet ectopic expression of Cdk5 in mammalian cells or yeast does not promote cell cycle progression (Meyerson et al., 1992; van den Heuvel and Harlow, 1993). Nonetheless, Cdk5 phosphorylates many of the same substrates as Cdk2 or Cdk1, making the absence of a role in cell cycle control a curious observation.

The Regulation of Cdk5

Like the traditional CDKs, the activation of Cdk5 requires its association with a regulatory partner. It can bind several of the normal cyclins, but it is not activated by them (Xiong et al., 1992; Zhang et al., 1993; Miyajima et al., 1995). It is activated instead by its specific partners—p35 and p39. These two proteins are structurally similar to cyclins, yet they share no homology to cyclins at the amino acid level. And since the phenotype of $p35^{-/-}$ and $p39^{-/-}$ double knockout mice is virtually identical to the Cdk5 knockout itself (Ko et al., 2001), the suggestion is that p35 and p39 are the only two relevant activators of Cdk5 during development.

A high level of Cdk5 kinase activity is primarily detected in the nervous system (Lew et al., 1994; Tsai et al., 1994; Tang et al., 1995) as this is where the levels of p35 and p39 are the highest. Nonetheless, the Cdk5 protein itself is found in all tissues. Besides the brain, low levels of Cdk5 kinase activity are also present in adult mouse prostate and embryonic limb buds (Zhang et al., 1997). The expression of p35 and p39 in brain is found in overlapping, but distinct, sub-cellular compartments including growth cones and synapses (Humbert et al., 2000a, b; Niethammer et al., 2000; Fu et al., 2001). Evidence points to the fact that p39 can compensate for most of the functions of p35 since, in $p35^{-/-}$ mice, Cdk5 kinase activity is present, albeit reduced (Hallows et al., 2003). The reverse is true as well; there are no obvious abnormalities in p39-deficient mice (Ko et al., 2001).

As discussed above, full activation of mitotic CDKs depends on the phosphorylation of a conserved residue in its activating T loop. Phosphorylation of the homologous loop of Cdk5, however, is not required for its maximal activation (Poon et al., 1997). One study shows that the structure of the Cdk5–p25 complex (p25 is a carboxy-terminal proteolytic product of p35) is in a fully active conformation which is indistinguishable from phosphorylated Cdk2 bound to cyclin A. This offers a hint as to why Cdk5 does not need to be phosphorylated to achieve full activity (Tarricone et al., 2001). As mentioned above, the mitotic CDKs are also regulated by the phosphorylation of Thr14 and Thr15 by kinases Wee1 and Myt1 (which serves to inhibit their activity). In

Cdk5, these two sites are conserved, but are not phosphorylated by Wee1 *in vitro* (Poon et al., 1997). Phosphorylation of Thr14 on Cdk5 by an inhibitory protein kinase purified from bovine thymus cytosol inactivates Cdk5 (Matsushita et al., 1996; Matsuura and Wang, 1996). Thr15 on Cdk5 is phosphorylated by c-Abl, but rather than blocking Cdk5 activity, this phosphorylation increases Cdk5 kinase activity (Zukerberg et al., 2000).

Function of Cdk5

The Roles of Cdk5 in the Adult

Synapse Function

Cdk5 and its activators p35 and p39 are present in sub-cellular fractions that are enriched for synaptic membranes (Humbert et al., 2000a; Niethammer et al., 2000), where it appears to play an important role in synaptic functions, including the modulation of neurotransmitter release by the phosphorylation of various pre- and post-synaptic proteins (Tomizawa et al., 2002; Wang et al., 2003) and by modulating CNS dopamine signaling (Bibb et al., 1999; Chergui et al., 2004; Moy and Tsai, 2004). Cdk5 also functions in the synapse to regulate endocytosis, an important step in synaptic transmission (Tomizawa et al., 2002; Tan et al., 2003).

Neuronal Survival

In human AD, it has been reported that there is an unusual accumulation of p25 due to the calpain-mediated cleavage of p35. This accumulated p25 results in elevated Cdk5 activity (Grynspan et al., 1997; Lee et al., 2000). Since Cdk5 can function as a tau kinase and since hyperphosphorylated tau is the major constituent of the neurofibrillary tangle, a pathological hallmark of AD, it has been proposed that elevated Cdk5 activity contributes to AD pathology, including neuronal degeneration. Cdk5 interacts closely with early neurofibrillary tangles in AD patients, further supporting this idea. Further, in two different lines, transgenic mice that overexpress p25, hyperphosphorylated tau, neuorofibrillary tangles, and neurodegeneration are found (Ahlijanian et al., 2000; Van den Haute et al., 2001; Bian et al., 2002; Cruz et al., 2003; Noble et al., 2003). It should be noted, however, that the activity of other kinases such as Gsk3β is also increased (Noble et al., 2003), leaving a level of uncertainty as to which kinase plays the main role in Aβ-related tau hyperphosphorylation and neuronal death (Bian et al., 2002; Hallows et al., 2003; Tandon et al., 2003; Giese et al., 2005; Hallows et al., 2006; Plattner et al., 2006).

In addition to AD, elevated levels of Cdk5 and p35 are observed in the cell body of apoptotic cells following ischemia (Hayashi et al., 1999) and in models of PD (Neystat et al., 2001). Elevated p25 and Cdk5 kinase activity are also found in the mouse model of ALS as well as in human ALS patients

(Nguyen et al., 2001; Nguyen and Julien, 2003). Together, these observations suggest an important and continuous role of Cdk5 in the physiology of the adult neurons.

The Roles of Cdk5 in Development

Migration

Cdk5-deficient mice reveal an important role of Cdk5 in CNS development. $Cdk5^{-/-}$ homozygotes die around birth with severe disruptions in the neuronal layering of cerebral cortex, hippocampus, cerebellum, and olfactory bulb (Ohshima et al., 1996, 1999; Gilmore et al., 1998). In the cerebral cortex of $Cdk5^{-/-}$ mice, after the early-arriving neurons of layer VI split the preplate, the later-born neurons cannot migrate past their predecessors to their appropriate superficial positions. Instead, they stall underneath the subplate.

Neuronal migration depends on the regulation of actin, microtubule, and intermediate-filament cytoskeletal components, and modulation of the cell adhesion. The kinase activity of Cdk5 is involved in all these processes. Cdk5 can phosphorylate and facilitate the degradation of nestin, an intermediate filament protein found in progenitor cells and downregulated after neuronal differentiation begins (Sahlgren et al., 2003). Nestin continues to be expressed in $Cdk5$-deficient cortex. Consistent with the delayed nestin degradation, there is also reduced expression of neuron-specific class III β-tubulin protein (an early neuronal marker) and a very low level of expression of Map2 (a mature neuronal marker). These data suggest that Cdk5 is involved in cytoskeletal maturation and neuronal differentiation (Cicero and Herrup, 2005). Cdk5 is also involved in the post-translational modification of microtubule-associated proteins such as tau (Patrick et al., 1999). In Cdk5-deficient neurons, the basal level of tau phosphorylation is reduced compared with the wild-type neurons (Cicero, Wang et al., and unpublished observations). Recent studies have identified several novel Cdk5 substrates that are involved in neuronal migration: focal adhesion kinase (FAK) and doublecortin (DCX). Phosphorylation of FAK by Cdk5 at Ser732 is involved in the regulation of a centrosome-associated microtubule structure to promote nuclear translocation, which is important for neuronal positioning (Xie et al., 2003). Phosphorylation of DCX by Cdk5 at Ser297 decreases its ability to bind and stabilize microtubules, and this phosphorylation is necessary for DCX-induced migration (Tanaka et al., 2004). These studies indicate that Cdk5 governs neuronal migration in the developing neocortex by phosphorylation of various proteins involved in these processes.

Survival

Several lines of evidence indicate that Cdk5 activity is also needed for neuronal survival. In $Cdk5^{-/-}$ mice, swelling of cell soma and nuclear marginalization is observed in brainstem and spinal cord neurons (Ohshima et al., 1996). In Cdk5 chimeric mice, the percentage of $Cdk5^{-/-}$ neurons present in the cerebral cortex is much lower than would be predicted if Cdk5 deficiency had no effect on cell survival (Gilmore and Herrup, 2001). Further evidence comes from large number of cells that are stained with either TUNEL or activated caspase-3 in the cortical plate of $Cdk5^{-/-}$ brains. This indicates substantial neuronal loss during the final maturation of even those neurons that migrate as far as cortical plate (Cicero and Herrup, 2005).

The activators of Cdk5, p35 and p39, are cleaved, respectively, into p25 and p29 by calpain (Lee et al., 2000; Patzke and Tsai, 2002), and Cdk5–p25 has increased kinase activity relative to Cdk5–p35. The generation of the p25 fragment from p35 often accompanies apoptosis upon challenge by a variety of apoptotic stimuli, like neurotoxicity, ischemia, and oxidative stress (Lee et al., 2000; Nath et al., 2000; Fu et al., 2002; Zhu et al., 2002). The phenotype of p25 overexpressing mice also suggests that too much Cdk5 activity may be as bad as too little. Further, the mislocalization of Cdk5, as well as its hyperactivation in neurons, may alter substrate specificity and trigger various events that induce neurodegeneration. Indeed, introduction of p25 into cultured primary neurons induces neurite retraction, microtubule collapse, and apoptosis (Patrick et al., 1999).

Cdk5 and the Cell Cycle Regulation in Neurons

As stated earlier Cdk5 is typically viewed as having no role in cell cycle regulation. This is somewhat of a mystery, given the high level of structural homology of Cdk5 with the traditional CDKs, and the intimate association of Cdk5 with many of the more traditional proteins typically associated with cell cycle regulation. It has been reported, for example, that Cdk5 binds normal cyclins such as cyclin D and cyclin E. It can even bind the DNA polymerase component, PCNA (Xiong et al., 1992; Zhang et al., 1993; Miyajima et al., 1995); however, no Cdk5 kinase activity is detected in those complexes.

Normal mammalian cells enter a quiescent stage (referred to as G0) when mitosis is arrested by serum withdrawal. When serum is restored, mitotic activity resumes, and when this happens the first step in executing a cell cycle is to shift from G0 to G1 (Fig. 2). In considering the process of cell cycle–related neuronal death, we are struck by the similarities between the stimuli that induce it and the consequences of giving back serum to normal cultured cells. In pursuing this analogy, several important concepts have come to light that make the relationship between Cdk5 and neuronal cell cycle regulation seem much less distant than is typically assumed.

The role of p27 is an example of one such concept. It has been reported that p27 levels are high while CDK activity is low in G0 cells. This is believed to be due to the binding of p27 to the CDK–cyclin complex. During cell cycle re-entry, p27 becomes phosphorylated on Thr187, which targets the protein for

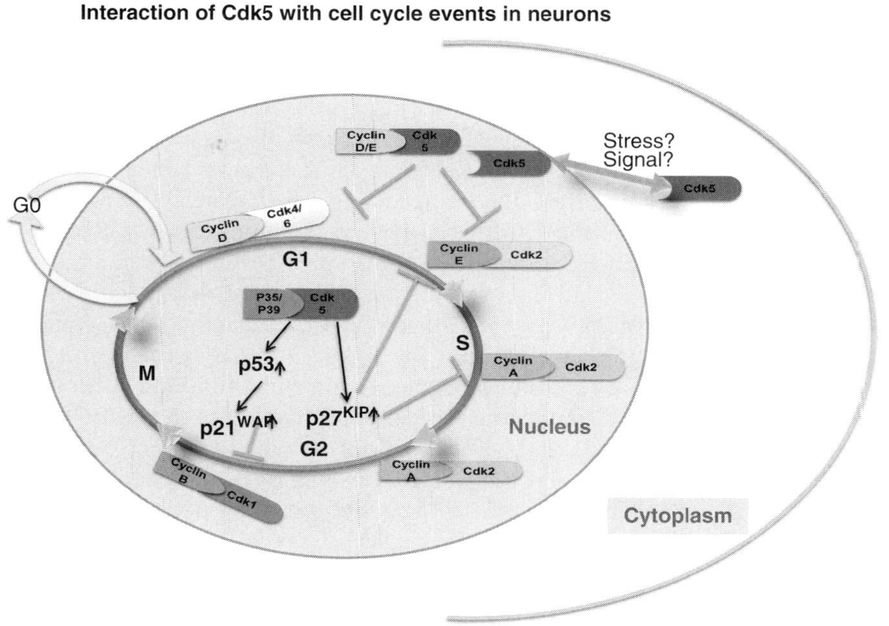

Interaction of Cdk5 with cell cycle events in neurons

Fig. 2 A model for the actions of Cdk5 in its role as a cell cycle inhibitor. Several possible modes of action are highlighted including the shuttling of the Cdk5 protein between nucleus and cytoplasm in response to stress as well as its effects in altering the levels of cell cycle active agents such as p21, p27, and p53 (*See* Color Insert)

degradation. In G0-arrested cells, p27 levels are high. But, as there is no detectable Cdk1–4 or 6 activity, it is hard to image how p27 can be phosphorylated. This suggests the existence of another kinase dedicated to this modification (Kaldis, 2007). In neurons, as in cell lines, high levels of p27 are essential for G0 arrest, and we presume that any re-entrance into a cell cycle would require that these levels be reduced. We propose that this is done by phosphorylation and that the "other" kinase is Cdk5 (Fig. 2). It has recently been reported that the Cdk5–p35 complex can phosphorylate p27 at both Ser10 and Thr187 (Kawauchi et al., 2006), The latter modification would de-stabilize the protein, but the Ser10 phosphorylation is believed to stabilize it. The suggestion is that Cdk5 can both upregulate and downregulate the levels of p27, and thus is in an excellent position to control both cell cycle arrest and re-entry.

Another well-known CKI protein is p21$^{WAF1/CIP1}$. Once transcriptionally activated by p53, p21 induces cell cycle arrest by interfering with the activity of Cdk2–cyclin E and Cdk1–cyclin B complexes (Levine, 1997; Cheng et al., 1999; Sherr and Roberts, 1999; Vogelstein et al., 2000). Cdk5 can phosphorylate p53, thus increasing its transcriptional activity and stimulating the production of p21 (Zhang et al., 2002). Thus, through pathways already known to be sensitive to its function, Cdk5 may regulate the cell cycle arrest through several different pathways.

One final note of interest involves the fact that Cdk5 is activated by the non-cyclin protein p35/p39. While unusual, this behavior is not unique; other CDKs can be activated by non-cyclin activator proteins that also lack primary amino acid sequence similarity to cyclins. These non-cyclin activators include the viral-based cyclin, cyclin H, and the RINGO/Speedy proteins, which were originally identified as regulators of the meiotic cell cycle in Xenopus oocytes. Recently, five different mammalian RINGO/Speedy family members have been reported, all of which can bind to and directly activate Cdk1 and Cdk2 (Zhang et al., 2002). Cdk2 is the CDK most closely related to Cdk5, having 60% sequence identity. This opens the possibility that the RINGO/Speedy proteins bind to and alter the activity of Cdk5. It is important to remember that the phenotypes of $p35^{-/-}$ and $p39^{-/-}$ double knockout mice are quite similar to those of $Cdk5^{-/-}$ mice. This suggests that p35 and p39 are the only two activators of Cdk5 (Ko et al., 2001). Yet these mice die before birth, leaving open the possibility that some of the postnatal functions of Cdk5 depend on other activator proteins including RINGO/Speedy protein (Dinarina et al., 2005). This is an intriguing possibility as RINGO/Speedy-activated CDKs have been shown to have a different rank order of substrate preference than cyclin-activated CDKs. If Cdk5 fits this pattern, it may be that these proteins target Cdk5 toward different substrates than the ones found to date. Given all of these interactions, the proposition that Cdk5 is not involved in any way in cell cycle regulation is more than a little mysterious.

The mystery may be beginning to be solved, however, and Cdk5 may yet take its place on the list of CDKs that help to regulate the cell cycle. Cicero et al. (2005) report a novel and unexpected role of Cdk5 in neuronal cell cycle control. Analysis of $Cdk5^{-/-}$ mice found that loss of Cdk5 leads to loss of cell cycle control in cortical plate cells of E16.5 mouse neocortex. This includes the abnormal expression of cell cycle proteins such as cyclin D, cyclin A, and PCNA as well as BrdU incorporation. These unexpected cell cycle events are found in multiple layers of $Cdk5^{-/-}$ neocortex. Double labeling with cell death markers reveals that the "cycling" neurons are dying, as indicated by their staining positive for DNA breakdown (TUNEL) and activated caspase-3. In vitro, cultures of primary $Cdk5^{-/-}$ neurons continue to express cell cycle–related proteins such as PCNA and continue to incorporate BrdU, consistent with the in vivo data. More importantly, introduction of expression plasmids encoding wild-type Cdk5 into $Cdk5^{-/-}$ neurons stops cell cycle re-entry. Further, both in vitro and in vivo, the abnormal cell cycle events are coupled with a failure of neuronal differentiation in Cdk5-deficient neurons. In the aggregate, therefore, these data strongly suggest that Cdk5 plays a dual role as a cell cycle suppressor and as a facilitator of cell differentiation during neuronal development.

Recent data from our laboratory expand upon this study and further suggest that localization of Cdk5 plays an important role in this cell cycle deregulation. As mentioned above, both Cdk5 levels and normal cell cycle control are deregulated in neurons of the human AD brain. As Cdk5 plays the role of a cell cycle suppressor in the developing neurons, we asked whether Cdk5 plays a role in the aberrant cell cycle re-entry in AD. Support for this hypothesis was first developed *in vitro* where we found BrdU incorporation and cell cycle protein re-expression in Aβ-treated cultured primary neurons. The level of Cdk5 in the treated cultures did not change. However, this deregulated cell cycle event was closely correlated with a shift in the localization of Cdk5 in the "cycling" neurons. In neurons that incorporated BrdU, the levels of Cdk5 were dramatically reduced in the nucleus, and increased in the cytoplasm. If we blocked Cdk5 translocation using the nuclear transport inhibitor leptomycin B, this abnormal cell cycle event was blocked. The same relationship between Cdk5 localization and cell cycle re-entry was also found in human AD brain and in AD mouse models (Cicero, Wang et al., and unpublished observations). Significantly, although Cdk5 localization changed, p35/25 protein did not change. It remained predominantly nuclear both *in vivo* and *in vitro*. This is consistent with the findings of others who have shown that Cdk5 and p35/25 can move in and out of the nucleus independently (Fu et al., 2006). These findings, if confirmed, suggest that the role of Cdk5 in the suppression of cell cycle in both young and mature neurons is related to its sub-cellular localization. The mechanisms involved in both the localization change and the cell cycle suppression remain to be investigated. Our hypothesis is that stress signals, in some yet to be defined way, alert the cell to export Cdk5 from the nucleus. When this occurs, the cell cycle inhibition achieved by the nuclear Cdk5 is lost and the cell is released to re-enter a cell cycle–like process, paving the way to its death.

Conclusion and Perspectives

Overall, Cdk5 is turning out to be a unique member in CDK family. Unlike the traditional cell cycle–dependent kinases, which facilitate cell cycle progression through different stages of the cell cycle, Cdk5 appears to act as a cell cycle suppressor in both young and adult neurons. This function of Cdk5 is important for young neurons exiting from the cell cycle and beginning their differentiation program. During this period, loss of Cdk5 leads to loss of control of the cell cycle and blocks neuronal differentiation. The result is neuronal death. In stressed neurons, Cdk5 moves from the nucleus to the cytoplasm and this is accompanied by aberrant cell cycle re-entry.

This shifting perspective on the role of Cdk5 in the health and well-being of mature neurons cell cycle is an issue with considerable clinical significance. In the development of optimal strategies for the treatment of neurodegenerative disease, it is crucial to define the consequences of the deregulation of Cdk5 activity. Early studies suggested that its activity was part of the pathogenic process. This suggests that the path to therapy is through Cdk5 inhibition. But the preliminary picture that emerges from our data indicates that Cdk5 may play a positive role as a cell cycle inhibitor, thus protecting a neuron from a cell cycle–related cell death. In this context, it would seem desirable to *increase* Cdk5 activity. Finally, the key to the positive or negative consequences of Cdk5 activity may lie in its localization rather than its absolute level of activity. Little is known about pharmacological means of altering this property of the kinase. In the end, we believe that caution is needed as treatments based on this atypical kinase are developed.

Acknowledgments All three authors wish to acknowledge support during the writing of this review by grants from the NIH (NS20591 and AG24494).

References

Ahlijanian MK, Barrezueta NX, Williams RD, Jakowski A, Kowsz KP, McCarthy S, Coskran T, Carlo A, Seymour PA, Burkhardt JE, Nelson RB, McNeish JD (2000) Hyperphosphorylated tau and neurofilament and cytoskeletal disruptions in mice over-expressing human p25, an activator of cdk5. Proc Natl Acad Sci U S A 97:2910–2915.

Akoulitchev S, Chuikov S, Reinberg D (2000) TFIIH is negatively regulated by cdk8-containing mediator complexes. Nature 407:102–106.

Al-Ubaidi MR, Font RL, Quiambao AB, Keener MJ, Liou GI, Overbeek PA, Baehr W (1992) Bilateral retinal and brain tumors in transgenic mice expressing simian virus 40 large T antigen under control of the human interphotoreceptor retinoid-binding protein promoter. J Cell Biol 119:1681–1687.

Al-Ubaidi MR, Mangini NJ, Quiambao AB, Myers KM, Abler AS, Chang CJ, Tso MO, Butel JS, Hollyfield JG (1997) Unscheduled DNA replication precedes apoptosis of photoreceptors expressing SV40 T antigen. Exp Eye Res 64:573–585.

Appert-Collin A, Hugel B, Levy R, Niederhoffer N, Coupin G, Lombard Y, Andre P, Poindron P, Gies JP (2006) Cyclin dependent kinase inhibitors prevent apoptosis of postmitotic mouse motoneurons. Life Sci 79:484–490.

Bian F, Nath R, Sobocinski G, Booher RN, Lipinski WJ, Callahan MJ, Pack A, Wang KK, Walker LC (2002) Axonopathy, tau abnormalities, and dyskinesia, but no neurofibrillary tangles in p25-transgenic mice. J Comp Neurol 446:257–266.

Bibb JA, Snyder GL, Nishi A, Yan Z, Meijer L, Fienberg AA, Tsai LH, Kwon YT, Girault JA, Czernik AJ, Huganir RL, Hemmings HC, Jr., Nairn AC, Greengard P (1999) Phosphorylation of DARPP-32 by Cdk5 modulates dopamine signalling in neurons. Nature 402:669–671.

Bloom J, Cross FR (2007) Multiple levels of cyclin specificity in cell-cycle control. Nat Rev Mol Cell Biol 8:149–160.

Busser J, Geldmacher DS, Herrup K (1998) Ectopic cell cycle proteins predict the sites of neuronal cell death in Alzheimer's disease brain. J Neurosci 18:2801–2807.

Chen P, Zindy F, Abdala C, Liu F, Li X, Roussel MF, Segil N (2003) Progressive hearing loss in mice lacking the cyclin-dependent kinase inhibitor Ink4d. Nat Cell Biol 5:422–426.

Cheng M, Olivier P, Diehl JA, Fero M, Roussel MF, Roberts JM, Sherr CJ (1999) The p21(Cip1) and p27(Kip1) CDK 'inhibitors' are essential activators of cyclin D-dependent kinases in murine fibroblasts. EMBO J 18:1571–1583.

Chergui K, Svenningsson P, Greengard P (2004) Cyclin-dependent kinase 5 regulates dopaminergic and glutamatergic transmission in the striatum. Proc Natl Acad Sci U S A 101:2191–2196.

Cicero S, Herrup K (2005) Cyclin-dependent kinase 5 is essential for neuronal cell cycle arrest and differentiation. J Neurosci 25:9658–9668.

Clarke AR, Maandag ER, van Roon M, van der Lugt NM, van der Valk M, Hooper ML, Berns A, te Riele H (1992) Requirement for a functional Rb-1 gene in murine development. Nature 359:328–330.

Cruz JC, Tseng HC, Goldman JA, Shih H, Tsai LH (2003) Aberrant Cdk5 activation by p25 triggers pathological events leading to neurodegeneration and neurofibrillary tangles. Neuron 40:471–483.

Desai D, Wessling HC, Fisher RP, Morgan DO (1995) Effects of phosphorylation by CAK on cyclin binding by CDC2 and CDK2. Mol Cell Biol 15:345–350.

Dinarina A, Perez LH, Davila A, Schwab M, Hunt T, Nebreda AR (2005) Characterization of a new family of cyclin-dependent kinase activators. Biochem J 386:349–355.

Ding XL, Husseman J, Tomashevski A, Nochlin D, Jin LW, Vincent I (2000) The cell cycle Cdc25A tyrosine phosphatase is activated in degenerating postmitotic neurons in Alzheimer's disease. Am J Pathol 157:1983–1990.

Farinelli SE, Park DS, Greene LA (1996) Nitric oxide delays the death of trophic factor-deprived PC12 cells and sympathetic neurons by a cGMP-mediated mechanism. J Neurosci 16:2325–2334.

Feddersen RM, Clark HB, Yunis WS, Orr HT (1995) In vivo viability of postmitotic Purkinje neurons requires pRb family member function. Mol Cell Neurosci 6: 153–167.

Feddersen RM, Ehlenfeldt R, Yunis WS, Clark HB, Orr HT (1992) Disrupted cerebellar cortical development and progressive degeneration of Purkinje cells in SV40 T antigen transgenic mice. Neuron 9:955–966.

Fu AK, Fu WY, Cheung J, Tsim KW, Ip FC, Wang JH, Ip NY (2001) Cdk5 is involved in neuregulin-induced AChR expression at the neuromuscular junction. Nat Neurosci 4:374–381.

Fu WY, Fu AK, Lok KC, Ip FC, Ip NY (2002) Induction of Cdk5 activity in rat skeletal muscle after nerve injury. Neuroreport 13:243–247.

Fu X, Choi YK, Qu D, Yu Y, Cheung NS, Qi RZ (2006) Identification of nuclear import mechanisms for the neuronal Cdk5 activator. J Biol Chem 281(51):39017–39021.

Giese KP, Ris L, Plattner F (2005) Is there a role of the cyclin-dependent kinase 5 activator p25 in Alzheimer's disease? Neuroreport 16:1725–1730.

Gilmore EC, Herrup K (2001) Neocortical cell migration: GABAergic neurons and cells in layers I and VI move in a cyclin-dependent kinase 5-independent manner. J Neurosci 21:9690–9700.

Gilmore EC, Ohshima T, Goffinet AM, Kulkarni AB, Herrup K (1998) Cyclin-dependent kinase 5-deficient mice demonstrate novel developmental arrest in cerebral cortex. J Neurosci 18:6370–6377.

Giovanni A, Wirtz-Brugger F, Keramaris E, Slack R, Park DS (1999) Involvement of cell cycle elements, cyclin-dependent kinases, pRb, and E2F x DP, in B-amyloid-induced neuronal death. J Biol Chem 274:19011–19016.

Grynspan F, Griffin WR, Cataldo A, Katayama S, Nixon RA (1997) Active site-directed antibodies identify calpain II as an early-appearing and pervasive component of neurofibrillary pathology in Alzheimer's disease. Brain Res 763:145–158.

Hallows JL, Chen K, DePinho RA, Vincent I (2003) Decreased cyclin-dependent kinase 5 (cdk5) activity is accompanied by redistribution of cdk5 and cytoskeletal proteins and increased cytoskeletal protein phosphorylation in p35 null mice. J Neurosci 23:10633–10644.

Hallows JL, Iosif RE, Biasell RD, Vincent I (2006) p35/p25 is not essential for tau and cytoskeletal pathology or neuronal loss in Niemann-Pick type C disease. J Neurosci 26:2738–2744.

Harper JW, Adami GR, Wei N, Keyomarsi K, Elledge SJ (1993) The p21 Cdk-interacting protein Cip1 is a potent inhibitor of G1 cyclin-dependent kinases. Cell 75:805–816.

Harper JW, Elledge SJ, Keyomarsi K, Dynlacht B, Tsai LH, Zhang P, Dobrowolski S, Bai C, Connell-Crowley L, Swindell E, et al. (1995) Inhibition of cyclin-dependent kinases by p21. Mol Biol Cell 6:387–400.

Hayashi T, Warita H, Abe K, Itoyama Y (1999) Expression of cyclin-dependent kinase 5 and its activator p35 in rat brain after middle cerebral artery occlusion. Neurosci Lett 265:37–40.

Herrup K, Busser JC (1995) The induction of multiple cell cycle events precedes target-related neuronal death. Development 121:2385–2395.

Herrup K, Yang Y (2007) Cell cycle regulation in the postmitotic neuron: oxymoron or new biology? Nat Rev Neurosci 8:368–378.

Höglinger G, et al. (2007) The pRb/E2F cell-cycle pathway mediates cell death in Parkinson's disease. Proc Natl Acad Sci U S A 104, 3585–3590.

Humbert S, Lanier LM, Tsai LH (2000a) Synaptic localization of p39, a neuronal activator of cdk5. Neuroreport 11:2213–2216.

Humbert S, Dhavan R, Tsai L (2000b) p39 activates cdk5 in neurons, and is associated with the actin cytoskeleton. J Cell Sci 113(Pt 6):975–983.

Husseman JW, Nochlin D, Vincent I (2000) Mitotic activation: a convergent mechanism for a cohort of neurodegenerative diseases. Neurobiol Aging 21:815–828.

Jacks T, Fazeli A, Schmitt EM, Bronson RT, Goodell MA, Weinberg RA (1992) Effects of an Rb mutation in the mouse. Nature 359:295–300.

Kaldis P (2007) Another piece of the p27Kip1 puzzle. Cell 128:241–244.

Kasten M, Giordano A (2001) Cdk10, a Cdc2-related kinase, associates with the Ets2 transcription factor and modulates its transactivation activity. Oncogene 20:1832–1838.

Kawauchi T, Chihama K, Nabeshima Y, Hoshino M (2006) Cdk5 phosphorylates and stabilizes p27kip1 contributing to actin organization and cortical neuronal migration. Nat Cell Biol 8:17–26.

Ko J, Humbert S, Bronson RT, Takahashi S, Kulkarni AB, Li E, Tsai LH (2001) p35 and p39 are essential for cyclin-dependent kinase 5 function during neurodevelopment. J Neurosci 21:6758–6771.

Lee EY, Chang CY, Hu N, Wang YC, Lai CC, Herrup K, Lee WH, Bradley A (1992) Mice deficient for Rb are nonviable and show defects in neurogenesis and haematopoiesis. Nature 359:288–294.

Lee MS, Kwon YT, Li M, Peng J, Friedlander RM, Tsai LH (2000) Neurotoxicity induces cleavage of p35 to p25 by calpain. Nature 405:360–364.

Lees E (1995) Cyclin dependent kinase regulation. Curr Opin Cell Biol 7:773–780.

Levine AJ (1997) p53, the cellular gatekeeper for growth and division. Cell 88:323–331.

Lew J, Beaudette K, Litwin CM, Wang JH (1992) Purification and characterization of a novel proline-directed protein kinase from bovine brain. J Biol Chem 267:13383–13390.

Lew J, Huang QQ, Qi Z, Winkfein RJ, Aebersold R, Hunt T, Wang JH (1994) A brain-specific activator of cyclin-dependent kinase 5. Nature 371:423–426.

Love S (2003) Neuronal expression of cell cycle-related proteins after brain ischaemia in man. Neurosci Lett 353:29–32.

Malumbres M, Barbacid M (2005) Mammalian cyclin-dependent kinases. Trends Biochem Sci 30:630–641.

Matsuoka S, Edwards MC, Bai C, Parker S, Zhang P, Baldini A, Harper JW, Elledge SJ (1995) p57KIP2, a structurally distinct member of the p21CIP1 Cdk inhibitor family, is a candidate tumor suppressor gene. Genes Dev 9:650–662.

Matsushita M, Tomizawa K, Lu YF, Moriwaki A, Tokuda M, Itano T, Wang JH, Hatase O, Matsui H (1996) Distinct cellular compartment of cyclin-dependent kinase 5 (Cdk5) and neuron-specific Cdk5 activator protein (p35nck5a) in the developing rat cerebellum. Brain Res 734:319–322.

Matsuura I, Wang JH (1996) Demonstration of cyclin-dependent kinase inhibitory serine/threonine kinase in bovine thymus. J Biol Chem 271:5443–5450.

McShea A, Lee HG, Petersen RB, Casadesus G, Vincent I, Linford NJ, Funk JO, Shapiro RA, Smith MA (2007) Neuronal cell cycle re-entry mediates Alzheimer disease-type changes. Biochim Biophys Acta 1772:467–472.

Meyerson M, Enders GH, Wu CL, Su LK, Gorka C, Nelson C, Harlow E, Tsai LH (1992) A family of human cdc2-related protein kinases. EMBO J 11:2909–2917.

Miyajima M, Nornes HO, Neuman T (1995) Cyclin E is expressed in neurons and forms complexes with cdk5. Neuroreport 6:1130–1132.

Morgan DO (1997) Cyclin-dependent kinases: engines, clocks, and microprocessors. Annu Rev Cell Dev Biol 13:261–291.

Moy LY, Tsai LH (2004) Cyclin-dependent kinase 5 phosphorylates serine 31 of tyrosine hydroxylase and regulates its stability. J Biol Chem 279:54487–54493.

Nagy Z, Esiri MM, Cato AM, Smith AD (1997) Cell cycle markers in the hippocampus in Alzheimer's disease. Acta Neuropathol (Berl) 94:6–15.

Nath R, Davis M, Probert AW, Kupina NC, Ren X, Schielke GP, Wang KK (2000) Processing of cdk5 activator p35 to its truncated form (p25) by calpain in acutely injured neuronal cells. Biochem Biophys Res Commun 274:16–21.

Neystat M, Rzhetskaya M, Oo TF, Kholodilov N, Yarygina O, Wilson A, El-Khodor BF, Burke RE (2001) Expression of cyclin-dependent kinase 5 and its activator p35 in models of induced apoptotic death in neurons of the substantia nigra in vivo. J Neurochem 77:1611–1625.

Nguyen MD, Julien JP (2003) Cyclin-dependent kinase 5 in amyotrophic lateral sclerosis. Neurosignals 12:215–220.

Nguyen MD, Lariviere RC, Julien JP (2001) Deregulation of Cdk5 in a mouse model of ALS: toxicity alleviated by perikaryal neurofilament inclusions. Neuron 30:135–147.

Nguyen MD, Boudreau M, Kriz J, Couillard-Despres S, Kaplan DR, Julien JP (2003) Cell cycle regulators in the neuronal death pathway of amyotrophic lateral sclerosis caused by mutant superoxide dismutase 1. J Neurosci 23:2131–2140.

Niethammer M, Smith DS, Ayala R, Peng J, Ko J, Lee MS, Morabito M, Tsai LH (2000) NUDEL is a novel Cdk5 substrate that associates with LIS1 and cytoplasmic dynein. Neuron 28:697–711.

Noble W, Olm V, Takata K, Casey E, Mary O, Meyerson J, Gaynor K, LaFrancois J, Wang L, Kondo T, Davies P, Burns M, Veeranna, Nixon R, Dickson D, Matsuoka Y, Ahlijanian M. Lau LF, Duff K (2003) Cdk5 is a key factor in tau aggregation and tangle formation in vivo. Neuron 38:555–565.

Ohshima T, Gilmore EC, Longenecker G, Jacobowitz DM, Brady RO, Herrup K, Kulkarni AB (1999) Migration defects of cdk5(–/–) neurons in the developing cerebellum is cell autonomous. J Neurosci 19:6017–6026.

Ohshima T, Ward JM, Huh CG, Longenecker G, Veeranna, Pant HC, Brady RO, Martin LJ, Kulkarni AB (1996) Targeted disruption of the cyclin-dependent kinase 5 gene results in abnormal corticogenesis, neuronal pathology and perinatal death. Proc Natl Acad Sci U S A 93:11173–11178.

Park DS, Farinelli SE, Greene LA (1996) Inhibitors of cyclin-dependent kinases promote survival of post-mitotic neuronally differentiated PC12 cells and sympathetic neurons. J Biol Chem 271:8161–8169.

Park KH, Hallows JL, Chakrabarty P, Davies P, Vincent I (2007) Conditional neuronal simian virus 40 T antigen expression induces Alzheimer-like tau and amyloid pathology in mice. J Neurosci 27:2969–2978.

Park DS, Levine B, Ferrari G, Greene LA (1997a) Cyclin dependent kinase inhibitors and dominant negative cyclin dependent kinase 4 and 6 promote survival of NGF-deprived sympathetic neurons. J Neurosci 17:8975–8983.

Park DS, Morris EJ, Greene LA, Geller HM (1997b) G1/S cell cycle blockers and inhibitors of cyclin-dependent kinases suppress camptothecin-induced neuronal apoptosis. J Neurosci 17:1256–1270.

Patrick GN, Zukerberg L, Nikolic M, de la Monte S, Dikkes P, Tsai LH (1999) Conversion of p35 to p25 deregulates Cdk5 activity and promotes neurodegeneration. Nature 402:615–622.

Patzke H, Tsai LH (2002) Calpain-mediated cleavage of the cyclin-dependent kinase-5 activator p39 to p29. J Biol Chem 277:8054–8060.

Plattner F, Angelo M, Giese KP (2006) The roles of cyclin-dependent kinase 5 and glycogen synthase kinase 3 in tau hyperphosphorylation. J Biol Chem 281:25457–25465.

Poon RY, Lew J, Hunter T (1997) Identification of functional domains in the neuronal Cdk5 activator protein. J Biol Chem 272:5703–5708.

Ranganathan S, Bowser R (2003) Alterations in G(1) to S phase cell-cycle regulators during amyotrophic lateral sclerosis. Am J Pathol 162:823–835.

Rashidian J, Iyirhiaro G, Aleyasin H, Rios M, Vincent I, Callaghan S, Bland RJ, Slack RS, During MJ, Park DS (2005) Multiple cyclin-dependent kinases signals are critical mediators of ischemia/hypoxic neuronal death in vitro and in vivo. Proc Natl Acad Sci U S A 102:14080–14085.

Ren S, Rollins BJ (2004) Cyclin C/cdk3 promotes Rb-dependent G0 exit. Cell 117:239–251.

Russo AA, Jeffrey PD, Pavletich NP (1996) Structural basis of cyclin-dependent kinase activation by phosphorylation. Nat Struct Biol 3:696–700.

Sahlgren CM, Mikhailov A, Vaittinen S, Pallari HM, Kalimo H, Pant HC, Eriksson JE (2003) Cdk5 regulates the organization of Nestin and its association with p35. Mol Cell Biol 23:5090–5106.

Sano M, Schneider MD (2003) Cyclins that don't cycle—cyclin T/cyclin-dependent kinase-9 determines cardiac muscle cell size. Cell Cycle 2:99–104.

Sherr CJ, Roberts JM (1999) CDK inhibitors: positive and negative regulators of G1-phase progression. Genes Dev 13:1501–1512.

Shuttleworth J (1995) The regulation and functions of cdk7. Prog Cell Cycle Res 1:229–240.

Tan TC, Valova VA, Malladi CS, Graham ME, Berven LA, Jupp OJ, Hansra G, McClure SJ, Sarcevic B, Boadle RA, Larsen MR, Cousin MA, Robinson PJ (2003) Cdk5 is essential for synaptic vesicle endocytosis. Nat Cell Biol 5:701–710.

Tanaka T, Serneo FF, Tseng HC, Kulkarni AB, Tsai LH, Gleeson JG (2004) Cdk5 phosphorylation of doublecortin ser297 regulates its effect on neuronal migration. Neuron 41:215–227.

Tandon A, Yu H, Wang L, Rogaeva E, Sato C, Chishti MA, Kawarai T, Hasegawa H, Chen F, Davies P, Fraser PE, Westaway D, St George-Hyslop PH (2003) Brain levels of CDK5 activator p25 are not increased in Alzheimer's or other neurodegenerative diseases with neurofibrillary tangles. J Neurochem 86:572–581.

Tang D, Yeung J, Lee KY, Matsushita M, Matsui H, Tomizawa K, Hatase O, Wang JH (1995) An isoform of the neuronal cyclin-dependent kinase 5 (Cdk5) activator. J Biol Chem 270:26897–26903.

Tarricone C, Dhavan R, Peng J, Areces LB, Tsai LH, Musacchio A (2001) Structure and regulation of the CDK5-p25(nck5a) complex. Mol Cell 8:657–669.

Tomizawa K, Ohta J, Matsushita M, Moriwaki A, Li ST, Takei K, Matsui H (2002) Cdk5/p35 regulates neurotransmitter release through phosphorylation and downregulation of P/Q-type voltage-dependent calcium channel activity. J Neurosci 22: 2590–2597.

Tsai LH, Delalle I, Caviness VS, Jr., Chae T, Harlow E (1994) p35 is a neural-specific regulatory subunit of cyclin-dependent kinase 5. Nature 371:419–423.

Van den Haute C, Spittaels K, Van Dorpe J, Lasrado R, Vandezande K, Laenen I, Geerts H, Van Leuven F (2001) Coexpression of human cdk5 and its activator p35 with human protein tau in neurons in brain of triple transgenic mice. Neurobiol Dis 8:32–44.

van den Heuvel S, Harlow E (1993) Distinct roles for cyclin-dependent kinases in cell cycle control. Science 262:2050–2054.

Vincent I, Bu B, Hudson K, Husseman J, Nochlin D, Jin L (2001) Constitutive Cdc25B tyrosine phosphatase activity in adult brain neurons with M phase-type alterations in Alzheimer's disease. Neuroscience 105:639–650.

Vincent I, Jicha G, Rosado M, Dickson DW (1997) Aberrant expression of mitotic cdc2/ cyclin B1 kinase in degenerating neurons of Alzheimer's disease brain. J Neurosci 17:3588–3598.

Vogelstein B, Lane D, Levine AJ (2000) Surfing the p53 network. Nature 408:307–310.

Wang J, Liu S, Fu Y, Wang JH, Lu Y (2003) Cdk5 activation induces hippocampal CA1 cell death by directly phosphorylating NMDA receptors. Nat Neurosci 6:1039–1047.

Wen Y, Yang S, Liu R, Simpkins JW (2005) Cell-cycle regulators are involved in transient cerebral ischemia induced neuronal apoptosis in female rats. FEBS Lett 579:4591–4599.

Xie Z, Sanada K, Samuels BA, Shih H, Tsai LH (2003) Serine 732 phosphorylation of FAK by Cdk5 is important for microtubule organization, nuclear movement, and neuronal migration. Cell 114:469–482.

Xiong Y, Hannon GJ, Zhang H, Casso D, Kobayashi R, Beach D (1993) p21 is a universal inhibitor of cyclin kinases. Nature 366:701–704.

Xiong Y, Zhang H, Beach D (1992) D type cyclins associate with multiple protein kinases and the DNA replication and repair factor PCNA. Cell 71:505–514.

Yang Y, Herrup K (2005) Loss of neuronal cell cycle control in ataxia-telangiectasia: a unified disease mechanism. J Neurosci 25:2522–2529.

Yang Y, Geldmacher DS, Herrup K (2001) DNA replication precedes neuronal cell death in Alzheimer's disease. J Neurosci 21:2661–2668.

Yang Y, Mufson EJ, Herrup K (2003) Neuronal cell death is preceded by cell cycle events at all stages of Alzheimer's disease. J Neurosci 23:2557–2563.

Yang Y, Varvel NH, Lamb BT, Herrup K (2006) Ectopic cell cycle events link human Alzheimer's disease and amyloid precursor protein transgenic mouse models. J Neurosci 26:775–784.

Zhang Q, Ahuja HS, Zakeri ZF, Wolgemuth DJ (1997) Cyclin-dependent kinase 5 is associated with apoptotic cell death during development and tissue remodeling. Dev Biol 183:222–233.

Zhang J, Krishnamurthy PK, Johnson GV (2002) Cdk5 phosphorylates p53 and regulates its activity. J Neurochem 81:307–313.

Zhang H, Xiong Y, Beach D (1993) Proliferating cell nuclear antigen and p21 are components of multiple cell cycle kinase complexes. Mol Biol Cell 4:897–906.

Zhu Y, Lin L, Kim S, Quaglino D, Lockshin RA, Zakeri Z (2002) Cyclin dependent kinase 5 and its interacting proteins in cell death induced in vivo by cyclophosphamide in developing mouse embryos. Cell Death Differ 9:421–430.

Zukerberg LR, Patrick GN, Nikolic M, Humbert S, Wu CL, Lanier LM, Gertler FB, Vidal M, Van Etten RA, Tsai LH (2000) Cables links Cdk5 and c-Abl and facilitates Cdk5 tyrosine phosphorylation, kinase upregulation, and neurite outgrowth. Neuron 26:633–646.

Cdk5 and Neuregulin-1 Signaling

Yi Wen, Haung Yu, and Karen Duff

Abstract Cyclin-dependent kinase 5 (cdk5) is implicated in many neurodegenerative diseases including Alzheimer's disease (AD), amyotrophic lateral sclerosis (ALS), and Parkinson's disease (PD). Overexpression of p25 in transgenic mice leads to enhanced cdk5 activity, together with aberrant phosphorylation of cytoskeletal components and the formation of hyperphosphorylated tau. Consistent with previous findings, we observed enhanced NRG-1/ErbB receptor signaling in the p25 overexpressing mice, together with increased PI3 kinase (P13 K)/Akt activity and GSK3β inhibition by S9 phosphorylation. Further, a specific cdk5 inhibitor CP-681301 reduces ErbB2 receptor tyrosine phosphorylation. These results imply that cdk5 is involved in neuregulin-dependent activation of the PI3 K/Akt neuronal survival pathway and potentially other NRG-1-related signaling pathways by regulating the phosphorylation of ErbB2/ErbB3.

Neurodegenerative tauopathies, including Alzheimer disease (AD), are characterized by abnormal hyperphosphorylation of the microtubule-associated protein tau, at proline-directed serine/threonine phosphorylation sites. Tau binds directly to microtubules and promotes microtubule polymerization [1,2]. The proline-directed serine/threonine kinases, cyclin-dependent kinase 5 (cdk5), and glycogen synthase kinase 3β (GSK3β), have been identified as prime candidates for aberrant tau hyperphosphorylation at disease-associated sites [3,4]. Cdk5 co-localizes with filamentous tau deposits in several tauopathies, including AD [5]. GSK3β generates disease-associated phospho-epitopes on tau and co-localizes with aggregates of hyperphosphorylated tau [6]. Cdk5 and GSK3β are considered as potential therapeutic targets in tauopathies. At present, their corresponding phosphorylation sites and specific functions have not been fully elucidated.

Cdk5 activation requires association with its neuron-specific activators, p35 and p39 [7]. Its kinase activity is essential for neuronal migration, neurite

K. Duff
Department of Pathology, Taub Institute for Alzheimer's Disease research, Columbia University; NYS Psychiatric Institute, New York, 10032, USA
e-mail: ked2115@columbia.edu

N.Y. Ip, L.-H. Tsai (eds.), *Cyclin Dependent Kinase 5 (Cdk5)*,
DOI: 10.1007/978-0-387-78887-6_10, © Springer Science+Business Media, LLC 2008

outgrowth, and synaptic transmission of the cerebral cortex [8,9]. Cdk5 and p35 are also known to regulate synaptic activity by phosphorylating Munc-18, amphiphysin, and the N-methyl-D-aspartate (NMDA) receptor [10,11]. In non-neuronal tissues, cdk5 regulates the expression of acetylcholine receptor clustering at the neuromuscular junction via activating ERK pathway [12]. On the other hand, p35 or p39 is cleaved by calpain to p25/p29, causing dysregulation of cdk5 [13,14]. Overexpression of p25 in transgenic mice leads to enhanced and deregulated cdk5 activity, aberrant phosphorylation of cytoskeletal components, and formation of hyperphosphorylated tau [15,16]. Postnatal induction of p25 leads to neurodegeneration and NFT formation reminiscent of that seen in AD [17]. Increased p25 expression in mice causes accelerated tau aggregation [18], and p25/cdk5-related tau pathology is induced in a triple transgenic animal model for AD (AChR) following administration of an inflammation modulator [19]. Such dysregulation of cdk5 is involved in many neurodegenerative diseases such as amyotrophic lateral sclerosis (ALS) [20] and Parkinson's disease (PD) [21].

Previous reports suggest that cdk5 is involved in the Neuregulin (NRGs) signaling system [12]. NRGs comprise a complex family of growth factor signaling molecules that include many membrane-associated and secreted proteins [22]. Most of these forms are generated by multiple transcription variants, and by alternative RNA splicing [23]. All NRG-1 isoforms share an epidermal growth factor (EGF)-like domain, and can be separated into several subgroups [23]. NRG-1 type I and II are transmembrane forms of NRG-1 that undergo proteolytic cleavage by proteases such as β-site APP cleavage enzyme 1 (BACE1) [24], or metalloproteinases [25] that act as paracrine signaling molecules. NRG-1 type III proteins are defined by their cysteine-rich domain (CRD) tethered to the cell surface after cleavage, and they function as a juxtacrine signal [26]. NRG-1 type III are associated with myelin formation under axonal guidance [27], through the interaction between axons and Schwann cells. Another important function of NRG-1 is that it functions as a transcription factor, promoting specific gene expression when it is cleaved and the C-terminal fragment (CTF) is released.

NRG-1 functions through its binding to ErbB receptors. NRG-1 binds to ErbB3 or ErbB4, and stimulates its dimerization to ErbB2. ErbB2-related signaling pathways are similar to many other growth factor pathways [28]. Activation of the ErbB receptor pathway robustly activates mitogen-activated protein kinase (MAPK) and PtdIns 3-kinase (PI3 K) pathways [29]. Cdk5 regulates NRG-induced AChR expression at the neuromuscular junction by phosphorylating the corresponding receptor ErbB2 [12]. In addition, cdk5 can phosphorylate Thr871 in ErbB2 and/or Ser1120 in ErbB3, and upregulate the ErbB2-related signaling pathway. Reduced PI3 K/Akt activity is reported in cdk5-deficient mice [30]. It is suggested that the phosphorylation does not directly regulate the receptor kinase activity, but primes the receptor for further activation [31].

In NRG-1 signaling pathways, ErbB3/4 is unique in that it has a kinase domain but has no activity [32,33]. ErbB2, which has a kinase domain and a corresponding tyrosine kinase activity, is essential for receptor activity with its hetero-dimerization with other ErbB family receptors. We find that cdk5 is involved in NRG-1 signaling by regulating ErbB receptor activity *in vivo*. In mice that overexpress the cdk5 activator p25, there is increased ErbB2 receptor activity and PI3 K/Akt activity, and enhanced GSK3β S9 phosphorylation (GSK3β inhibition) (Fig. 1). It has been demonstrated that activated ErbB receptors recruit PI3 K to the plasma membrane through its regulatory subunit [34] and generates the phosphoinositide phosphates inositol 1,4,5-diphosphate and inositol 1,4,5-trisphosphate, which activates PDK1. PDK1, in turn, phosphorylates Ser473 and Thr308 on Akt kinase. Activated Akt targets many different apoptotic substrate molecules including Fork-head transcription factors, caspase-9 and Bad, inhibiting their activity and, promoting cell survival [35]. Phosphorylation of ErbB receptors by cdk5 is essential to their function by priming the receptor for further activation. Thus, cdk5 is closely related to, and required for, NRG-ErbB receptor pathway. Ser/Thr phosphorylation in ErbB2/ErbB3 is reduced in cdk5-deficient mice cortical neuronal cultures. Similarly, ErbB2 receptor tyrosine phosphorylation is reduced in mice treated with cdk5 inhibitor, CP-681301 (Fig. 1). These results imply that cdk5 is

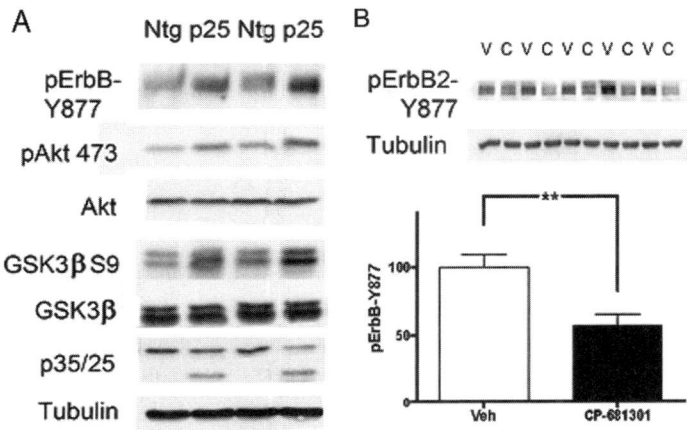

Fig. 1 (**Panel A**) Representative blots showing enhanced NRG-1/ErbB2 signaling pathway in brain extracts from p25 overexpressing and non-transgenic (Ntg) neonatal mice (postnatal day 4). pErbB2-Y877: Phospho-specific Her/ErbB2 at tyrosine 877; pAkt 473: Phospho-specific Akt at Ser473; Akt: Anti-Akt total protein; pGSK3β-S9: Phospho-specific GSK3β at Ser9; GSK3β: GSK3β total protein; p25: p25 protein. Tubulin: β-tubulin protein for loading control. (**Panel B**) *upper panel*: representative blots showing reduced ErbB2 signaling with cdk5 inhibitor CP-681301 treatment in adult mice. pErbB2-Y877: Phospho-specific Her/ErbB2 at tyrosine 877. V (Vehicle treatment); C: (CP-681301 treatment). *Lower panel*: statistical analysis of pErbB2-Y877 in Vehicle (Veh) and CP-681301-treated mice. Data were analyzed with Student's t-test. ** indicates there is a significant difference between the two sets of samples ($p < 0.01$)

Fig. 2 Schematic showing interaction of cdk5 and NRG-1 signaling pathways

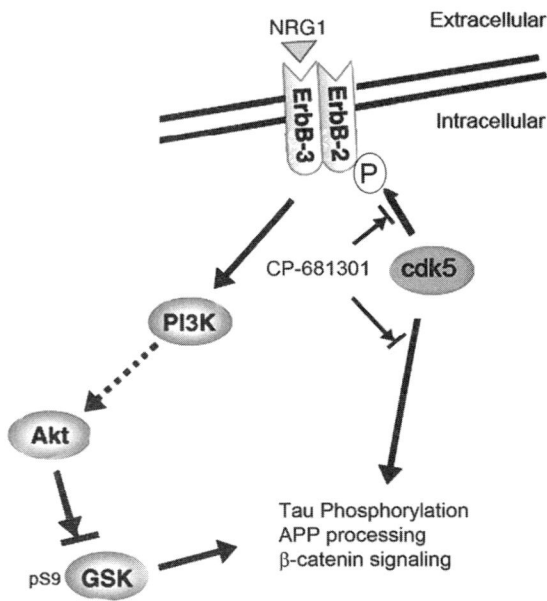

involved in NRG-dependent activation of the PI3 K/Akt neuronal survival pathway by regulating the phosphorylation of ErbB2/ErbB3. Our results are consistent with studies showing that cdk5 is involved in NRG-mediated regulation of neuromuscular junction development [12]. The proposed pathway schematic is shown in Fig. 2.

Our findings indicate that cdk5 activates NRG-1 signaling pathway *in vivo*, and may have wide biological implications. First, this observation supports the role that in neonatal mice, cdk5 inhibits apoptosis by activation of the PI3 K/Akt pathway, and promotes cell survival pathways. However, in some neurodegenerative disorders, it has been suggested that deregulation of cdk5 activity by p25 is pro-apoptotic and correlates with neuronal pathology in AD brains [36]. The versatility of cdk5 in development, function, and survival of the nervous system is probably due to a number of different environmental factors and proximity to different target substrates.

Regulation of NRG-1 signaling by cdk5 may also be important in other neurological diseases. NRG-1 has been identified as a strong susceptibility candidate gene for schizophrenia [37,38]. Furthermore, NRG-1 receptor (ErbB4) and the substrate Akt have been identified as weak susceptibility candidate genes for schizophrenia [39]. Biochemical experiments suggest NRG-1 signaling is enhanced in post-mortem schizophrenia patients, leading to suppressed NMDR receptor function [37,40]. As inhibition of cdk5 *in vivo* by pharmacological inhibitors leads to reduced NRG-1 signaling, cdk5 may be a potential therapeutic target for the treatment of schizophrenia, in addition to many neurodegenerative diseases.

References

1. Pelech, S. L. (1995) *Neurobiology of aging* **16**(3), 247–256; discussion 257–261
2. Michaelis, M. L., Dobrowsky, R. T., and Li, G. (2002) *J Mol Neurosci* **19**(3), 289–293
3. Hanger, D. P., Hughes, K., Woodgett, J. R., Brion, J. P., and Anderton, B. H. (1992) *Neurosci Lett* **147**(1), 58–62
4. Paudel, H. K., Lew, J., Ali, Z., and Wang, J. H. (1993) *J Biol Chem* **268**(31), 23512–23518
5. Shelton, S. B., and Johnson, G. V. (2004) *J Neurochem* **88**(6), 1313–1326
6. Ishizawa, T., Sahara, N., Ishiguro, K., Kersh, J., McGowan, E., Lewis, J., Hutton, M., Dickson, D. W., and Yen, S. H. (2003) *Am J Pathol* **163**(3), 1057–1067
7. Dhavan, R., and Tsai, L. H. (2001) *Nat Rev Mol Cell Biol* **2**(10), 749–759
8. Tsai, L. H., Delalle, I., Caviness, Jr., V. S., Chae, T., and Harlow, E. (1994) *Nature* **371**(6496), 419–423
9. Cruz, J. C., and Tsai, L. H. (2004) *Curr Opin Neurobiol* **14**(3), 390–394
10. Fletcher, A. I., Shuang, R., Giovannucci, D. R., Zhang, L., Bittner, M. A., and Stuenkel, E. L. (1999) *J Biol Chem* **274**(7), 4027–4035
11. Wang, J., Liu, S., Fu, Y., Wang, J. H., and Lu, Y. (2003) *Nature Neurosci* **6**(10), 1039–1047
12. Fu, A. K., Fu, W. Y., Cheung, J., Tsim, K. W., Ip, F. C., Wang, J. H., and Ip, N. Y. (2001) *Nat Neurosci.* **4**(4), 374–381
13. Lee, M. S., Kwon, Y. T., Li, M., Peng, J., Friedlander, R. M., and Tsai, L. H. (2000) *Nature* **405**(6784), 360–364
14. Patzke, H., and Tsai, L. H. (2002) *J Biol Chem* **277**(10), 8054–8060
15. Ahlijanian, M. K., Barrezueta, N. X., Williams, R. D., Jakowski, A., Kowsz, K. P., McCarthy, S., Coskran, T., Carlo, A., Seymour, P. A., Burkhardt, J. E., Nelson, R. B., and McNeish, J. D. (2000) *Proc Natl Acad Sci USA* **97**(6), 2910–2915
16. Bian, F., Nath, R., Sobocinski, G., Booher, R. N., Lipinski, W. J., Callahan, M. J., Pack, A., Wang, K. K., and Walker, L. C. (2002) *J Comp Neurol* **446**(3), 257–266
17. Cruz, J. C., Kim, D., Moy, L. Y., Dobbin, M. M., Sun, X., Bronson, R. T., and Tsai, L. H. (2006) *J Neurosci* **26**(41), 10536–10541
18. Noble, W., Olm, V., Takata, K., Casey, E., Mary, O., Meyerson, J., Gaynor, K., LaFrancois, J., Wang, L., Kondo, T., Davies, P., Burns, M., Veeranna, Nixon, R., Dickson, D., Matsuoka, Y., Ahlijanian, M., Lau, L. F., and Duff, K. (2003) *Neuron* **38**(4), 555–565
19. Kitazawa, M., Oddo, S., Yamasaki, T. R., Green, K. N., and LaFerla, F. M. (2005) *J Neurosci* **25**(39), 8843–8853
20. Nguyen, M. D., Boudreau, M., Kriz, J., Couillard-Despres, S., Kaplan, D. R., and Julien, J. P. (2003) *J Neurosci* **23**(6), 2131–2140
21. Smith, P. D., Crocker, S. J., Jackson-Lewis, V., Jordan-Sciutto, K. L., Hayley, S., Mount, M. P., O'Hare, M. J., Callaghan, S., Slack, R. S., Przedborski, S., Anisman, H., and Park, D. S. (2003) *Proc Natl Acad Sci USA* **100**(23), 13650–13655
22. Falls, D. L. (2003) *Exp Cell Res* **284**(1), 14–30
23. Law, A. J., Lipska, B. K., Weickert, C. S., Hyde, T. M., Straub, R. E., Hashimoto, R., Harrison, P. J., Kleinman, J. E., and Weinberger, D. R. (2006) *Proc Natl Acad Sci USA* **103**(17), 6747–6752
24. Willem, M., Garratt, A. N., Novak, B., Citron, M., Kaufmann, S., Rittger, A., DeStrooper, B., Saftig, P., Birchmeier, C., and Haass, C. (2006) *Science* (New York, NY) **314**(5799), 664–666
25. Horiuchi, K., Zhou, H. M., Kelly, K., Manova, K., and Blobel, C. P. (2005) *Dev Biol* **283**(2), 459–471
26. Nave, K. A., and Salzer, J. L. (2006) *Curr Opin Neurobiol* **16**(5), 492–500
27. Taveggia, C., Zanazzi, G., Petrylak, A., Yano, H., Rosenbluth, J., Einheber, S., Xu, X., Esper, R. M., Loeb, J. A., Shrager, P., Chao, M. V., Falls, D. L., Role, L., and Salzer, J. L. (2005) *Neuron* **47**(5), 681–694

28. Burgess, A. W., Cho, H. S., Eigenbrot, C., Ferguson, K. M., Garrett, T. P., Leahy, D. J., Lemmon, M. A., Sliwkowski, M. X., Ward, C. W., and Yokoyama, S. (2003) *Mol Cell* **12**(3), 541–552

29. Esper, R. M., Pankonin, M. S., and Loeb, J. A. (2006) *Brain Res Rev* **51**(2), 161–175

30. Li, B. S., Ma, W., Jaffe, H., Zheng, Y., Takahashi, S., Zhang, L., Kulkarni, A. B., and Pant, H. C. (2003) *J Biol Chem* **278**(37), 35702–35709

31. Lee, M. S., and Tsai, L. H. (2001) *Nat Neurosci* **4**(4), 340–342

32. Guy, P. M., Platko, J. V., Cantley, L. C., Cerione, R. A., and Carraway, K. L., III. (1994) *Proc Natl Acad Sci USA* **91**(17), 8132–8136

33. Sierke, S. L., Cheng, K., Kim, H. H., and Koland, J. G. (1997) *Biochem J* **322**(Pt 3), 757–763

34. Fukazawa, T., Reedquist, K. A., Panchamoorthy, G., Soltoff, S., Trub, T., Druker, B., Cantley, L., Shoelson, S. E., and Band, H. (1995) *J Biol Chem* **270**(34), 20177–20182

35. Datta, S. R., Brunet, A., and Greenberg, M. E. (1999) *Genes Dev* **13**(22), 2905–2927

36. Patrick, G. N., Zukerberg, L., Nikolic, M., de la Monte, S., Dikkes, P., and Tsai, L. H. (1999) *Nature* **402**(6762), 615–622

37. Norton, N., Moskvina, V., Morris, D. W., Bray, N. J., Zammit, S., Williams, N. M., Williams, H. J., Preece, A. C., Dwyer, S., Wilkinson, J. C., Spurlock, G., Kirov, G., Buckland, P., Waddington, J. L., Gill, M., Corvin, A. P., Owen, M. J., and O'Donovan, M. C. (2006) *Am J Med Genet B Neuropsychiatr Genet* **141**(1), 96–101

38. Lachman, H. M., Pedrosa, E., Nolan, K. A., Glass, M., Ye, K., and Saito, T. (2006) *Am J Med Genet B Neuropsychiatr Genet* **141**(1), 102–109

39. Straub, R. E., and Weinberger, D. R. (2006) *Biol Psychiatry* **60**(2), 81–83

40. Hahn, C. G., Wang, H. Y., Cho, D. S., Talbot, K., Gur, R. E., Berrettini, W. H., Bakshi, K., Kamins, J., Borgmann-Winter, K. E., Siegel, S. J., Gallop, R. J., and Arnold, S. E. (2006) *Nat Med* **12**(7), 824–828

Cyclin-Dependent Kinase 5 and Insulin Secretion

Christina Bark, Marjan Rupnik, Marko Jevsek, Slavena A. Mandic, and Per-Olof Berggren

Abstract Cyclin-dependent kinase 5 (Cdk5) is emerging as a multifunctional kinase involved in regulating numerous cellular processes. Lately, Cdk5 has also emerged as a key controller of regulated membrane fusion in secretory cells. The pancreatic β-cell is highly specialized to secrete insulin in response to elevated glucose concentrations in the blood. The final biochemical events leading to insulin release from the β-cell are governed by a secretion apparatus that is similar to the presynaptic machinery mediating synaptic transmission in neuronal networks. We now summarize recent developments in the field of Cdk5 and regulated exocytosis and also present some novel findings regarding Cdk5's effect on insulin secretion.

The small proline-directed serine/threonine cyclin-dependent kinase 5 (Cdk5) is most known for its multifunctional roles in postmitotic neurons (Dhavan and Tsai, 2001; Cheung et al., 2006). However, during recent years this kinase has also been demonstrated to play roles outside the nervous system including being a key factor during cellular differentiation (Lazaro et al., 1997; Philpott et al., 1997; Sahlgren et al., 2003) and as a regulator of exocytosis in endocrine cells (Lilja et al., 2001, 2004; Xin et al., 2004; Wei et al., 2005). Recently the function of Cdk5 in pancreatic β-cells has been in focus since a variance in the CDKAL1 gene, encoding the Cdk5 regulatory subunit associated protein 1-like1, is associated with the prevalence to develop type 2 diabetes (Saxena et al., 2007; Scott et al., 2007; Steinthorsdottir et al., 2007; Zeggini et al., 2007). Therefore, the search for novel pharmacological principles in the treatment of type 2 diabetes will most likely include studies of Cdk5 and Cdk5 regulatory proteins in insulin exocytosis and β-cell physiology.

C. Bark
The Rolf Luft Research Center for Diabetes and Endocrinology, Karolinska Institutet, Stockholm, Sweden
e-mail: christina.bark@ki.se

N.Y. Ip, L.-H. Tsai (eds.), *Cyclin Dependent Kinase 5 (Cdk5)*,
DOI: 10.1007/978-0-387-78887-6_11, © Springer Science+Business Media, LLC 2008

Glucose Stimulation/Insulin Secretion Coupling in the Pancreatic β-Cell

Regulated secretion of insulin has to be tightly controlled to ensure glucose homeostasis in the human body (MacDonald et al., 2005). Insulin is released from β-cells present in the islets of Langerhans in the pancreas. The pancreatic β-cell is the key player in maintaining physiological levels of blood glucose as it is the only cell in the body that has the ability to synthesize, store and after sensing the appropriate stimuli, release the right dose of insulin into the blood stream (Leibiger et al., 2002; Berggren and Leibiger, 2006). Blood glucose is effectively taken up into the β-cell by specific glucose transporters GLUT1 and GLUT2 in order to precisely monitor changes in plasma levels of glucose (Fig. 1). Intracellular increase in glucose results in increased glucose metabolism and ATP generation in glycolysis and the mitochondrial Krebs cycle, leading to elevated levels of intracellular ATP. This increase in cellular ATP leads to closure of ATP-sensitive K^+-channels (K_{ATP}-channels) in the plasma membrane (for recent account see, Speier et al., 2005). K_{ATP}-channel closure results in membrane depolarization of the β-cell. Depolarization opens L-type voltage-dependent Ca^{2+}-channels, and the influx of Ca^{2+} triggers exocytosis of insulin-filled secretory granules (SGs; Fig. 1) (Yang and Berggren, 2006). The capacity of β-cells to respond to extracellular glucose stimulation is also modulated by direct electrical communication between cells (Speier et al., 2007) and by autocrine and paracrine crosstalk between the different cell types in the islets of Langerhans, as well as different actions of a multitude of other factors released from other body organs or nerve endings innervating the pancreas.

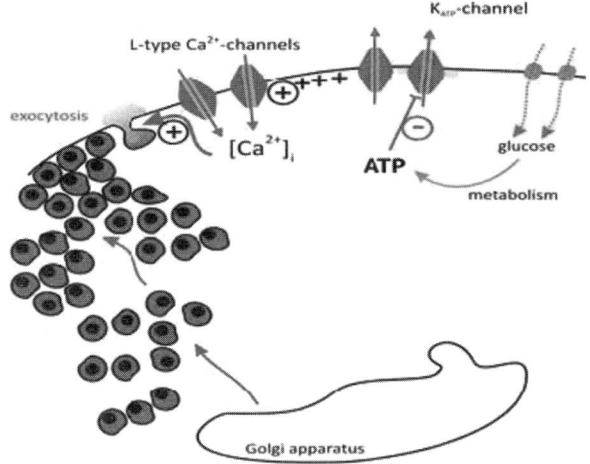

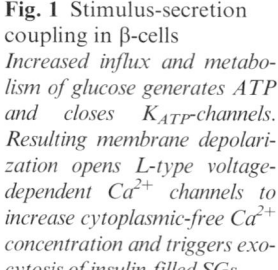

Fig. 1 Stimulus-secretion coupling in β-cells
Increased influx and metabolism of glucose generates ATP and closes K_{ATP}-channels. Resulting membrane depolarization opens L-type voltage-dependent Ca^{2+} channels to increase cytoplasmic-free Ca^{2+} concentration and triggers exocytosis of insulin-filled SGs

Regulated Exocytosis of Insulin

The pancreatic β-cell is highly specialized to secrete insulin in response to elevated glucose concentrations in the blood. However, the final biochemical events leading to the release of insulin from SGs in β-cells are governed by an exocytotic apparatus that, in principle, resembles the well-studied presynaptic machinery mediating exocytosis from synaptic vesicles (SVs) in neurons (Fig. 2A,B) (Südhof, 2004; Jackson and Chapman, 2006; Jahn and Scheller, 2006). These last steps of vesicle trafficking are usually referred to as translocation, docking, priming, and finally fusion with the plasma membrane (Fig. 2B). The docking of the vesicles at the plasma membrane is usually defined biochemically when a protein connection is created between vesicle and plasma membrane. In order to become releasable, the vesicles need to undergo a process referred to as priming that involves Ca^{2+}- and ATP-dependent steps, and renders the vesicles into a fusion-competent state. During the priming process, which also involves several protein rearrangements, a transmembrane soluble N-ethylmaleimide-sensitive factor (NSF) attachment protein receptor (SNARE) complex (Söllner et al., 1993a,b) is formed. The core trans-SNARE

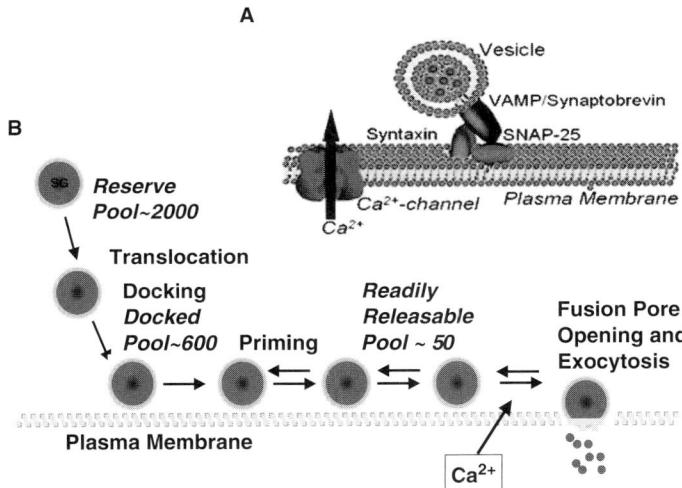

Fig. 2 Schematic illustration of SNARE-complex formation and regulated exocytosis (**A**) *According to the soluble N-ethylmaleimide-sensitive factor attachment protein receptor (SNARE) hypothesis, one of the last steps prior to vesicle fusion and secretion is the formation of a transmembrane SNARE complex consisting of vesicular-SNARE, vesicle-associated membrane protein VAMP/synaptobrevin, and target-SNAREs: syntaxin and synaptosomal-associated protein of 25 kDa, SNAP-25.* (**B**) *Regulated exocytosis is a highly controlled process occurring in all eukaryotic cells and required for both neurotransmission and release of hormones. This pathway involves functionally defined stages and distinct populations of vesicles that need to be translocated from the reserve pool, docked, primed, and finally fused with the plasma membrane in order to release their content. (See Color Insert)*

complex consists of three compartmentally defined SNARE proteins: syntaxin 1, VAMP2/synaptobrevin, and synaptosomal-associated protein of 25 kD, SNAP-25 (Fig. 2A). The three SNARE proteins are held together by strong protein–protein interactions, whereby the cytoplasmic domains form a four α-helix coiled-coiled bundle (Sutton et al., 1998). In the triggering of membrane fusion, the family of Ca^{2+}-sensors named synaptotagmins that are associated with the SNARE complex also plays important roles. When binding incoming Ca^{2+}-ions, they have the ability to promote fusion by changing the curvature of lipid membranes (Martens et al., 2007). The current hypothesis is that the SNARE complexes also take part in the actual fusion event and have intrinsic, albeit slow, capability to perform the merging of membranes (Giraudo et al., 2006; Pobbati et al., 2006).

The different processes described above, except maybe fusion, are also reversible, resulting in vesicles residing in different vesicle pools in the cell. Pancreatic β-cells hold a higher amount of secretory vesicles compared to neurons; however, only a fraction of those are fusion competent. A single β-cell contains more than 13,000 insulin-containing SGs that are residing in different functional vesicle pools. Approximately 600 are docked at the plasma membrane (Dean, 1973; Rorsman and Renström, 2003), but of those, on average only 50 are available for immediate release (Barg et al., 2002). However, a larger pool of vesicles comprising approximately 2,000 of the total vesicles in the β-cell is present in the vicinity of the plasma membrane and referred to as the reserve pool (Rorsman and Renström, 2003). The remaining 10,000 insulin granules are present in different cytosolic storage pools. The sizes of SGs and SVs differ, with insulin granules being 5–10-fold larger in diameter, approximately 200–350 nm (Barg et al., 2002). Another feature that distinguishes secretion of insulin from neurotransmission is the speed of fusion. Triggered exocytosis of SVs is 10–200-fold faster than insulin secretion from SGs (Martin, 2003). When using capacitance recordings to detect the dynamic changes in cell surface area due to exocytosis, SVs fuse within 0.1–1 ms, whereas the fusion time for SG in pancreatic β-cells is as slow as 85–2,000 ms (Rose et al., 2007).

Despite the fact that the SNARE complex in itself has fusogenic capacity, accessory proteins are required in order to maintain accuracy and speed of fusion and membrane trafficking. Among those is the family of Sec1/Munc18 (SM) proteins, first described as membrane-trafficking mutants in yeast (sec1) and *C. elegans* (unc18) (Brenner, 1974; Novick et al., 1980). There are three different Munc18 proteins present in mammals: Munc18-1, Munc18-2, and Munc18-3 (Garcia et al., 1994; Hata et al., 1993; Hata and Südhof, 1995; Pevsner et al., 1994; Tellam et al., 1995). All Munc18 proteins interact with different syntaxin isoforms during distinct steps of vesicle trafficking, but for the last biochemical stages of regulated exocytosis, maybe the Munc18-1 isoform is the most interesting as it specifically interacts with syntaxin 1, both during a step when vesicles are docked at the plasma membrane (Dulubova et al., 1999; Misura et al., 2000) and as recently demonstrated when syntaxin 1 is present in the fusogenic trans-SNARE complex (Dulubova et al., 2007; Shen et al., 2007). Furthermore, it has been suggested that Munc18-1 also influences the dynamics of the formed fusion pore (Fisher et al., 2001). Thus, Munc18-1 appears to be a key player both in several of the intermediate protein complexes leading to fusion and in the fusion event itself.

Cdk5 in Pancreatic β-Cells

A powerful way to regulate function of the different proteins participating in the exocytotic machinery is by phosphorylation (Snyder et al., 2006). The adding of a phosphate group to the side chains of amino acids occurs typically on threonins, serines, or tyrosines. Protein kinases A (PKA) and C (PKC) are two well-known kinases regulating exocytosis, but recently Cdk5 has also been identified as an important modulator of the fusion of vesicles and in the process of vesicle retrieval, termed endocytosis (Shuang et al., 1998; Fletcher et al., 1999; Barclay et al., 2004; Lilja et al., 2004; Floyd et al., 2001; Tan et al., 2003; Tomizawa et al., 2003; Lee et al., 2004). A role of Cdk5 in insulin secretion from pancreatic β-cells was first reported in 2001 (Lilja et al., 2001). The presence of Cdk5 in primary β-cells and pancreatic islets was demonstrated by RT-PCR, immunocytochemistry, and immunoblotting. By using sucrose-density fractionation of unstimulated and glucose-stimulated insulin-secreting mouse pancreatic β-cell homogenates, a glucose-induced translocation of membrane-bound Cdk5 to soluble fractions was revealed, suggesting that the kinase was involved in mechanisms related to insulin exocytosis. Roscovitine is a powerful inhibitor of Cdk5 activity; however, this potent chemical also acts on cell cycle Cdks in the same micromolar concentration range and thus is instrumental in preventing the progression through the cell cycle (Bach et al., 2005). Therefore, caution is needed to interpret the effect of roscovitine on glucose-induced insulin secretion by insulinoma cell lines. Terminally differentiated primary mouse β-cells can be isolated, subjected to roscovitine, and their capacity to release insulin can be readily assessed. The inhibition of Cdk5 with roscovitine in isolated mouse β-cells reduces glucose-stimulated secretion by approximately 35% compared with control, and when the cells are depolarized with both glucose and KCl, roscovitine decreases the amount of released insulin by 65% compared to control β-cells. The involvement of Cdk5 in insulin secretion has also been analyzed by whole-cell capacitance recordings in primary mouse β-cells (Lilja et al., 2001, 2004). Overexpression of wild-type (WT) Cdk5 is not significantly increasing membrane capacitance unless cAMP is excluded from the pipette solution during electrophysiological recordings. This implies that the effects of Cdk5 on insulin secretion can possibly be masked by other kinases activated by cAMP. Measurements of cytoplasmic-free Ca^{2+}-concentrations prior to membrane fusion in roscovitine-treated β-cells demonstrates no difference in total Ca^{2+}-influx after stimulation, although the kinetics of Ca^{2+}-influx appeared slightly different

(Lilja et al., 2001). This suggests that the effect of roscovitine treatment in primary mouse β-cells affects biochemical steps close to secretion, after Ca^{2+}-influx. However, in pituarity cells, roscovitine treatment causes reorganization of the actin cytoskeleton and thus impairs the ability of SGs to traffic to the plasma membrane (Xin et al., 2004). A reduction in docked and fusion-competent vesicles at the site of fusion will also result in a decreased release. Similar results have also been observed during roscovitine treatment of other secretory cells such as neutrophils, anterior pituarity cells, and chromaffin cells, where in all instances inhibition of Cdk5 decreases secretion (Fletcher et al., 1999; Rosales et al., 2004; Xin et al., 2004). In contrast, inhibition of Cdk5 in combination with the p35 activator increases insulin secretion from β-cells during high-glucose conditions (Wei et al., 2005). These apparently conflicting results regarding Cdk5 and regulated insulin exocytosis might be explained by differences in experimental conditions as the numbers of potential Cdk5 substrates in the cell make it possible that Cdk5 can act both as a positive and as a negative regulator of secretion dependent on what biochemical step is affected and analyzed. A second possibility is that Cdk5/p35 and Cdk5/p39 by recognizing different substrates are actors in an intimate interplay tuning regulated exocytosis toward different modes of secretion. Tomizawa and colleagues have demonstrated that Cdk5/p35 has a negative effect on insulin secretion during high-glucose conditions (Wei et al., 2005), whereas Lilja et al. reported a positive effect of specifically Cdk5/p39 in a biochemical step close to vesicle fusion in β-cells (Lilja et al., 2004). Indeed, the p35 activator cannot substitute for p39 function, neither when analyzed with electrophysiological recordings of membrane increase using capacitance measurements in primary β-cells nor when using a glucose-triggered human growth hormone release assay in insulinoma cell lines (Lilja et al., 2004). The findings that Cdk5/p39 participates in regulatory steps close to insulin secretion were further confirmed with identification of the SNARE-interacting protein Munc18-1 as a substrate for Cdk5/p39 in primary β-cells (Lilja et al., 2004). This has been proven using a series of intact and mutated templates for Munc18-1 (Fig. 3). In Fig. 3 it is shown that overexpression of both Munc18-1 WT or a Munc18-1 template mutated in all potential PKC phosphorylation sites in primary β-cells has a positive effect on capacitance increase. This analysis, performed in the absence of cAMP in the pipette solution, demonstrates that the experimental design did not activate PKC, which is a kinase that has previously been demonstrated to phosphorylate Munc18-1 and mediate enhanced stimulated exocytosis (Morgan et al., 2005). Interestingly, using the same experimental approach, no increase in cell capacitance is observed when a Munc18-1 template with a destroyed Cdk5 phosphorylation site is analyzed. Co-transfection of Munc18-1 WT or a Munc18-1 template mutated in the PKC phosphorylation sites together with Cdk5 or p39 increases insulin exocytosis significantly. However, if the Munc18-1 template mutated in the Cdk5 phosphorylation site is co-transfected with Cdk5 or p39, no significant capacitance increase is observed. Also co-transfection of p35 and Munc18-1 WT did not augment secretion.

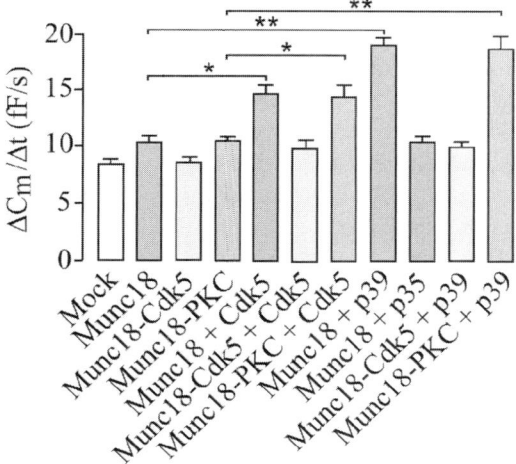

Fig. 3 *Histogram of the mean rate of exocytosis in single primary β-cells as measured by whole-cell capacitance. Primary β-cells were transiently transfected with expression constructs encoding Munc18-1, Cdk5, p35, and p39. The histogram represents mean rates of increase in cell capacitance measured in either mock-transfected cells or cells transiently transfected with WT Munc18-1 (Munc18), a Munc18-1 Cdk5 phosphorylation mutant (Munc18-Cdk5), a Munc18-1 PKC phosphorylation mutant (Munc18-PKC), or the different Munc18 templates co-transfected with Cdk5, p35, or p39. Data are mean values of 10–12 experiments from 3 to 4 individual cell preparations. *, $p < 0.01$ versus mock-transfected cells. **, $p < 0.001$ versus mock-transfected cells. Figure modified from Lilja et al., 2004*

Cdk5/p39 in Insulin Exocytosis and β-cell Physiology

Cdk5 and its regulator p39 seem to play an important role in the mechanism driving exocytosis of secretory vesicles in the endocrine β-cell. Classical approaches, using patch-clamp-based membrane capacitance measurements specifically connected Cdk5/p39 to the exocytotic machinery in β-cells using Munc18-1 as a mediator to stimulate secretion (Lilja et al., 2004; Fig. 3). Several gene targeted mouse mutants have been generated for Cdk5 and the Cdk5 activators, primarily to investigate the role Cdk5 plays in nervous system development and function. Targeting of the Cdk5 gene in mice leads to neonatal death and widespread disruption of neuronal layering in the brain (Ohshima et al., 1996). Ablation of the p35 gene results in aberrant layering of neocortex. but the mice are viable (Chae et al., 1997). Targeting of the p39 gene does not give an obvious neuronal phenotype; however, generating a double knockout by crossing p35- and p39 null mouse mutants leads to a phenotype similar to the Cdk5 knockout mouse (Ko et al., 2001). The findings from the different animal models suggest that p35 is the most important Cdk5 activator in brain and can mask most of the p39 protein function. The functional role of the p39 protein in neuronal cells is more elusive since it can rescue some, but not all, of the roles of p35. Furthermore, the results from the animal models suggest that p35 and p39 are the two most important, and probably the only, activators of Cdk5.

Ablation of p39 in mouse β-cells was tested using fresh slices from whole pancreas. Gene-targeted p39 mutants were kindly provided by Prof. Tsai LH. This revealed significant differences in the physiology of p39 null β-cells compared to WT and heterozygous littermates. In p39 null mutants fed *ad libitum,* hyperglycemia was observed (data not shown), a finding that appeared to be dependent on feeding, but consistent in mice used in the recent patch-clamp experiments. Other obvious differences in cell physiology were the profile of voltage-activated Ca^{2+}-channels, where appearance of the low-voltage-activated (LVA) channels was increased in p39 null β-cells. Consistently, β-cells from these animals show significantly larger excitability induced by glucose perifusion. A perifusion of β-cells with 6 mM glucose reproducibly induced electrical activity of the WT β-cell syncytium approximately 700 s after the start of perifusion (Fig. 4). This response latency is characteristic for a given electrically coupled cluster of β-cells and can be induced repetitively by decreasing or increasing the glucose concentration in the perifusion chamber. In p39-ablated β-cell syncytia, a significantly shorter response latency time to elicit the electrical activity was found: average less than 300 s (Fig. 4).

To elicit inward Ca^{2+}-currents, elevate cytosolic Ca^{2+}-concentration and trigger Ca^{2+}-dependent secretion, trains of depolarizing pulses were applied. Similar to what has been demonstrated previously using membrane capacitance measurements (Speier and Rupnik, 2003; Rose et al., 2007), about one-third of the β-cells did not show evoked exocytosis.

To unequivocally tell responsive β-cells from unresponsive β-cells, we used a stimulation consisting of 50 depolarizations of 40 ms duration from –90 to 0 mV

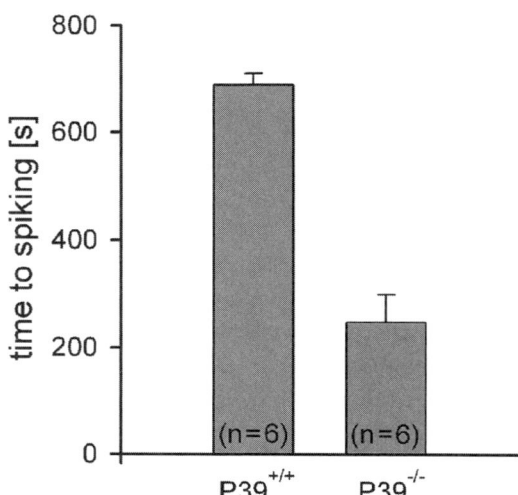

Fig. 4 *Time to onset of electrical activity in β-cell clusters stimulated with 6 mM glucose in WT and p39-ablated animals, respectively*

Fig. 5 *Exocytosis during trains of depolarizing pulses is impaired in β-cells with ablated p39*

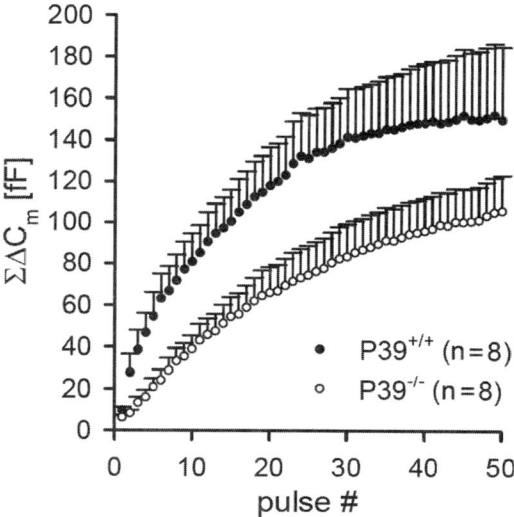

at a frequency of 10 Hz (Rose et al., 2007). When WT control littermate p39 β-cells were stimulated with a train of depolarizing pulses, there was an increase in C_m reaching on average 151 fF at the end of the stimulation (Fig. 5). This corresponded to the release of approximately 43 SGs, assuming a unitary C_m of 3.5 fF for a β-cell SG ($C_m = \varepsilon \times \pi \times d2$, where ε is 9 fF μm^{-2} and $d = 348$ nm; MacDonald et al., 2005). The β-cell of p39-ablated animals reached a total $\Sigma\Delta C_m$ of 106 fF (~30 SGs). Quantitative analysis of the data showed that there was no significant difference between the β-cells of both genotypes in the absolute $\Sigma\Delta C_m$ reached at the end of the train stimulation (Fig. 5). However, the kinetics of the C_m increase was significantly different: Whereas the average release rate in controls stayed high from the 2 nd to the 11th depolarization (80 ± 10 fF s^{-1}, $n = 8$), $p39^{-/-}$ β-cells showed a significantly slower initial response (40 ± 8 fF s^{-1}, $n = 8$, $p < 0.01$). These recent findings are in agreement with the findings by Lilja et al. showing that Cdk5/p39 regulates late biochemical steps in secretion of insulin from pancreatic β-cells (Lilja et al., 2001, 2004).

Is Cdk5 Positively or Negatively Regulating Insulin Secretion?

Cdk5 was initially identified as a neuronal kinase, active only in postmitotic neurons. However, lately Cdk5 has emerged as a critical regulator also in many non-neuronal cells, including the insulin-secreting endocrine β-cell (for review, see Rosales and Lee, 2006). There are many potential Cdk5 substrates participating in the fine-tuned release of insulin. An examination of steps immediately upstream of vesicle fusion, biochemical events that β-cells share with neuronal and neuroendocrine cells, reveal that Cdk5/p39 together with Munc18-1 augments insulin exocytosis (Lilja et al., 2004). The fact that insulin secretion is regulated by Cdk5/p39 is also supported by the recent findings using electrophysiological recordings on slices from p39 null mutant mice (Fig. 5). Furthermore, as β-cells from p39 null mice demonstrate larger excitability induced by glucose, the glucose-triggering pathway of insulin secretion is also regulated by p39 (Fig. 4). A positive effect by p39 on glucose-stimulated release of endogenously added growth hormone from insulinoma cells also demonstrated a positive effect (Lilja et al., 2004). However, there are also recent data demonstrating a negative effect of Cdk5/p35 on insulin exocytosis (Wei et al., 2005). At first glance, those data seem to contradict the findings that Cdk5/p39 plays a positive role in insulin secretion, but this is not necessarily the case. Insulin release, as presented in the Wei et al. paper (Wei et al., 2005), is measured after 1 h of high-glucose treatment, a time point where hormone release represents the sum of both triggering and augmenting pathways and p35 gene expression, Cdk5 activation, and cytoskeletal rearrangements are stimulated. The secretion deficiencies observed in the p39 null mutants may be hidden during extended high-glucose incubations. In the well-established diabetic GK rat model with a prominent deficiency in short-term glucose-triggered insulin release, both upregulation of VDCC and augmented mobilization of SG during prolonged stimulations can be found (Rose et al., 2007). It is clear that both Cdk5/p39 and Cdk5/p35 play key roles in the secretion of insulin from pancreatic β-cells. The p39 activator has been specifically documented to be a positively acting factor involved in insulin exocytosis via specified substrates. On the other hand, the p35 activator appears fatal for glucose-stimulated insulin secretion. A likely explanation for this puzzling scenario is that during extended periods of glucose stimulation, resembling cell stress, part of the effect of Cdk5/p35 can be attributed to the fact that the β-cells respond by activating several signal-transduction pathways. It has been demonstrated that glucose induces expression of the p35 gene and after 1 h of glucose exposure, increased Cdk5/p35 activity can be detected (Ubeda et al., 2004). Chronic exposure to high glucose also decreases insulin mRNA levels, but inhibition of Cdk5 prevents this reduction in gene expression (Ubeda et al., 2006). In neuronal systems, hyperactivation of Cdk5/p35 leads to neurodegeneration and apoptosis (Cheung et al., 2006), and in the neuroendocrine PC12 cells, this has been shown to be mediated by phosphorylation of Map kinase kinase-1 (MEK1), in the MAP kinase pathway (Sharma et al., 2007). Envisioning a similar scenario in β-cells, inhibition of high-glucose-induced detrimental Cdk5/p35 activity, would prevent the cascade of events unfavorable for β-cell survival and function and results in restoration of normal insulin secretion.

In conclusion, during the last decade, Cdk5 has proven to be a powerful kinase, playing important regulatory roles in many cellular processes spanning from neuronal migration and differentiation to membrane trafficking and synaptic plasticity. However, in parallel, Cdk5 has demonstrated a ferocious side by its involvement in the progression of degenerative and destructive processes in neuronal systems. Thus, a similar scenario may be envisioned for the endocrine β-cell. Under normal circumstances, Cdk5/p39 is beneficial for insulin secretion and β-cell physiology, but the kinase can be malicious when the balanced activity between Cdk5/p39 and Cdk5/p35 is disturbed.

Acknowledgments Our own work presented here has been supported by funding to one or more of the authors by the following grant agencies, to which we are sincerely thankful: The Swedish Research Council, The Family Erling-Persson Foundation, The Novo Nordisk Foundation, Berth von Kantzow's Foundation, Funds from Karolinska Institutet, The Swedish Diabetes Association, EFSD, Eurodia and The Slovenian Research Agency.

References

Bach S, Knockaert M, Reinhardt J, Lozach O, Schmitt S, et al. (2005) Roscovitine targets, protein kinases and pyridoxal kinase. J Biol Chem 280:31208–31219

Barclay JW, Aldea M, Craig TJ, Morgan A, Burgoyne RD (2004) Regulation of the fusion pore conductance during exocytosis by cyclin-dependent kinase 5. J Biol Chem 279:41495–41503

Barg S, Eliasson L, Renstrom E, Rorsman P (2002) A subset of 50 secretory granules in close contact with L-type Ca2+ channels accounts for first-phase insulin secretion in mouse β-cells. Diabetes 51 Suppl 1:S74–S82

Berggren PO, Leibiger IB (2006) Novel aspects on signal-transduction in the pancreatic β-cell. Nutr Metab Cardiovasc Dis.16 Suppl 1:S7–S10

Brenner S (1974) The genetics of *Caenorhabditis elegans*. Genetics 77:71–94

Chae T, Kwon YT, Bronson R, Dikkes P, Li E, Tsai LH (1997) Mice lacking p35, a neuronal specific activator of Cdk5, display cortical lamination defects, seizures, and adult lethality. Neuron 18:29–42

Cheung ZH, Fu AKY, Ip NY (2006) Synaptic roles of Cdk5: Implications in higher cognitive functions and neurodegenerative diseases. Neuron 50:13–18

Dean PM (1973) Ultrastructural morphometry of the pancreatic β-cell. Diabetologia 9:115–119

Dhavan R, Tsai LH (2001) A decade of CDK5. Nat Rev Mol Cell Biol 2:749–759

Dulubova I, Sugita S, Hill S, Hosaka M, Fernandez I, Südhof TC, Rizo J (1999) A conformational switch in syntaxin during exocytosis: role of munc18. EMBO J 18:4372–4382

Dulubova I, Khvotchev M, Liu S, Huryeva I, Südhof TC, Rizo J (2007) Munc18-1 binds directly to the neuronal SNARE complex. Proc Natl Acad Sci USA 104:2697–2702

Fisher RJ, Pevsner J, Burgoyne RD (2001) Control of Fusion Pore Dynamics During Exocytosis by Munc18. Science 291:875–878

Fletcher AI, Shuang R, Giovannucci DR, Zhang L, Bittner MA, et al. (1999) Regulation of exocytosis by cyclin-dependent kinase 5 via phosphorylation of Munc18. J Biol Chem 274:4027–4035

Floyd SR, Porro EB, Slepnev VI, Ochoa GC, Tsai LH, De Camilli P (2001) Amphiphysin 1 binds the cyclin-dependent kinase (cdk) 5 regulatory subunit p35 and is phosphorylated by cdk5 and cdc2. J Biol Chem 276:8104–8110

Garcia EP, Gatti E, Butler M, Burton J, De Camilli P (1994) A rat brain Sec1 homologue related to Rop and UNC18 interacts with syntaxin. Proc Natl Acad Sci USA 91:2003–2007

Giraudo CG, Eng WS, Melia TJ, Rothman JE (2006) A clamping mechanism involved in SNARE-dependent exocytosis. Science 313:676–680

Hata Y, Slaughter CA, Sudhof TC (1993) Synaptic vesicle fusion complex contains unc-18 homologue bound to syntaxin. Nature 366:347–351

Hata Y and Südhof TC (1995) A novel ubiquitous form of Munc-18 interacts with multiple syntaxins. Use of the yeast two-hybrid system to study interactions between proteins involved in membrane traffic. J Biol Chem 270:13022–13028

Jahn R, Scheller RH (2006) SNAREs – engines for membrane fusion. Nat Rev Mol Cell Biol 7:631–643

Jackson MB, Chapman ER (2006) Fusion pores and fusion machines in Ca2+-triggered exocytosis. Annu Rev Biophys Biomol Struct 35:135–160

Ko J, Humbert S, Bronson RT, Takahashi S, Kulkarni AB, et al. (2001) p35 and p39 are essential for cyclin-dependent kinase 5 function during neurodevelopment. J Neurosci 21:6758–6771

Lazaro JB, Kitzmann M, Poul MA, Vandromme M, Lamb NJ, et al. (1997) Cyclin dependent kinase 5, cdk5, is a positive regulator of myogenesis in mouse C2 cells. J Cell Sci 110:1251–1260

Lee SY, Wenk MR, Kim Y, Nairn AC, De Camilli P (2004) Regulation of synaptojanin 1 by cyclin dependent kinase 5 at synapses. Proc Natl Acad Sci USA 101:546–551

Leibiger IB, Leibiger B, Berggren PO (2002) Insulin feedback action on pancreatic β-cell function. FEBS Lett 532:1–6

Lilja L, Yang SN, Webb DL, Juntti-Berggren L, Berggren PO, Bark C (2001) Cyclin-dependent kinase 5 promotes insulin exocytosis. J Biol Chem 276:34199–34205

Lilja L, Johansson JU, Gromada J, Mandic SA, Fried G, Berggren PO, Bark C (2004) Cyclin-dependent kinase 5 associated with p39 promotes Munc18-1 phosphorylation and Ca^{2+}-dependent exocytosis. J Biol Chem 279:29534–29541

MacDonald PE, Joseph JW, Rorsman P (2005) Glucose-sensing mechanisms in pancreatic β-cells. Phil Trans R Soc B 360:2211–2225

Martens S, Kozlov MM, McMahon HT (2007) How Synaptotagmin promotes membrane fusion. Science 316:1205–1208

Martin TFJ (2003) Tuning exocytosis for speed: fast and slow modes. Biochem Biophy Acta 1641:157–165

Misura KMS, Scheller RH, Weis WI (2000) Three-dimensional structure of the neuronal-Sec1-syntaxin 1a complex. Nature 404:355–362

Morgan A, Burgoyne RD, Barclay JW, Craig TJ, Prescott GR, Ciufo LF, Evans GJ, Graham ME (2005) Regulation of exocytosis by protein kinase C. Biochem Soc Trans 33:1341–1344

Novick P, Field C, Schekman R (1980) Identification of 23 complementation groups required for post-translational events in the yeast secretory pathway. Cell 21:205–215

Ohshima T, Ward JM, Huh CG, Longenecker G, Veeranna, Pant HC, Brady RO, Martin LJ, Kulkarni AB (1996) Targeted disruption of the cyclin-dependent kinase 5 gene results in abnormal corticogenesis, neuronal pathology and perinatal death. Proc Natl Acad Sci USA 93:11173–11178

Pevsner J, Hsu SC, Scheller RH (1994) n-Sec1: a neural-specific syntaxin-binding protein. Proc Natl Acad Sci USA 91:1445–1449

Philpott A, Porro EB, Kirschner MW, Tsai LH (1997) The role of cyclin dependent kinase 5 and a novel regulatory subunit in regulating muscle differentiation and patterning. Genes Dev 11:1409–1421

Pobbati AV, Stein A, Fasshauer D (2006) N- to C-terminal SNARE complex assembly promotes rapid membrane fusion. Science 313:673–676

Rorsman P, Renström E (2003) Insulin granule dynamics in pancreatic β-cells. Diabetologia 46:1029–1045

Rosales JL, Ernst JD, Hallows J, Lee KY (2004) GTP-dependent secretion from neutrophils is regulated by Cdk5. J Biol Chem 279:53932–53936

Rosales JL, Lee KY (2006) Extraneuronal roles of cyclin-dependent kinase 5. BioEssays 28:1023–1034

Rose T, Efendic S, Rupnik M (2007) Ca^{2+}-secretion coupling is impaired in diabetic Goto Kakizaki rats. J Gen Physiol 129(6):493–508

Sahlgren CM, Mikhailov A, Vaittinen S, Pallari HM, Kalimo H, et al. (2003) Cdk5 regulates the organization of Nestin and its association with p35. Mol Cell Biol 23:5090–5106

Saxena R, Voight BF, Lyssenko V,. Burtt NP, de Bakker PIW, et al. (2007) Genome-Wide Association Analysis Identifies Loci for Type 2 Diabetes and Triglyceride Levels. Published Online April 26, 2007 Science DOI: 10.1126/science.1142358

Scott LJ, Mohlke KL, Bonnycastle LL, Willer CJ, Li Y, et al. (2007) A Genome-Wide Association Study of Type 2 Diabetes in Finns Detects Multiple Susceptibility Variants. Published Online April 26, 2007 Science DOI: 10.1126/science.1142382

Sharma M, Hanchate NK, Tyagi RK, Sharma P (2007) Cyclin dependent kinase 5 (Cdk5) mediated inhibition of the MAP kinase pathway results in CREB down regulation and apoptosis in PC12 cells. Biochem Biophys Res Commun May 4; [Epub ahead of print]

Shen S, Tareste DC, Paumet F, Rothman JE, Melia TJ (2007) Selective activation of cognate SNAREpins by Sec1/Munc18 proteins. Cell 128:183–195

Shuang R, Zhang L, Fletcher A, Groblewski GE, Pevsner J, et al. (1998) Regulation of Munc-18/syntaxin 1A interaction by cyclin-dependent kinase 5 in nerve endings. J Biol Chem 273:4957–4966

Smith DS, Tsai LH (2002) Cdk5 behind the wheel: a role in trafficking and transport? Trends Cell Biol 12:28–36

Snyder DA; Kelly ML, Woodbury DJ (2006) SNARE complex regulation by phosphorylation. Cell Biochem Biophys 45:111–123

Speier S, Rupnik M (2003) A novel approach to in situ characterization of pancreatic β-cells. Pflugers Arch 446:553–558

Speier S, Yang SB, Sroka K, Rose T, Rupnik M (2005) KATP-channels in β-cells in tissue slices are directly modulated by millimolar ATP. Mol Cell Endocrinol 230:51–58

Speier S, Gjinovci A, Charollais A, Meda P, Rupnik M (2007) Cx36-Mediated Coupling Reduces {beta}-Cell Heterogeneity, Confines the Stimulating Glucose Concentration Range, and Affects Insulin Release Kinetics. Diabetes 56:1078–1086

Steinthorsdottir V, Thorleifsson G, Reynisdottir I, Benediktsson R, Jonsdottir T, et al. (2007) A variant in *CDKAL1* influences insulin response and risk of type 2 diabetes. Nature Genetics Published online: 26 April 2007 | doi:10.1038/ng2043

Sutton RB, Fasshauer D, Jahn R, Brunger AT (1998) Crystal structure of a SNARE complex involved in synaptic exocytosis at 2.4 A resolution. Nature 395:347–353

Südhof TC (2004) The synaptic vesicle cycle. Annu Rev Neurosci 27:509–547

Söllner T, Whiteheart SW, Brunner M, Erdjument-Bromage H, Geromanos S, Tempst P, Rothman JE (1993a) SNAP receptors implicated in vesicle targeting and fusion. Nature 362:318–324

Söllner T, Bennett MK, Whiteheart SW, Scheller RH, Rothman JE (1993b) A protein assembly disassembly pathway in vitro that may correspond to sequential steps of synaptic vesicle docking, activation, and fusion. Cell 75:409–418

Tan TC, Valova VA, Malladi CS, Graham ME, Berven LA, Jupp OJ, Hansra G, McClure SJ, Sarcevic B, Boadle RA, Larsen MR, Cousin MA, Robinson PJ (2003) Cdk5 is essential for synaptic vesicle endocytosis. Nat Cell Biol 5:701–710

Tellam JT, McIntosh S, James DE (1995) Molecular identification of two novel Munc-18 isoforms expressed in non-neuronal tissues. J Biol Chem 270:5857–58631

Tomizawa K, Sunada S, Lu YF, Oda Y, Kinuta M, Ohshima T, Saito T, Wei FY, Matsushita M, Li ST, Tsutsui K, Hisanaga S, Mikoshiba K, Takei K, Matsui H. (2003) Cophosphorylation of amphiphysin I and dynamin I by Cdk5 regulates clathrin-mediated endocytosis of synaptic vesicles. J Cell Biol 163:813–824

Ubeda M, Kemp DM, Habener JF (2004) Glucose-induced expression of the cyclin-dependent protein kinase 5 activator p35 involved in Alzheimer's disease regulates insulin gene transcription in pancreatic β-cells. Endocrinology 145:3023–3031

Ubeda M, Rukstalis JM, Habener JF (2006) Inhibition of cyclin-dependent protein kinase 5 activity protects pancreatic β cells from glucotoxicity. J Biol Chem 281:28858–28864

Wei FY, Nagashima K, Ohshima T, Saheki Y, Lu YF, et al. (2005) Cdk5-dependent regulation of glucose-stimulated insulin secretion. Nat Med 11:1104–1108

Xin X, Ferraro F, Back N, Eipper BA, Mains RE (2004) Cdk5 and Trio modulate endocrine cell exocytosis. J Cell Sci 117:4739–4748

Yang SN, Berggren PO (2006) The role of voltage-gated calcium channels in pancreatic β-cell physiology and pathophysiology. Endocr Rev 27:621–676

Zeggini E, Weedon MN, Lindgren CM, Frayling TM, Elliott KS, et al. (2007) Replication of Genome-Wide Association Signals in U.K. Samples Reveals Risk Loci for Type 2 Diabetes. Published Online April 26, 2007 Science DOI: 10.1126/science.1142364

Protein–Protein Interactions Involving the N-Terminus of p35

Gary Kar Ho Ng, Lisheng He, and Robert Z. Qi

Abstract As a major regulator of Cdk5, the protein p35 exerts various effects on neuronal cells. Under certain conditions, p35 undergoes truncation of its N-terminus to form the protein p25. Although p25 is equally potent in activating Cdk5 as p35, it displays a number of differences from the latter, indicating an indispensable role of the N-terminal region for carrying out many of the Cdk5-p35 activities. A number of proteins have been identified that interact with p35 via its N-terminus. Such binding confers p35 and Cdk5 a multitude of properties, including regulation of Cdk5 activity by interacting with the proteins SET, CK2, importins β, 5 and 7, as well as the microtubule cytoskeleton; mediation of nuclear import of p35 through binding with the importin proteins; and participation in the control of protein biogenesis by regulating ribosomal protein S6 kinase 1. p35 itself also functions as a microtubule-associated protein through its N-terminus to regulate microtubule dynamics which is required for neurite outgrowth. Therefore, the discovery of p35–N-terminal binding partners has facilitated the elucidation of p35 and Cdk5 functions and regulations.

Introduction

Cdk5 is a proline-directed serine/threonine kinase that belongs to the cyclin-dependent kinase (CDK) family. Despite its sequence homology to Cdk1 and other CDK family members, Cdk5 does not participate in cell cycle control. Although Cdk5 is expressed in all tissues, its kinase activity is primarily observed in brain due to the almost exclusive expression of its activators, p35 and p39, in neurons of the central nervous system [1–4]. p39 is a homologue of p35 discovered during the screening of a human hippocampus library using a bovine p35 cDNA [2]. p35 and p39 specifically activate Cdk5 but not other

R.Z. Qi
Department of Biochemistry, The Hong Kong University of Science and Technology,
Clear Water Bay, Kowloon, Hong Kong, China
e-mail: qirz@ust.hk

N.Y. Ip, L.-H. Tsai (eds.), *Cyclin Dependent Kinase 5 (Cdk5)*,   159
DOI: 10.1007/978-0-387-78887-6_12, © Springer Science+Business Media, LLC 2008

CDK members, and cyclins cannot activate Cdk5, indicating that p35 and p39 are specific activators of Cdk5 [2,5].

Gene-targeting studies further confirm that these two proteins are neuronal activators of Cdk5. In $cdk5^{-/-}$ mice, the neocortex layers are inverted: neurons born later on appear to pile up under earlier waves of migrating cells; these mice die just before or after birth [6]. In $p35^{-/-}$ mice, the layering pattern is inverted, but is less severe and widespread than that observed in $cdk5^{-/-}$ mice; the $p35^{-/-}$ mice are viable and fertile, and can reach adulthood, although they have increased susceptibility to develop seizures [7]. While $p39^{-/-}$ mice do not show any noticeable defects, the p35/p39 double-null mice display the same brain abnormalities as seen in $cdk5^{-/-}$ mice [8], implying that p35 and p39 together are necessary and sufficient for regulating Cdk5 kinase activity during nervous system development, and that p35 is the major neuronal activator of Cdk5.

p25 is an N-terminal truncation of the p35 protein created as a result of neurotoxic stress and neuronal injury. Further investigation showed that p35 can be cleaved into p25, a process mediated directly by calpain in a Ca^{2+}-dependent manner [9,10]. Despite that p25 contains sufficient information to bind to and activate Cdk5, there are a number of differences between p25 and p35. p25 is more stable than p35 with substantially longer half-life [11]. In addition to that, there is significant difference in their subcellular localizations. Due to the lack of the N-terminal 98 amino acids of p35, designated p10 which contains the myristoylation signal motif, p25 is enriched in the nuclear and perinuclear regions of the cell, whereas p35 appears in the cell peripheries [11]. It was postulated that such proteolytic truncation disrupts the normal regulation of Cdk5, causing prolonged activation and mislocalization of Cdk5.

The Cdk5–p35 kinase complex plays a pivotal role in a multitude of neuronal functions, ranging from neuronal migration and differentiation to synaptic plasticity, through exerting its kinase activity on various substrates. It is therefore important for physiological neuronal activities to properly control Cdk5 kinase activity, as evidence has indicated that the deregulation of Cdk5 activity is involved in several neurodegenerative diseases, such as Alzheimer's disease and amytrophic lateral sclerosis [11–13]. In brain, Cdk5–p35 exists in large multiprotein complexes [14]. Unlike Cdk5–p35, Cdk5–p25 exists only as a heterodimer, suggesting that protein–protein interactions may link Cdk5–p35 to its regulators and effectors, and that the N-terminus of p35 is required for association with a myriad of proteins. In this chapter, we will review several proteins isolated to interact with the N-terminus of p35, and the functional and regulatory properties revealed by these protein–protein interactions.

N-Terminus of p35 Interacts with Various Proteins

In neuronal cells, Cdk5 is known to exist in three forms: a free monomeric protein that can be activated, Cdk5–p35 that associates with a number of proteins and exists as large multiprotein complexes, and a Cdk5–p25

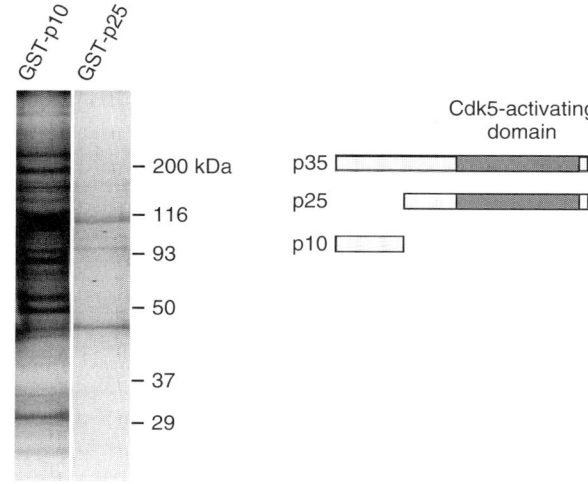

Fig. 1 Biochemical isolation of proteins interacting with p10 and p25 of p35. Note that the GST-p10 column pulled down copious amounts of proteins, whereas the GST-p25 column only pulled down a few, revealing an important role of p10 in association with cellular targets. Below: diagrams showing the location of p25 and p10 on p35

heterodimer [14]. The fact that p25, the truncated form of p35 without the N-terminal 98 amino acids, exists mostly as a heterodimer implies that the N-terminus of p35 is required for Cdk5–p35 to associate with its target proteins. The proteins p35 associates with and the role the N-terminus of p35 plays in the interactions were explored. To do that, biochemical affinity isolation was employed together with mass spectrometry. Interestingly, a number of proteins were found to selectively interact with p10 and few proteins were pulled down with p25 (Fig. 1). Subsequent characterization of the identified interactions revealed proteins carrying different functions which were not found to associate with Cdk5 and p35 previously. In the following sections, we will discuss how the N-terminus of p35 interacts with these proteins and the effects they exert on cellular functions.

Regulation of Cdk5 Activation

The activities of CDKs are regulated in several ways: by binding to cyclins and inhibitor proteins, and by its phosphorylation/dephosphorylation. Although the level of Cdk5 is much more abundant than p35 and p39 in neurons, Cdk5 needs to interact with p35 or p39 to get activated, implying that the levels of p35 and p39 are major determinants in the activity of Cdk5. Unlike the authentic CDKs, Cdk5 can be highly activated by p35 or p25 without its phosphorylation [1,15]. In addition, phosphorylation of Cdk5 at Tyr15, a conserved residue within the CDK family, further enhances its kinase activity; such phosphorylation is mediated by the cytoplasmic kinase c-Abl in conjunction with Cables as well as by the Src family member Fyn [16,17]. In contrast, phosphorylation of the same residue in other CDKs inhibits their activities. Also, a mode of CDK regulation involves CDK inhibitory proteins (CKIs) that bind to CDKs or CDK–cyclin complexes and inhibit their activity through stoichiometric interferences. This includes members of the Cip/Kip family of inhibitors. Intriguingly, despite the fact that Cdk5–cyclin D is a target of Cip/Kip proteins, Cdk5–p35 complex is not recognized by these inhibitors [18]. The observations above suggest distinct regulatory properties of Cdk5–p35.

As mentioned previously, Cdk5 needs to associate with p35 in order to form multiprotein complexes and exert its kinase activity. It has been identified that the N-terminus of p35 is responsible for interaction with several proteins such as importin-β (Impβ), importin-5 (Imp5), importin-7 (Imp7), protein kinase CK2, and cellular microtubules, blocking p35 association with Cdk5 and thus Cdk5 activation. A class of proteins called importins has been identified which mediate nuclear transport by associating with protein cargoes. Impβ, Imp5, and Imp7 directly associate with a basic region at the N-terminus of p35, thus inhibiting the binding of p35 to Cdk5 [19]. Similarly, protein kinase CK2 is also found to inhibit the activity of Cdk5–p35 [20]. CK2 holoenzyme co-exists with Cdk5 and p35 in the brain, and the alpha-subunit of CK2 (abbreviated CK2α) physically interacts with p35 and Cdk5. Investigation of the interactions between Cdk5, p35, and microtubules in a binding assay revealed that microtubules disrupted the interaction between p35 and Cdk5 in a dose-dependent manner, indicating that microtubule polymers inhibit the activity of Cdk5–p35 [21]. The association between the aforementioned proteins and p35 N-terminus deactivates the activity of Cdk5. As the Cdk5-binding and -activating domain is located in p25, such association with the N-terminus perhaps causes p35 conformational change to hinder the interaction between Cdk5 and p35.

In addition to the inhibitors, the nuclear protein SET was found as a positive regulator of Cdk5 activity via interaction with the N-terminus of p35 [22]. SET co-exists with Cdk5–p35 in the nuclear compartment of neuronal cells where they form a trimolecular complex; such binding was found to enhance the Cdk5–p35 kinase activity. Together, the regulation of Cdk5 activity is conferred by the N-terminus, in addition to the Cdk5-binding domain of p35.

Nuclear Import of p35

In addition to the presence of Cdk5 and p35 in the cell soma and neurites, a population of Cdk5 and p35 exists in the nucleus [22–24]. Some of the Cdk5–p35-associating proteins and substrates exhibit nucleus-localizing properties, such as SET and MEF2, which are a modulator and an effector of the kinase, respectively [22,23]. These observations suggest that Cdk5–p35 may exert its nuclear function by phosphorylating its substrates. Interestingly, p35 appears to localize to the nucleus in a signal-mediated manner, as neuregulin induces the nuclear accumulation of p35 but not Cdk5.

Protein transport between the cytoplasm and the nucleus of a cell occurs at the nuclear pore complex, a structure which allows diffusion of small molecules and active transport of large molecules [25]. Ran-GTP, existing primarily in the nucleus, associates with import carriers to mediate cargo release. The nuclear import properties of p35 were investigated by employing nuclear import essays, in which recombinant p35 proteins were incubated with intact nuclei of semi-permeabilized cells, cytoplasmic import factors supplied as rabbit reticulocyte lysate, and an energy-regenerating system [19]. After the reaction, the import was visualized by p35 immunostaining. p35 import happens in an energy-dependent and Ran-GTP/GDP-mediated manner, and requires soluble import factors, as the import is inhibited under each of the following conditions: (1) either cytosolic import factors or the energy system is omitted in the assay; (2) the assay is performed at low temperature to inhibit active protein transport but not passive diffusion; (3) GTP-loaded RanQ69L, a GTPase-deficient Ran mutant, is included [19].

Importins are a family of proteins that mediate active nuclear transport through association with protein cargoes [25]. We isolated three importin family members, Impβ, Imp5, and Imp7, that interact with p10 in rat brain extracts [19]. Subsequent binding characterization revealed that Impβ, Imp5, and Imp7 bind to the amino acid 31–98 region, rich in basic amino acid residues, a feature commonly found in the binding sites of Impβ/5/7 substrates. The effect of Ran-GTP/GDP on p35 interaction with Impβ/5/7 was also examined. Impβ, Imp5, and Imp7 did not interact with p35 in the presence of RanQ69L-GTP, in contrast to the result when Ran-GDP was used, meaning that GTP-bound Ran abrogates p35 association with Impβ/5/7 [19]. This is consistent with the RanQ69L-GTP effect observed from the p35 nuclear import assay using rabbit reticulocyte lysate.

The import activities of Impβ/5/7 with p35 were then tested *in vitro* using the nuclear import assay. This time, cytoplasmic import factors were supplied from purified recombinant proteins of Impβ/5/7 and other factors required facilitating nuclear import instead of rabbit reticulocyte lysate. Impβ, Imp5, and Imp7 can individually stimulate nuclear import of p35 in the assay; among them, Impβ displays the strongest import activity [19]. Given that Impβ/5/7 binds to amino acid 31–98, the role of this region in p35 nuclear import was assessed in the following ways. First, when expressed in fusion with GFP and GST, 31–98 targeted the chimera to the nuclei, showing the nucleus-localizing function of this sequence [19]. Second, Ala substitution for three consecutive Lys residues within this region, [61]KKK[63], abolished p35 interaction with Impβ and reduced the p35 content in the nuclei of transfected cells [19]. These results demonstrate that the N-terminal region plays an important role in p35 targeting to the nucleus via association with certain nuclear import carriers.

To explore the potential role of Cdk5 in p35 nuclear import, we created two p35 mutants, p35(L151,152 N) and p35(D288A/L289A), that lack the binding activity for Cdk5 [19]. When transiently expressed, these mutants appeared in both the cytoplasm and the nuclei with the nucleocytoplasmic distributions similar to that of the wild type [19]. This shows that Cdk5 association is dispensable for nuclear localization of p35. Moreover, we found that Cdk5 and the importins above are mutually exclusive in forming complexes with p35 [19], suggesting that p35 and Cdk5 may undertake separate pathways for nuclear import. As the kinase activity of Cdk5 is drastically stimulated by associating with p35, which causes phosphorylation of p35 and subsequent p35 degradation, employing different modes of transport for Cdk5 and p35 may prevent Cdk5 activation and hence p35 phosphorylation and degradation during transport. Also, our results raise the possibility that the controlled nuclear import of p35 may regulate Cdk5 function in the nucleus.

Cdk5–p35 in the Control of S6 Kinase 1 Activation

The 40S ribosomal protein S6 kinase 1 (S6K1) exists as p70 (α_2 isoform) and the alternative translation variant p85 (α_1 isoform); both of them have an acidic N-terminus, a Ser/Thr kinase catalytic domain, and a regulatory C-terminal tail [26,27]. S6K1 is perceived as a critical component in the control of cell growth and animal size. Activation of this kinase is controlled by a complex mechanism involving at least eight Ser/Thr residues (Thr229, Ser371, Thr389, Ser404, Ser411, Ser418, Thr421, and Ser424), located at the catalytic domain and the C-terminus. Multiple phosphorylation of S6K1 occurs in a step-by-step manner starting at the Ser/Thr-Pro sites located within the autoinhibitory domain adjacent to the C-terminus [26,27]. Such process mediates Thr389 phosphorylation in the linker region, which relieves the autoinhibition and releases the kinase from the translation preinitiation complex. The resulting phosphorylated S6K1 then binds to and gets phosphorylated by phosphoinositide-dependent protein kinase 1 (PDK1) at Thr229 in the catalytic activation loop for activation. Among the multiple phosphorylation events, Thr389 phosphorylation, a rapamycin-sensitive process mediated by mTOR/raptor, appears to be a crucial and rate-limiting step in S6K1 activation.

From the isolation of p35-associating proteins, p70 S6K1 was found specifically in the pull-downs of p10 from rat brain extracts. Further characterization of the interactions between Cdk5–p35 and S6K1 using immuno- precipitation and *in vitro* binding assays revealed that S6K1 forms a ternary complex with Cdk5–p35 via physical association with the p10 region of p35 [28]. To explore the function of the association, we tested whether S6K1 is a substrate of Cdk5 in an *in vitro* phosphorylation experiment. A recombinant S6K1 protein got phosphorylated readily when subjected to Cdk5–p35 phosphory-lation [28]. Scanning of the S6K1 sequence revealed that Ser411 and Ser424 conformed to the consensus motif for Cdk5 phosphorylation. In order to determine whether Ser411 and Ser424 were targeted by the Cdk5–p35 phosphorylation above, mutational studies were carried out and found that the phosphorylation occurs primarily at Ser411 and Ser424 with preference for Ser411. In HEK293T cells, activation of Cdk5 by p35 expression enhances S6K1 phosphorylation at Ser411 but not at Ser424 [28], suggesting the mediation of Ser411 phosphorylation by Cdk5–p35 *in vivo*.

Ser411 is located in the conserved RSPRR sequence which confers an inhibitory effect on mTOR-mediated S6K1 activation [29]. To test the impact of Ser411 phosphorylation on S6K1 activity, a Ser411-non-phosphorylatable mutant and a Ser411-phosphomimetic mutant of S6K1 were created. In HEK293T cells, the phosphomimetic mutant was phosphorylated at Thr389 and activated to a similar extent as the wild type upon insulin stimulation; the same stimulation failed to induce Thr389 phosphorylation for the non-phosphorylatable mutant [28]. In addition, the non-phosphorylatable mutation S411A does not affect the kinase activity of the Thr389-phosphomimetic mutant. Together, these observations show that Ser411 phosphorylation mediates S6K1 activation via the control of Thr389 phosphorylation. As the RSPRR sequence is located in the autoinhibitory domain at the C-terminus of S6K1, Ser411 phosphorylation may disrupt the interaction between the N-and C-terminus of S6K1, rendering the kinase to adopt a partially open conformation and therefore relieving part of the negative modulation imparted by the RSPRR segment.

In cultured neurons, S6K1 activation is inhibited by blocking S6K1 phosphorylation at Ser411 and Thr389 through suppression of Cdk5 expression by RNA interference or through inhibition of Cdk5 activity using pharmacological inhibitor [28], meaning that Cdk5 activity is required by nervous system neurons for S6K1 phosphorylation and activation. It has been known that long-term synaptic plasticity involves the production of proteins, especially local protein synthesis in the dendrites [30,31]. The mTOR signaling pathway is implicated in regulating this type of synaptic plasticity, including LTP and LTD [32]. Since S6K1 is a downstream effector of mTOR, this shows that S6K1 plays an important role in the control of LTP, LTD, and thus synaptic plasticity. The fact that S6K1 activation initiates the encoding of protein translation components leads to the belief that Cdk5 participates in protein synthesis in response to synaptic demands.

Regulation of Microtubule Organization

Microtubules are a major cytoskeletal element in eukaryotic cells which intimately regulate intracellular transport, cell mobility, and morphology. In living cells, microtubule-associated proteins (MAPs), which bind microtubule polymers and promote microtubule assembly by stabilizing the polymer structure, are found to modulate the dynamic properties of microtubules. MAPs often exist as phosphoproteins *in vivo*. Phosphorylation of MAPs is site specific, and phosphorylation at different sites may exert variable effects on microtubule association and function. Cdk5 has been implicated to participate in the regulation of microtubule dynamics via phosphorylating several MAPs, such as tau and doublecortin, which interferes with their microtubule-binding and stabilizing activities [33–35].

Recently, direct association of p35 with tubulin and microtubules and the functioning of p35 as a MAP were identified [21]. The binding was mapped to N-terminal regions of p35 without any recognizable microtubule-binding motif, implicating that p35 may contain novel microtubule- and tubulin-binding domains. p35 co-localizes with microtubule structures in neurites, including growth cones of cortical neurons [24]. When p35 distribution was assessed in cultured neurons by extracting cytoplasmic and microtubule proteins, about one-third of p35 was found to associate with microtubules. However, Cdk5 was not detectable in the microtubule fraction, implying that p35 does not bind to Cdk5 when associated with microtubules [21]. Indeed, p35 does not confer Cdk5 attachment to microtubules. Instead, microtubules segregate p35 from Cdk5, acting as an inhibitor of Cdk5 activity [21]. To further analyze the microtubule association, microtubules were isolated from rat brain through three cycles of temperature-induced assembly and disassembly. In each cycle, p35 co-assembled robustly with both cold-labile and cold-stable microtubules, while Cdk5 and p25 appeared in the cold-labile microtubule fractions [21]. Therefore, p35 exhibits prominent association with microtubules including cold-stable microtubules, hinting that p35 may be one of the microtubule cold stabilizers.

The aforementioned interaction of p35 with microtubules and tubulin prompted us to investigate if p35 alters microtubule assembly characteristics. Addition of p35 to purified tubulin in a microtubule assembly assay stimulated the polymerization of tubulin into microtubule polymers; such activity is specific for p35, as p10 and p25 were unable to induce microtubule assembly [21]. Furthermore, microtubules polymerized with p35 were resistant to cold-induced disassembly. When samples from these assembly assays were analyzed using electron microscopy, most of the p35-assembled microtubules were observed to exist in bundles [21]. These results reveal that p35 cross-bridges microtubules in addition to promoting their assembly, and stabilizes them in the form of bundles.

Recall that p35 is a substrate of Cdk5 that gets phosphorylated in brain [15,36,37]. Whether p35 phosphorylation affects its microtubule-binding and polymerizing properties was investigated using microtubule sedimentation and assembly assays, respectively. The assays were conducted with phosphorylated and unphosphorylated forms of p35. Phosphorylation of p35 inhibited its microtubule-binding and assembly-promoting activities [38]. p35 gets phosphorylated by Cdk5 primarily at residues Ser8 and Thr138 *in vivo*, which exhibit different phosphorylation patterns during brain development [36]. To probe the effects of these phosphorylations on the microtubule-polymerizing activity, two mutants were generated, each carrying phosphomimetic mutation of one of the residues. The S8E mutant displayed activity similar to the wild type in polymerizing microtubules, whereas the T138E mutant lost such activity, indicating that the residue Thr138 is paramount in the control of the microtubule-polymerizing function [38]. The microtubule association was tested in HEK293T cells using p35 wild type and non-phosphorylatable mutants co-expressed with Cdk5 to catalyze p35 phosphorylation. Results showed that phosphorylation at either Ser8 or Thr138 impair p35 interaction with microtubules. The phosphorylation at Thr138 appears to be under tight regulation in adult brain by unidentified mechanisms involving protein phosphatases [36]. During embryonic development, however, such activity may serve to maintain a high level of dynamic microtubule cytoskeleton during rapid brain development.

The role of p35 in neurite formation was explored in PC12 cultures. Application of a specific Cdk5 inhibitor, GW8510 [28], effectively abolished nerve growth factor (NGF)-induced neurite outgrowth, but expression of p35 significantly reversed the outgrowth inhibition. Thus, p35 can promote NGF-induced neurite outgrowth independent of Cdk5 activity. As Cdk5 activation by p35 has been linked to neurite outgrowth [24,39,40], these observations suggest that p35 plays a multifunctional role in neurite formation (Fig. 2). Recall that phosphorylation of p35 at Thr138 interferes with the MAP function of p35. A Thr138-phosphomimetic mutant was used to investigate the potential effect of p35 phosphorylation in neurite outgrowth promotion. Mutation T138E impaired the outgrowth-promoting activity, whereas the non-phosphorylatable T138A mutation did not show any effect (Fig. 2).

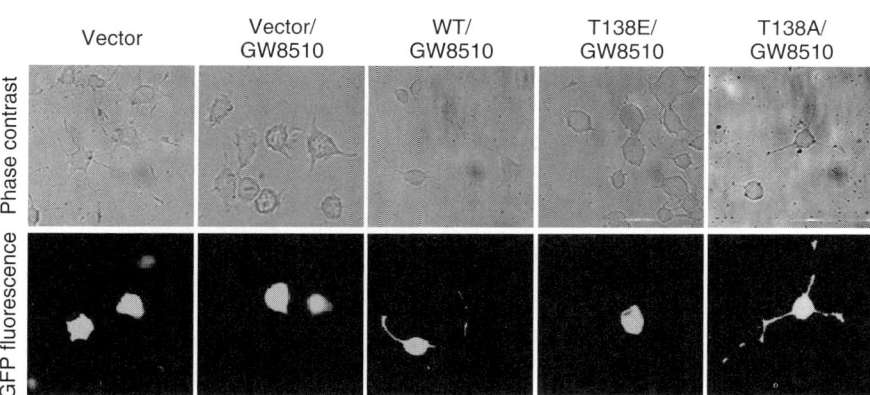

Fig. 2 p35 promotes NGF-induced neurite formation of PC12 cells under Cdk5-inhibitory condition. PC12 cells, transfected with GFP (Vector), GFP-p35 wild type (WT), or mutants, were induced with NGF for differentiation. Micrographs are shown in phase contrast and with GFP fluorescence. Administration of Cdk5 inhibitor GW8510 (2 μM) effectively inhibited NGF-induced neurite outgrowth. The expression of p35 WT promoted neurite outgrowth in the presence of the Cdk5 inhibitor. This p35 activity was impaired by the mutation T138E but not by T138A (*See* Color Insert)

Together, our results suggest that p35 exerts its MAP function in neurite outgrowth, and that phosphorylation of the Thr138 residue of p35 plays a critical role in the control of p35 function. As both Cdk5 and p35 are present in the perinuclear soma, neurites, and axonal growth cones of developing neurons [24], areas which are consistent with their functions in neuronal migration, neurite extension, and axonal growth, it is likely that Cdk5 exerts its regulatory function of the microtubule network via phosphorylation of several cytoskeletal proteins, including p35.

Conclusion

Although Cdk5 is a member of the CDK family, it displays functional activities that are completely distinct from its counterparts. In brain, p35 is the major component in regulating Cdk5 kinase activity and function. Through the N-terminal 98 amino acid region, p35 interacts with a number of proteins to carry out various functions within neuronal cells. Intriguingly, some of these functions are independent of Cdk5 association, revealing that p35 is not merely a regulator of Cdk5. Also, this p35 region confers regulation of Cdk5 activity via interaction with certain protein factors or subcellular structures, which sequester p35 from Cdk5. As the N-terminal 98 amino acids are absent in the proteolytically truncated p25 form, such truncation abrogates many of the p35 functional properties, and the resulting Cdk5-p25 escapes the Cdk5 activation control exerted by the N-terminus-binding factors of p35. These and other differences between p35 and p25 provide a molecular basis for their distinct functions in nervous system neurons. The potential of p35 N-terminus in interacting with proteins will lead to more regulators and effectors for Cdk5 and p35 to be discovered, and their additional functions to be explored.

Acknowledgments This work was supported by the Earmarked Research Grant from the Research Grants Council and the Area of Excellence Scheme under the University Grants Committee of Hong Kong.

References

1. Lew, J., Q. Q. Huang, Z. Qi, R. J. Winkfein, R. Aebersold, T. Hunt, and J. H. Wang (1994) A brain-specific activator of cyclin-dependent kinase 5. Nature 371:423–426
2. Tang, D., A. C. Chun, M. Zhang, and J. H. Wang (1997) Cyclin-dependent kinase 5 (Cdk5) activation domain of neuronal Cdk5 activator: Evidence of the existence of cyclin fold in neuronal Cdk5a activator. J. Biol. Chem. 272:12318–12327
3. Tsai, L. H., I. Delalle, V. S. Caviness, Jr., T. Chae, and E. Harlow (1994) p35 is a neural-specific regulatory subunit of cyclin-dependent kinase 5. Nature 371:419–423

4. Tsai, L. H., T. Takahashi, V. S. Caviness, Jr., and E. Harlow (1993) Activity and expression pattern of cyclin-dependent kinase 5 in the embryonic mouse nervous system. Development 119:1029–1040

5. Poon, R. Y., J. Lew, and T. Hunter (1997) Identification of functional domains in the neuronal Cdk5 activator protein. J. Biol. Chem. 272:5703–5708

6. Ohshima, T., J. M. Ward, C. G. Huh, G. Longenecker, Veeranna, H. C. Pant, R. O. Brady, L. J. Martin, and A. B. Kulkarni (1996) Targeted disruption of the cyclin-dependent kinase 5 gene results in abnormal corticogenesis, neuronal pathology and perinatal death. Proc. Natl. Acad. Sci. USA 93:11173–11178

7. Chae, T., Y. T. Kwon, R. Bronson, P. Dikkes, E. Li, and L. H. Tsai (1997) Mice lacking p35, a neuronal specific activator of Cdk5, display cortical lamination defects, seizures, and adult lethality. Neuron 18:29–42

8. Ko, J., S. Humbert, R. T. Bronson, S. Takahashi, A. B. Kulkarni, E. Li, and L. H. Tsai (2001) p35 and p39 are essential for cyclin-dependent kinase 5 function during neurode-velopment. J. Neurosci. 21:6758–6771

9. Kusakawa, G., T. Saito, R. Onuki, K. Ishiguro, T. Kishimoto, and S. Hisanaga (2000) Calpain-dependent proteolytic cleavage of the p35 cyclin-dependent kinase 5 activator to p25. J. Biol. Chem. 275:17166–17172

10. Lee, M. S., Y. T. Kwon, M. Li, J. Peng, R. M. Friedlander, and L. H. Tsai (2000) Neurotoxicity induces cleavage of p35 to p25 by calpain. Nature 405: 360–364

11. Patrick, G. N., L. Zukerberg, M. Nikolic, M. S. de la, P. Dikkes, and L. H. Tsai (1999) Conversion of p35 to p25 deregulates Cdk5 activity and promotes neurodegeneration. Nature 402:615–622

12. Cruz, J. C., H. C. Tseng, J. A. Goldman, H. Shih, and L. H. Tsai (2003) Aberrant Cdk5 activation by p25 triggers pathological events leading to neurodegeneration and neurofi-brillary tangles. Neuron 40:471–483

13. Nguyen, M. D., R. C. Lariviere, and J. P. Julien (2001) Deregulation of Cdk5 in a mouse model of ALS: toxicity alleviated by perikaryal neurofilament inclusions. Neuron 30:135–147

14. Lee, K. Y., J. L. Rosales, D. Tang, and J. H. Wang (1996) Interaction of cyclin-dependent kinase 5 (Cdk5) and neuronal Cdk5 activator in bovine brain. J. Biol. Chem. 271:1538–1543

15. Qi, Z., Q. Q. Huang, K. Y. Lee, J. Lew, and J. H. Wang (1995) Reconstitution of neuronal Cdc2-like kinase from bacteria-expressed Cdk5 and an active fragment of the brain-specific activator. Kinase activation in the absence of Cdk5 phosphorylation. J. Biol. Chem. 270:10847–10854

16. Sasaki, Y., C. Cheng, Y. Uchida, O. Nakajima, T. Ohshima, T. Yagi, M. Tani-guchi, T. Nakayama, R. Kishida, Y. Kudo, S. Ohno, F. Nakamura, and Y. Goshima (2002) Fyn and Cdk5 mediate semaphorin-3A signaling, which is involved in regulation of dendrite orientation in cerebral cortex. Neuron 35:907–920

17. Zukerberg, L. R., G. N. Patrick, M. Nikolic, S. Humbert, C. L. Wu, L. M. Lanier, F. B. Gertler, M. Vidal, R. A. Van Etten, and L. H. Tsai (2000) Cables links Cdk5 and c-Abl and facilitates Cdk5 tyrosine phosphorylation, kinase upregulation, and neurite outgrowth. Neuron 26:633–646

18. Lee, M. H., M. Nikolic, C. A. Baptista, E. Lai, L. H. Tsai, and J. Massague (1996) The brain-specific activator p35 allows Cdk5 to escape inhibition by p27Kip1 in neurons. Proc. Natl. Acad. Sci. USA 93:3259–3263

19. Fu, X., Y. K. Choi, D. Qu, Y. Yu, N. S. Cheung, and R. Z. Qi (2006) Identification of nuclear import mechanisms for the neuronal CDK5 activator. J. Biol. Chem. 281:39014–39021

20. Lim, A. C., Z. Hou, C. P. Goh, and R. Z. Qi (2004) Protein kinase CK2 is an inhibitor of the neuronal Cdk5 kinase. J. Biol. Chem. 279:46668–46673

21. Hou, Z., Q. Li, L. He, H. Y. Lim, X. Fu, N. S. Cheung, D. X. Qi, and R. Z. Qi (2007) Microtubule association of the neuronal p35 activator of Cdk5. J. Biol. Chem. 282:18666–18670

22. Qu, D., Q. Li, H. Y. Lim, N. S. Cheung, R. Li, J. H. Wang, and R. Z. Qi (2002) The protein SET binds the neuronal Cdk5 activator p35nck5a and modulates Cdk5/p35nck5a activity. J. Biol. Chem. 277:7324–7332

23. Gong, X., X. Tang, M. Wiedmann, X. Wang, J. Peng, D. Zheng, L. A. Blair, J. Marshall, and Z. Mao (2003) Cdk5-mediated inhibition of the protective effects of transcription factor MEF2 in neurotoxicity-induced apoptosis. Neuron 38:33–46

24. Nikolic, M., H. Dudek, Y. T. Kwon, Y. F. Ramos, and L. H. Tsai (1996) The cdk5/p35 kinase is essential for neurite outgrowth during neuronal differentiation. Genes Dev. 10:816–825

25. Gorlich, D. and U. Kutay (1999) Transport between the cell nucleus and the cytoplasm. Annu. Rev. Cell Dev. Biol. 15:607–660

26. Dufner, A. and G. Thomas (1999) Ribosomal S6 kinase signaling and the control of translation. Exp. Cell Res. 253:100–109

27. Martin, K. A. and J. Blenis (2002) Coordinate regulation of translation by the PI 3-kinase and mTOR pathways. Adv. Cancer Res. 86:1–39

28. Hou, Z., L. He, and R. Z. Qi (2007) Regulation of s6 kinase 1 activation by phosphorylation at ser-411. J. Biol. Chem. 282:6922–6928

29. Schalm, S. S., A. R. Tee, and J. Blenis (2005) Characterization of a conserved C-terminal motif (RSPRR) in ribosomal protein S6 kinase 1 required for its mammalian target of rapamycin-dependent regulation. J. Biol. Chem. 280:11101–11106

30. Kandel, E. R. (2001) The molecular biology of memory storage: a dialogue between genes and synapses. Science 294:1030–1038

31. Steward, O. and E. M. Schuman (2001) Protein synthesis at synaptic sites on dendrites. Annu. Rev. Neurosci. 24:299–325

32. Hay, N. and N. Sonenberg (2004) Upstream and downstream of mTOR. Genes Dev. 18:1926–1945

33. Baumann, K., E. M. Mandelkow, J. Biernat, H. Piwnica-Worms, and E. Mandelkow (1993) Abnormal Alzheimer-like phosphorylation of tau-protein by cyclin-dependent kinases cdk2 and cdk5. FEBS Lett. 336:417–424

34. Paudel, H. K., J. Lew, Z. Ali, and J. H. Wang (1993) Brain proline-directed protein kinase phosphorylates tau on sites that are abnormally phosphorylated in tau associated with Alzheimer's paired helical filaments. J. Biol. Chem. 268:23512–23518

35. Tanaka, T., F. F. Serneo, H. C. Tseng, A. B. Kulkarni, L. H. Tsai, and J. G. Gleeson (2004) Cdk5 phosphorylation of doublecortin Ser297 regulates its effect on neuronal migration. Neuron 41:215–227

36. Kamei, H., T. Saito, M. Ozawa, Y. Fujita, A. Asada, J. A. Bibb, T. C. Saido, H. Sorimachi, and S. Hisanaga (2007) Suppression of calpain-dependent cleavage of the CDK5 activator p35 to p25 by site-specific phosphorylation. J. Biol. Chem. 282:1687–1694

37. Patrick, G. N., P. Zhou, Y. T. Kwon, P. M. Howley, and L. H. Tsai (1998) p35, the neuronal-specific activator of cyclin-dependent kinase 5 (Cdk5) is degraded by the ubiquitin-proteasome pathway. J. Biol. Chem. 273:24057–24064

38. He, L., Z. Hou, and R. Z. Qi; unpublished data

39. Harada, T., T. Morooka, S. Ogawa, and E. Nishida (2001) ERK induces p35, a neuron-specific activator of Cdk5, through induction of Egr1. Nat. Cell Biol. 3:453–459

40. Paglini, G., G. Pigino, P. Kunda, G. Morfini, R. Maccioni, S. Quiroga, A. Ferreira, and A. Caceres (1998) Evidence for the participation of the neuron-specific CDK5 activator P35 during laminin-enhanced axonal growth. J. Neurosci. 18:9858–9869

The Kinase Activity of Cdk5 and Its Regulation

Shin-ichi Hisanaga and Koichi Ishiguro

Abstract Cdk5 is a member of cyclin-dependent kinase (Cdk) family and is uniquely activated by binding to its neuron-specific non-cyclin activator p35 or p39. In this chapter we describe the enzymatic properties and the regulation mechanisms of Cdk5 activity in post-mitotic neurons in comparison to cell cycle Cdks. The regulation mechanisms are synthesis and degradation of p35, the interaction with p35- or Cdk5-binding proteins, phosphorylation of Cdk5 or p35, and the association with membranes. In addition to these physiological regulations, deregulation of the activity is induced by the cleavage of p35 to p25 by calpain, leading to hyperactivation and eventually neuronal cell death. The kinase activity and stability of Cdk5–p39, which still remains to be characterized, are also described here. The biochemical information will be useful in studying Cdk5 and elucidating its functions.

Introduction

Cdk5 is a member of the cyclin-dependent kinase (Cdk) family that requires the regulatory subunit p35 or p39 for activation. Cdk5 activity is predominantly detected in post-mitotic neurons where p35 and p39 are mainly expressed (Tang and Wang 1996; Dhavan and Tsai 2002; Hisanaga and Saito 2003). Cdk5 plays a role in a variety of neuronal functions, including neuronal migration and neurite extension during brain development, synaptic activities in mature neurons, and neuronal cell death in neurodegenerative diseases. The kinase activity of Cdk5 should be tightly coupled to these neuronal functions, as the activities of cell cycle Cdks in proliferating cells are closely connected to cell cycle events. To understand Cdk5 functions in neurons, it is important to know the regulatory mechanism of the kinase activity. Whereas the activation mechanism of Cdks in proliferating cells has been extensively investigated, that of Cdk5

S.-i. Hisanaga
Molecular Neuroscience, Department of Biological Sciences, Graduate School
of Science, Tokyo Metropolitan University, Hachioji, Tokyo 192-0397, Japan
e-mail: hisanaga-shinichi@c.metro-u.ac.jp

N.Y. Ip, L.-H. Tsai (eds.), *Cyclin Dependent Kinase 5 (Cdk5)*,   171
DOI: 10.1007/978-0-387-78887-6_13, © Springer Science+Business Media, LLC 2008

remains to be elucidated. In this chapter, we describe the current understanding of the kinase properties and the activation mechanisms of Cdk5. Although most properties of Cdk5 have been obtained with Cdk5 activated by p35 or its N-terminal truncated form, p25, the kinase activity of Cdk5–p39 is here compared with that of Cdk5–p35.

Kinase Properties of Cdk5

Cdk5 Activated by p35 or its N-Terminal Truncated Form p25

Cdk5 is a catalytic subunit of the active kinase complex with a regulatory subunit p35, or its N-terminal truncated form (known as p25) (Fig. 1). The kinase activities of Cdk5 have been studied with several different Cdk5 preparations. The first is Cdk5–p25 purified from bovine, porcine, or rat brains by column chromatography (Ishiguro et al. 1992; Lew et al. 1992; Shetty et al. 1993; Hisanaga et al. 1995). Unfortunately, Cdk5–p35 has not yet been purified from brain tissues, so the following alternative methods have been used: Cdk5–p35 isolated from rat or mouse brains by immunoprecipitation; the Cdk5–p35 or p25 complex prepared from transfected cultured cells by immunoprecipitation; Cdk5–p35 (or p25) synthesized and purified from Sf9 cells with Ni-beads (Tarricone et al. 2001; Hashiguchi et al. 2002; Saito et al. 2003: Yamada et al. 2007); or *in vitro* reconstitution. The Cdk5–p25 complex is

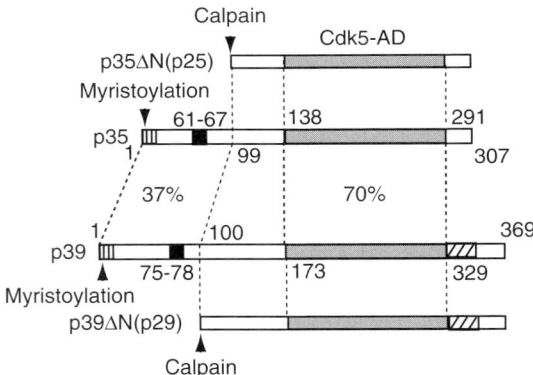

Fig. 1 Molecular structure of p35 and p39 Cdk5 activators, and their C-terminal fragments generated by calpain cleavage. There are 307 amino acids composing p35 and 369 amino acids for p39. Both p35 and p39 contain a myristoylation site at Gly2 in the N-terminal common 9 amino acids (*stripe*), and a lysine cluster at amino acids 61–67 in p35 and at amino acids 75–85 in p39 (*black*). Calpain cleaves p35 between amino acids 98 and 99 to generate p25, and p39 cleaves between amino acids 99 and 100 to generate p29. The Cdk5 activation domain is present in the C-terminal two-thirds encompassing about 150 amino acids (Cdk5-AD, *gray*). A specific insertion after the Cdk5 activation domain (*hatched*) is present in p39

formed *in vitro* by incubating Cdk5 and p25, both of which are separately synthesized and prepared from *Escherichia coli* (Qi et al. 1995; Tang et al. 1997; Amin et al. 2002). In some cases, monomeric Cdk5 from brain extracts is used for reconstruction with p25 prepared from bacteria (Lee et al. 1996a). However, because it is hard to prepare a soluble form of p35 from bacteria, only a few articles describe the kinase activity of Cdk5–p35 reconstructed *in vitro* (Amin et al. 2002).

The properties of kinase activity have not been compared in detail for Cdk5 complexes from different preparations. The basic kinase properties have mainly been determined in earlier Cdk5 studies using the Cdk5–p25 complex purified from mammalian brains by column chromatography (Ishiguro et al. 1992; Lew et al. 1992; Shetty et al. 1992; Hisanaga et al. 1993, 1995) or using Cdk5–p25 reconstituted *in vitro* (Qi et al. 1995; Tang et al. 1997; Amin et al. 2002). The purified Cdk5–p25 was highly active for histone H1 with specific activities of about 0.3 µmol/min/mg (Lew et al. 1992), tau with 0.02 µmol/min/mg (Ishiguro et al. 1992), neurofilament H subunit (NH-H) with 0.003 µmol/min/mg (Hisanaga et al. 1995), and NF-H peptide with 0.5 µmol/min/mg (Shetty et al. 1993). A K_m of 16 µM was measured for tau protein (Ishiguro et al. 1992).

The kinase activity of Cdk5 (named tau protein kinase II) was originally assayed in microtubule assembly buffer solution with ATP (Ishiguro et al. 1992). This is because strong activity was obtained in the *in vitro* microtubule assembly conditions. Cdk5 may have adapted its kinase activity in a cellular environment favorable for microtubule formation, such as in neurites. The optimum kinase activity of Cdk5–p25 is obtained in the reaction mixture at a slightly acidic pH, low ionic strength, and about 1 mM concentrations of Mg^{2+}–ATP with 0.1–1% nonionic detergent. Cdk5–p25 shows the highest kinase activity between pH 5.5 and 6.0 and decreases to about one-third at pH 7.4. The kinase activity is suppressed as NaCl concentration is increased in the assay mixture; the half-inhibition is obtained between 40 and 100 mM, and the activity is reduced to one-quarter of the maximum at 150 mM. The optimal Mg^{2+} concentration depends on ATP because the complex of Mg^{2+} and ATP is used as a substrate. Excess free ATP inhibits the activity. K_m for ATP is 33 µM in the presence of 0.5 mM Mg^{2+}. Stronger kinase activity is obtained when a nonionic detergent such as Triton X-100 or Nonidet P-40 is included in the assay mixture (Fig. 2A). Detergent appears to have triple effects on the kinase activity. Cdk5–p35 is a sticky protein complex, and its adsorption to tube walls is suppressed by detergent. Further, nonionic detergent stimulates the kinase activity, probably through the conformational change of the complex and the removal of membrane fragments bound to the complex.

Cdk5–p35 and Cdk5–p25 are inhibited by chemical inhibitors for Cdks, such as roscovitine, olomoucine, and butyrolactone I at the concentrations required for inhibition of Cdk1–cyclin B (Hisanaga et al. 1995; Meijer et al. 1997; Gray et al. 1999). In general, Cdk5 shows an inhibitory profile against these inhibitors similar to those of Cdk1 and Cdk2 rather than Cdk4 and Cdk6. Because Cdk5 is the only active Cdk family kinase in post-mitotic neurons, these inhibitors have

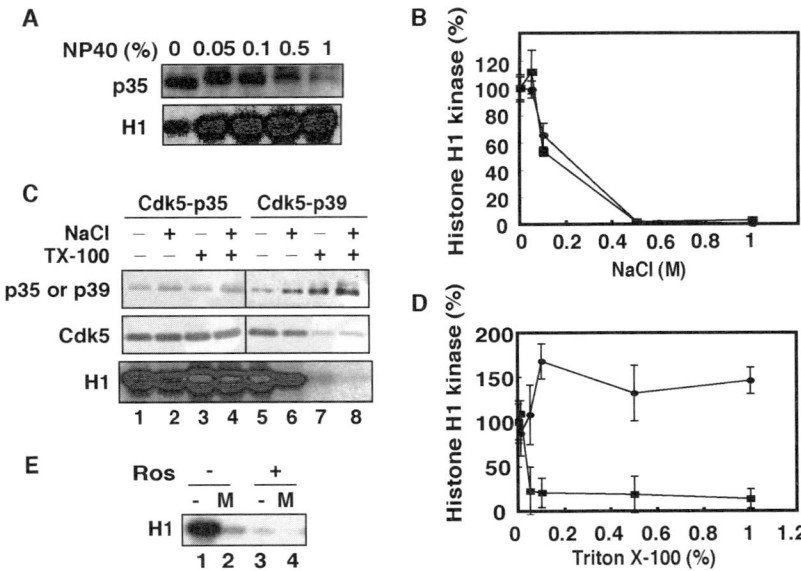

Fig. 2 Several kinase properties of Cdk5 activated by p35 or p39. (A) Activation of Cdk5–p35 by Nonidet P-40 (NP40). Cdk5–p35 eluted at the void volume fractions of Sepharose CL-4B gel filtration of rat brain extract was assayed for p35 or histone H1 phosphorylation in the presence of NP40 (0–1%). Phosphorylation was analyzed by the upward shift of p35 on immunoblotting after SDS-PAGE (*upper panel*) or autoradiography for histone H1 phos-phorylation (*lower panel*). (B) The effect of NaCl on the kinase activities of Cdk5–p35 (*circle*) and Cdk5–p39 (*square*). (C) Dissociation of Cdk5 and p39 by nonionic detergent. Ni-beads-bound Cdk5–p35his (lanes 1–4) and Cdk5–p39his (lanes 5–8) were washed with buffers containing 0.75 M NaCl and/or 1% Triton X-100 (TX-100), and Cdk5 protein and kinase activity bound to Ni-beads were examined on immunoblots with anti-Cdk5 antibody (*middle panel*) and by histone H1 phosphorylation (*lower panel*), respectively. Immunoblotting of p35 or p39 is shown in upper panel. (D) Effect of Triton X-100 on the kinase activity of Cdk5–p35 and Cdk5–p39. The histone H1 kinase activity of Cdk5–p35 (*circle*) and Cdk5–p39 (*square*) was measured as a function of TX100 concentration. Kinase activity is expressed as percen-tage of that in the absence of TX100. (E) Inhibition of Cdk5–p35 kinase activity by mem-branes. Cdk5–p35 immunoprecipitated from the void volume fraction of Sepharose CL-4B of rat brain extract was assayed for histone H1 phosphorylation in the absence or presence of the membrane fraction (M) prepared from rat liver and 40 µM roscovitine (Ros)

been frequently used for demonstrating the involvement of Cdk5 in particular neuronal activities. As a Cdk5–p25 selective inhibitor, Cdk5 inhibitory peptide (CIP), a fragment of p35 consisting of amino acids 154–279, is reported (Amin et al. 2002). CIP would provide a strategy to address the pathological activation of Cdk5–p25 without affecting normal Cdk5 activity (Zheng et al. 2005).

Cdk5 is a proline-directed serine/threonine kinase, targeting serine or threo-nine residues followed by proline with a preference for those with lysine, arginine, or histidine at the third position. The preferred consensus sequence is (S/T)PX(K/R/H) (Beaudette et al. 1993; Shetty et al. 1993). The number of known substrate proteins continues to increase (Dhavan and Tsai 2002; Shelton and Johnson 2004). However, histone H1 appears to be the best protein substrate for Cdk5, and this is frequently used for detecting Cdk5 activity, although several peptide substrates, including the (S/T)P sequence synthesized after the amino acid sequence of histone H1, bradykinin, pp60$^{c\text{-src}}$, tau, and neurofilament H subunit have also been used for the kinase assay (Lew et al. 1992; Beaudette et al. 1993; Ishiguro et al. 1992; Hisanaga et al. 1993; Shetty et al. 1993; Lee et al. 1996a; Ohshima et al. 1996).

Phosphorylation sites predicted from the consensus motifs have been identified as the actual phosphorylation sites for many substrate proteins. However, these sites are not determined solely by the primary sequence. For example, tau is one of the best substrates for Cdk5. There are 17 (S/T)P sites in human tau 441, but four of the sites, Ser202, Thr205, Ser235, and Ser404, are mainly phosphorylated *in vitro* (Ishiguro et al. 1991; Wada et al. 1998). Of the four sites, only Ser404 in the SPRH sequence conforms to the preferred consensus sequence, and the other three are in the minimum consensus of the SP or TP motif. It is not known why these sites are selected by Cdk5. Another example is mouse disable 1 (mDab1) protein. There are two closely positioned Ser residues, Ser491 and Ser515, in the identical SPSK consensus sequence in the C-terminal domain of mDab1, but Cdk5 phosphorylates only Ser491. On the other hand, Ser400 in the SP sequence is chosen as the phosphorylation site (Sato et al. 2007b). Even in crude brain extracts, preferential phosphorylation of certain proteins is observed. For example, strong phosphorylation was shown on microtubule-associated proteins such as tau and MAP2, but phosphorylation of other soluble proteins was barely detectable in bovine brain extracts (Ishiguro et al. 1992). Many of the phosphorylated proteins were associated with the macromolecular structures recovered in the pellet fraction after high-speed centrifugation. Other factors such as binding to macromolecular structures may operate on the selection of phosphorylation sites.

Cdk5 Activated by p39

p39, an isoform of p35, was isolated from hippocampus cDNAs by the polymerase chain reaction (PCR) using the DNA sequence of p35 as the primer (Tang et al. 1995). The p39 isoform has about 150 amino acids in the Cdk5 activation domain (Cdk5-AD) in its C-terminal two-thirds, with a high sequence identity (about 70%) to p35 (Tang et al. 1995; Zheng et al. 1998) (Fig. 1). In contrast, the N-terminal one-third shows a low sequence identity (37%) with p35 in the primary sequence, except for the extreme nine N-terminal amino acids, including a myristoylation signal (Patrick et al. 1999) and the lysine cluster in amino acids 75–85. At the C-terminus following the Cdk5-binding region, p39 has a specific insertion that is shown to bind muskelin, a protein-regulating cytoskeleton organization in adherent cells (Ledee et al. 2005). On SDS-PAGE, p39 migrates at about 45 kDa.

The physiological significance of Cdk5–p39 is shown by knockout studies. Together, p35 and p39 constitute all of the Cdk5 activators in neurons and appear to play distinct roles in neurons with some developmental and regional overlaps. Mice lacking Cdk5 die perinatally, displaying disrupted lamination patterns of neurons in many regions of the brain, including the cerebral cortex, hippocampus, and cerebellum (Ohshima et al. 1996). In contrast, p35-deficient mice survive, although they also show an inverted neuronal lamination pattern in the cerebral cortex (Chae et al. 1997). Mice deficient in p39 do not display a specific phenotype (Ko et al. 2001), but mice lacking both p35 and p39 show the perinatal lethality, the same phenotype as Cdk5-deficient mice (Ko et al. 2001). Studies of mRNA and protein expression show that p39 is more abundant in the posterior part of the brain and appears with a delay after p35 in the forebrain (Zheng et al. 1998; Takahashi et al. 2003). The striped expression of p39 in the cerebellum (Jeong et al. 2003) is interesting, but the significance of this is unknown.

Compared with Cdk5–p35 or Cdk5–p25, little is known about the kinase property of Cdk5 activated by p39. Several reports describe the kinase activity of Cdk5 complexed with the C-terminal recombinant fragment of p39 expressed in bacteria (Tang et al. 1995), and of Cdk5–p39 isolated by immunoprecipitation from brain or cultured neurons (Tang et al. 1997; Humbert et al. 2000; Ko et al. 2001; Takahashi et al. 2003; Lilja et al. 2004; Ohshima et al. 2005). However, detailed comparison with the activity of Cdk5–p35 (or Cdk5–p25) has not been reported. One reason for the small number of reports may be the labile property of the complexes. It has recently been shown that Cdk5–p39 is dissociated and inactivated in the presence of detergent (Yamada et al. 2007). Because Cdk5-acitivator complexes bind to membranes or cytoskeletons in brain, detergent has usually been used for extraction of Cdk5–p39 as well as Cdk5–p35. These procedures have been useful and perhaps even necessary for the isolation of Cdk5–p35, but do not appear to be appropriate for Cdk5–p39.

It is extremely difficult to purify Cdk5–p39 from brain tissues by column chromatography. Nor is it easy to isolate the active Cdk5–p39 from brain extracts or cultured cells in sufficient quantities and purity for biochemical studies using the immunoprecipitation procedure. Therefore, we have purified Cdk5–p39 as the kinase-active complex from Sf9 cells with Ni-beads to purity levels such that Cdk5 and p39 are the major components in the purified fraction (Yamada et al. 2007). The purified Cdk5–p39 still contained contaminating proteins, but it did not have other histone H1 kinase activity, because the kinase activity was completely inhibited by roscovitine or dominant-negative Cdk5. However, some limitations remained for biochemical analysis because the complex gradually loses kinase activity when it is eluted from Ni-beads with high concentrations of imidazole. Some of the kinase properties, however, can be characterized using this Cdk5–p39 complex. Cdk5–p39 shows specific activity for histone H1 similar to Cdk5–p35. The kinase activity of Cdk5–p39, as well as Cdk5–p35, decreased with NaCl concentrations in the assay mixture (Fig. 2B). Cdk5–p39 is inhibited by roscovitine at almost identical

Color Insert

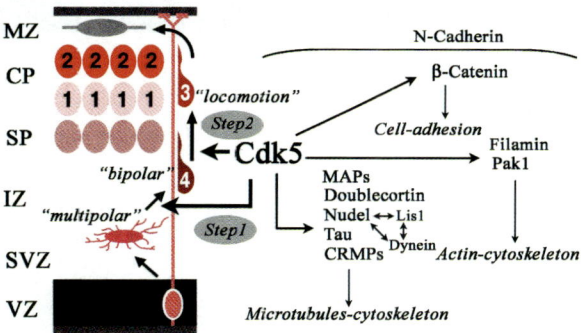

Chapter 1, Fig. 3 Cellular and molecular functions of Cdk5 in neuronal migration. In the cerebral cortex, Cdk5 is required for the radial migration of later-generated neurons. Cdk5 is necessary for multipolar-to-bipolar transition (*Step 1*) and locomotion through the regulation of nucleo-kinesis of migrating neurons (*Step 2*). For these steps, Cdk5 regulates the dynamics of actin-cytoskeleton and microtubules-cytoskeleton and cell adhesion through the phosphorylation of its substrate proteins

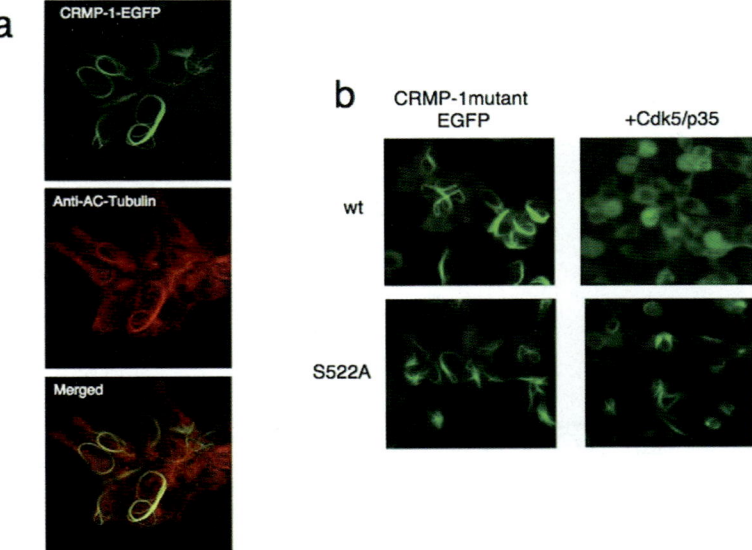

Chapter 2, Fig. 1 Distribution of CRMP1 and tubulin in COS-7 cells. CRMP1-EGFP or Cdk5/p35 was ectopically expressed in COS-7. In (**a**), the cells were immunostained with anti-α-tubulin antibody. The localization of wild-type CRMP1-EGFP showed filamentous structure, and this distribution pattern became diffusely distributed within the cytoplasm when co-transfected with Cdk5/p35 (**b**)

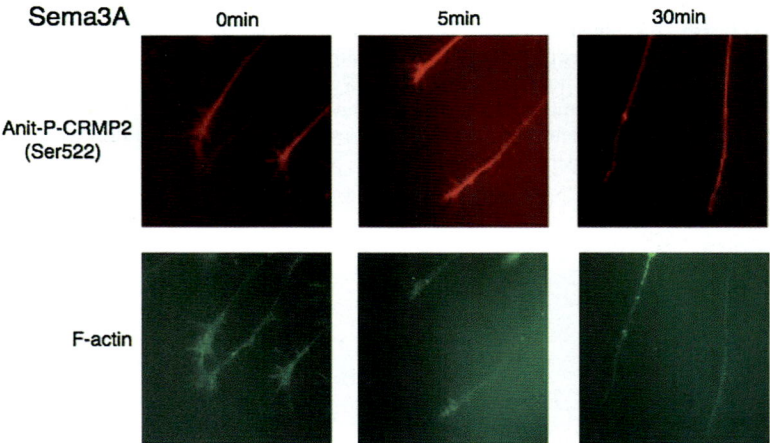

Chapter 2, Fig. 2 Sema3A induced growth cone collapse and increased the levels of CRMP2 phosphorylated at Ser522 in the growth cones. Cultured E7 chick DRG neurons were stimulated with Sema3A (1 nM), fixed and double stained with FITC-phalloidin and anti-P-CRMP1/2 antibody. Scale bar, 50 μm

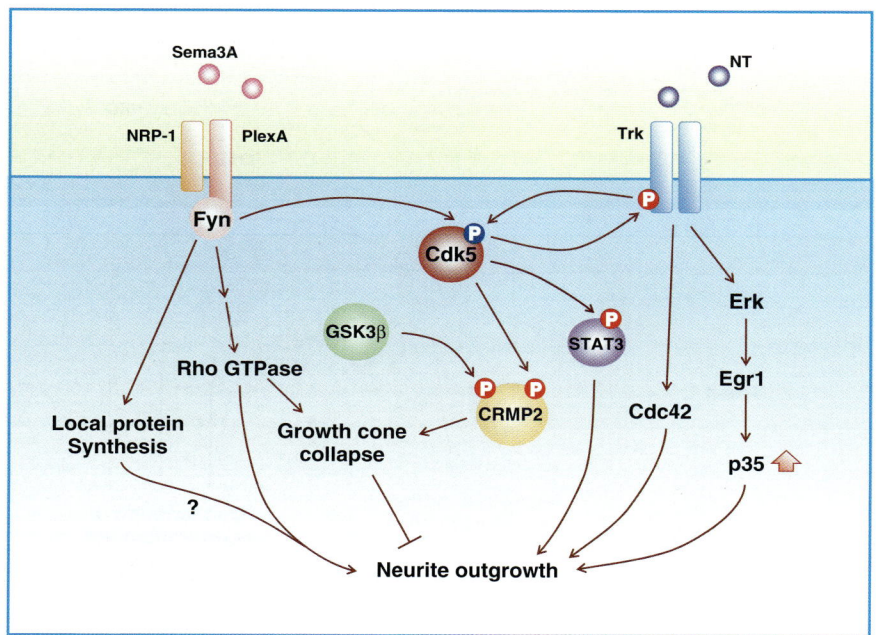

Chapter 5, Fig. 1 Regulation of neurite growth by Cdk5. Cdk5 was observed to regulate neurite development through modulating the downstream signaling of extracellular cues including semaphorins 3A (Sema 3A) and neurotrophins. Cdk5 was found to be required for Sema 3A–induced growth cone collapse via two mechanisms. Phosphorylation of Cdk5 by Fyn at Tyr15 leads to enhanced Cdk5 activity and recruitment of Cdk5 to the Sema 3A receptor complex of neuropilin 1 (NRP-1) and plexinA (PlexA). Both Fyn and Cdk5 are required for Sema 3A–induced growth cone collapse and induction of local protein synthesis, but whether they play a role in the regulation of Rho GTPase activity downstream of Sema 3A treatment remains to be explored. On the other hand, Cdk5 also modulates Sema 3A–induced growth cone collapse through phosphorylation of CRMP2. Dual phosphorylation of CRMP2 by GSK3β and Cdk5 is required for Sema 3A–triggered growth cone collapse. In addition to Sema 3A, Cdk5 also contributes to NGF-induced neurite outgrowth in PC12 cells. NGF treatment leads to sustained Erk1/2 activation, induction of transcription factor Egr-1, and elevated transcription of p35. This enhances Cdk5 activity, and reversal of this increase inhibits NGF-induced neurite outgrowth. Furthermore, Cdk5 may also affect neurite outgrowth through phosphorylation of STAT3, since STAT3 activation following NGF stimulation is required for neurite outgrowth. A recent study reveals that Cdk5 also contributes to BDNF-induced dendrite growth in hippocampal neurons. BDNF stimulation triggers phosphorylation of Cdk5 at Tyr15, resulting in enhanced Cdk5 activity. Cdk5 in turn phosphorylates TrkB, which is required for BDNF-stimulated increase in primary dendrites and activation of Rho GTPase Cdc42. Finally, Cdk5 also modulates Rho GTPase activity through phosphorylation of GEFs or their effectors. Tyrosine phosphorylation is depicted by blue circles, while red circles represent serine/threonine phosphorylation

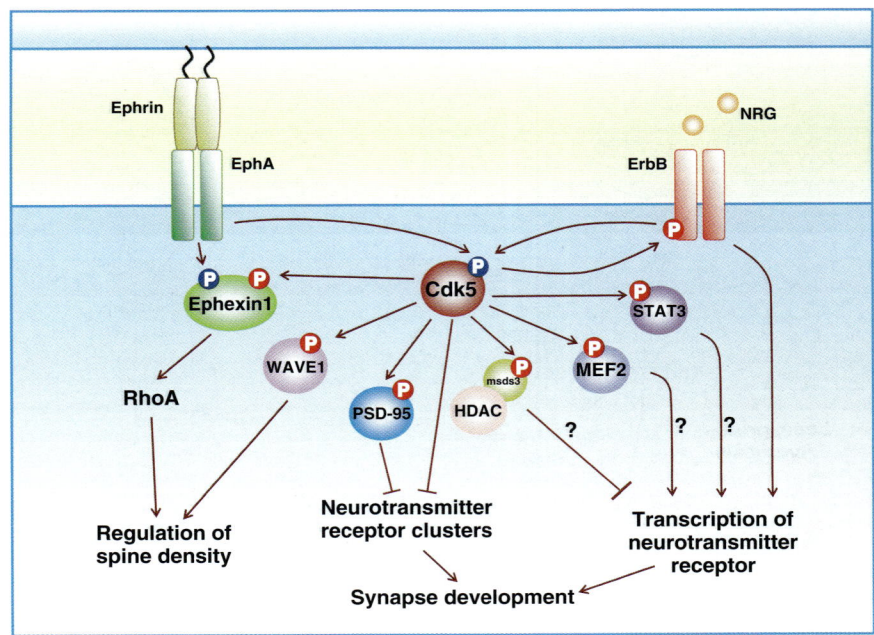

Chapter 5, Fig. 2 Regulation of spine morphogenesis and synapse formation by Cdk5. Cdk5 is recently implicated in spine morphogenesis through modulation of Rho GTPase activity. Through phosphorylation of Rho GEF ephexin, Cdk5 is required for ephrinA1-induced RhoA activation and spine retraction in hippocampal neurons. Cdk5 also inhibits activity of WAVE1, thereby affecting spine morphology. On the other hand, Cdk5 modulates transcription of neurotransmitter receptors at the synapse through regulation of neuregulin (NRG)/ErbB signaling. NRG-triggered increase in Cdk5 activity is required for NRG-induced transcription of acetylcholine receptors and GABA$_A$ receptors at the NMJ and central synapses, respectively. Cdk5-mediated phosphorylation of STAT3 also contributes to NRG-stimulated gene transcription. In addition, Cdk5 regulates gene transcription through inhibition of MEF2 and facilitation of mSds3-HDAC-dependent gene repression, although whether these events are implicated in the regulation of neurotransmitter receptor transcription remains to be determined. Finally, Cdk5 is an important regulator of neurotransmitter receptor clustering. Acetylcholine receptor endplate is broadened in Cdk5-deficient diaphragm. Cdk5 also phosphorylates PSD-95 to inhibit clustering of neurotransmitter receptors in the CNS. Tyrosine phosphorylation is depicted by blue circles, while red circles represent serine/threonine phosphorylation

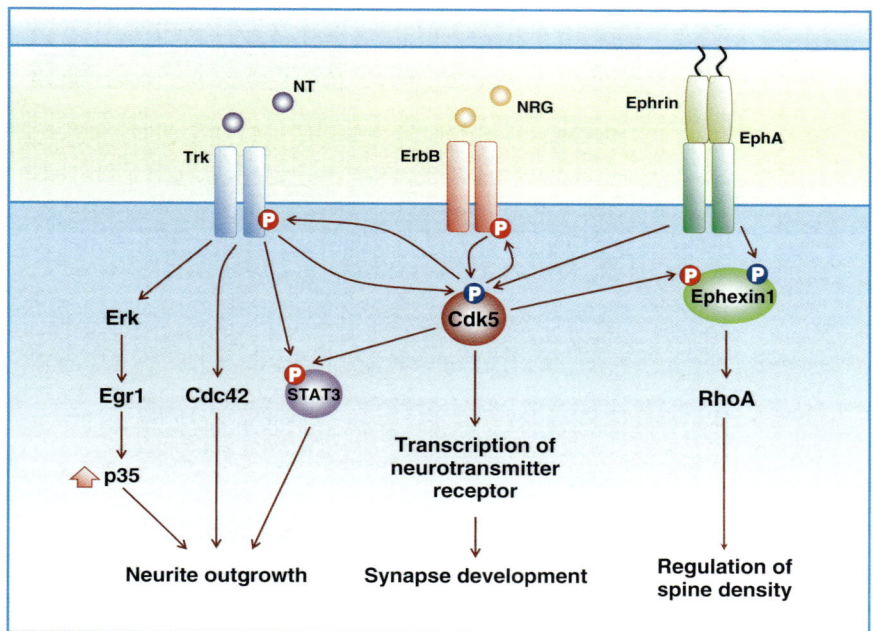

Chapter 5, Fig. 3 The emerging role of Cdk5 as a regulator of RTK signaling. Cdk5 was observed to modulate signaling and functions downstream of RTK activation, thereby affecting dendrite outgrowth, spine morphogenesis, and synapse formation. Cdk5 is required for BDNF-stimulated dendrite growth and Cdc42 activation in hippocampal neurons. On the other hand, Cdk5 is also required for the NGF-induced neurite outgrowth in PC12 cells. NGF stimulation triggers sustained Erk activation and induction of transcription factor Egr-1, leading to p35 expression. Inhibition of this increase attenuates NGF-induced neuronal differentiation. In addition, STAT3 activation downstream of NGF stimulation also contributes to neurite outgrowth in PC12 cells. Since Cdk5 phosphorylates STAT3 to enhance its transcriptional activity, Cdk5 may also affect neurite outgrowth through phosphorylation of STAT3. Cdk5 also regulates synapse formation by modulating the signaling of NRG. Cdk5-mediated phosphorylation of NRG receptor ErbB is required for NRG-stimulated upregulation of acetylcholine receptor expression. NRG also triggers phosphorylation of STAT3 by Cdk5, thereby regulating gene transcription at synapses. Finally, Cdk5 affects spine morphogenesis by modulating signaling downstream of EphA4 activation. Cdk5-mediated phosphorylation of Rho GEF ephexin1 is required for the activation of RhoA and the induction of spine retraction following ephrinA1 treatment. Tyrosine phosphorylation is depicted by blue circles, while red circles represent serine/threonine phosphorylation

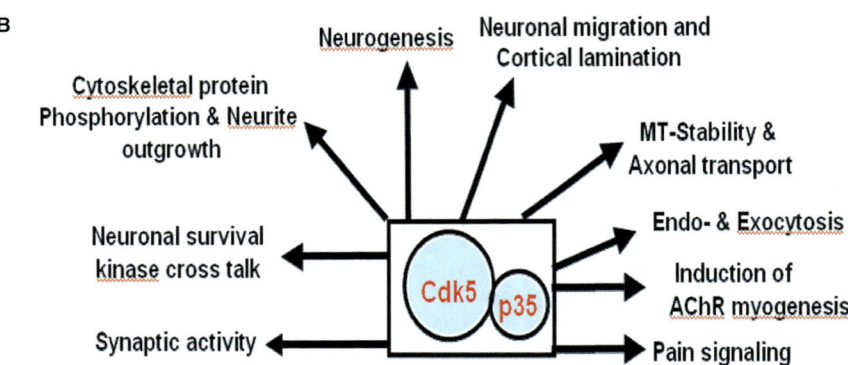

Chapter 6, Fig. 1 (**A**) Comparison of Cdk5 amino acid sequence among various species. Amino acids in black represent conserved residues and that in red represent non-conserved residues. (**B**) Schematic presentation shows involvement of Cdk5 in a wide range of neuronal functions

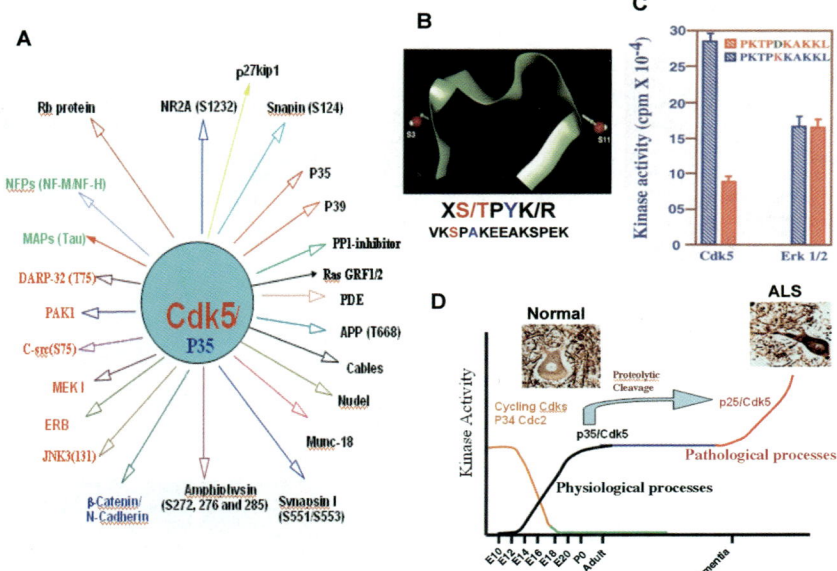

Chapter 6, Fig. 2 (**A**) Some of the Cdk5 substrates (**B**) Ribbon diagram of the best-fit average structure of the peptide VKSPAKEEAKSPEK repeat in rat NF-H, in which the first serine is phosphorylated by Cdk5. (**C**) Comparison of Cdk5 and Erk1/2 activity using peptide substrates, PKTPKKAKKL and PKTPDKAKKL. (**D**) Cdk5 activity is tightly regulated during nervous system development; dividing cells lack Cdk5 activity; during nervous system development, increased expression of neuron-specific p35 protein induces Cdk5 activity by binding to Cdk5. Upon neuronal insult, proteolytic cleavage of p35 produces p25 that deregulates Cdk5 activity

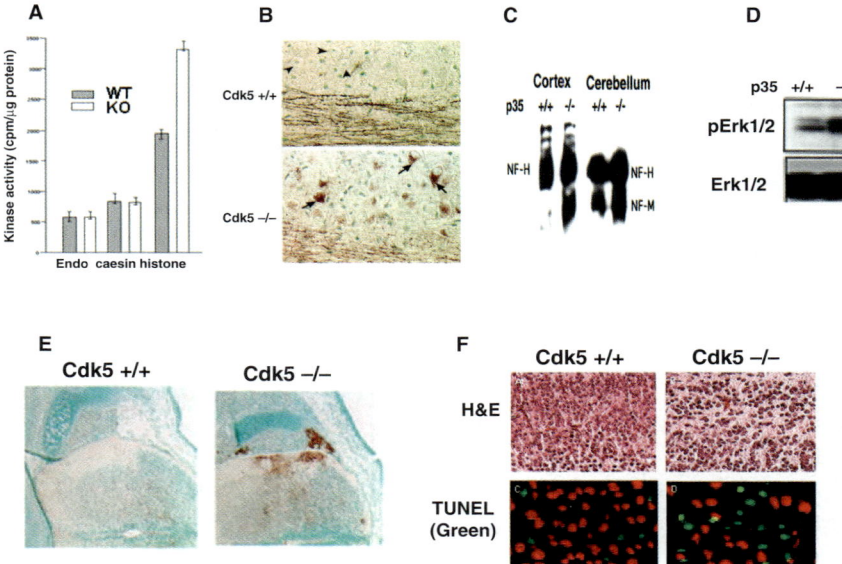

Chapter 6, Fig. 3 (**A**) Comparison of endogenous Cdk5 activity in the brain extracts isolated from wild-type (WT or Cdk5$^{+/+}$) and Cdk5$^{-/-}$ mice, using casein, histone H1 (type III) as substrates. (**B**) Immunohistochemical analyses of E18 brain stem neurons from WT and Cdk5$^{-/-}$ mice using SMI31 antibody. (**C**) Western blot analysis of NF-M/H using SMI31 antibody in the WT and p35$^{-/-}$ brain lysates. (**D**) Phospho-Erk1/2 and total Erk1/2 expression in the WT and p35$^{-/-}$ brain homogenate. (**E**) Caspase-3 activity in E14 spinal cord neurons from WT and CDk5$^{-/-}$ mice. (**F**) Analysis of apoptosis in E16.5 cortical plate from WT and Cdk5$^{-/-}$ mice using TUNEL assay to identify apoptotic cells. *Upper panel* shows the H&E staining of sections

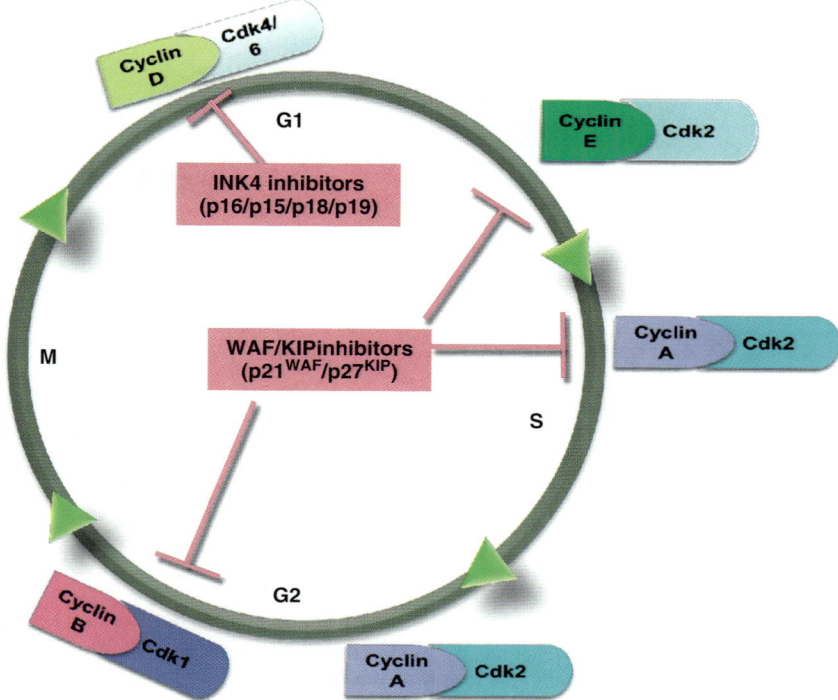

Chapter 9, Fig. 1 A highly simplified diagram of the proteins that drive a vertebrate cell cycle through its four phases, G1, S, G2, and M. We have emphasized the sequential action of the cyclin-dependent kinases, their activating cyclins, and the cyclin-dependent kinase inhibitors.

Interaction of Cdk5 with cell cycle events in neurons

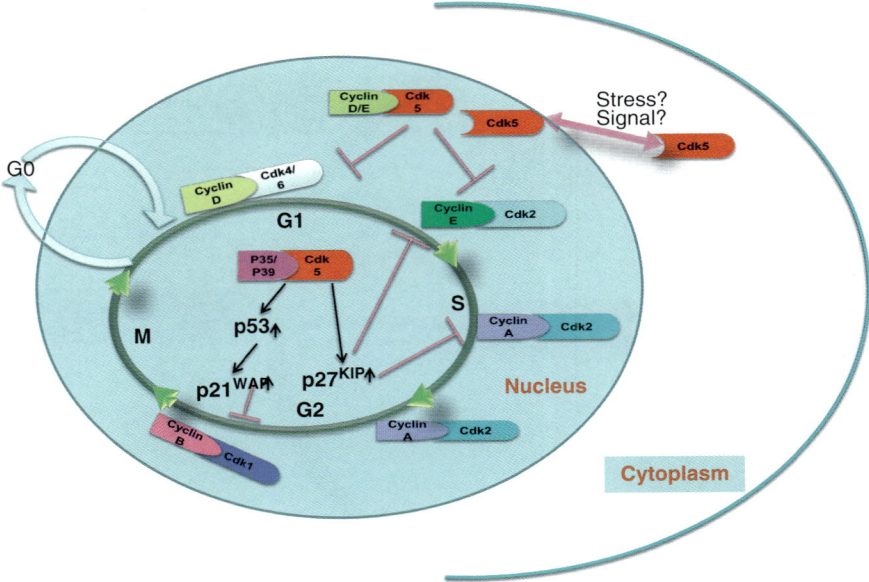

Chapter 9, Fig. 2 A model for the actions of Cdk5 in its role as a cell cycle inhibitor. Several possible modes of action are highlighted including the shuttling of the Cdk5 protein between nucleus and cytoplasm in response to stress as well as its effects in altering the levels of cell cycle active agents such as p21, p27, and p53

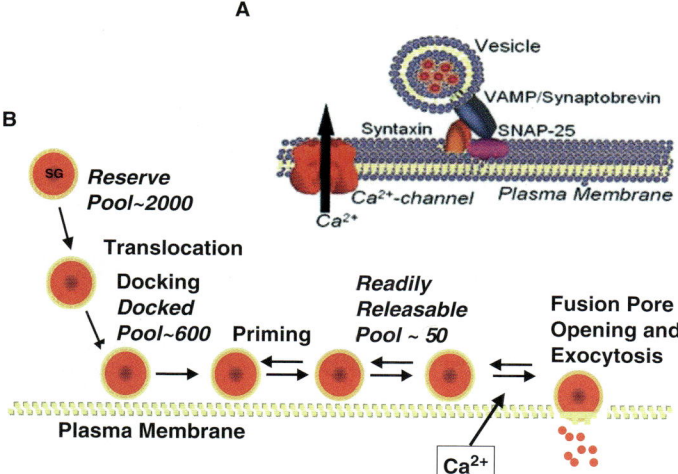

Chapter 11, Fig. 2 Schematic illustration of SNARE-complex formation and regulated exocytosis (**A**) *According to the soluble N-ethylmaleimide-sensitive factor attachment protein receptor (SNARE) hypothesis, one of the last steps prior to vesicle fusion and secretion is the formation of a transmembrane SNARE complex consisting of vesicular-SNARE, vesicle-associated membrane protein VAMP/synaptobrevin, and target-SNAREs: syntaxin and synaptosomal-associated protein of 25 kDa, SNAP-25.* (**B**) *Regulated exocytosis is a highly controlled process occurring in all eukaryotic cells and required for both neurotransmission and release of hormones. This pathway involves functionally defined stages and distinct populations of vesicles that need to be translocated from the reserve pool, docked, primed, and finally fused with the plasma membrane in order to release their content.*

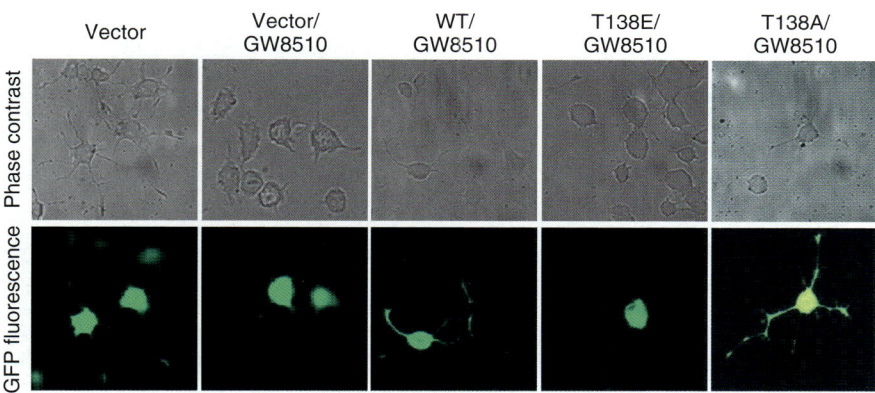

Chapter 12, Fig. 2 p35 promotes NGF-induced neurite formation of PC12 cells under Cdk5-inhibitory condition. PC12 cells, transfected with GFP (Vector), GFP-p35 wild type (WT), or mutants, were induced with NGF for differentiation. Micrographs are shown in phase contrast and with GFP fluorescence. Administration of Cdk5 inhibitor GW8510 (2 μM) effectively inhibited NGF-induced neurite outgrowth. The expression of p35 WT promoted neurite outgrowth in the presence of the Cdk5 inhibitor. This p35 activity was impaired by the mutation T138E but not by T138A

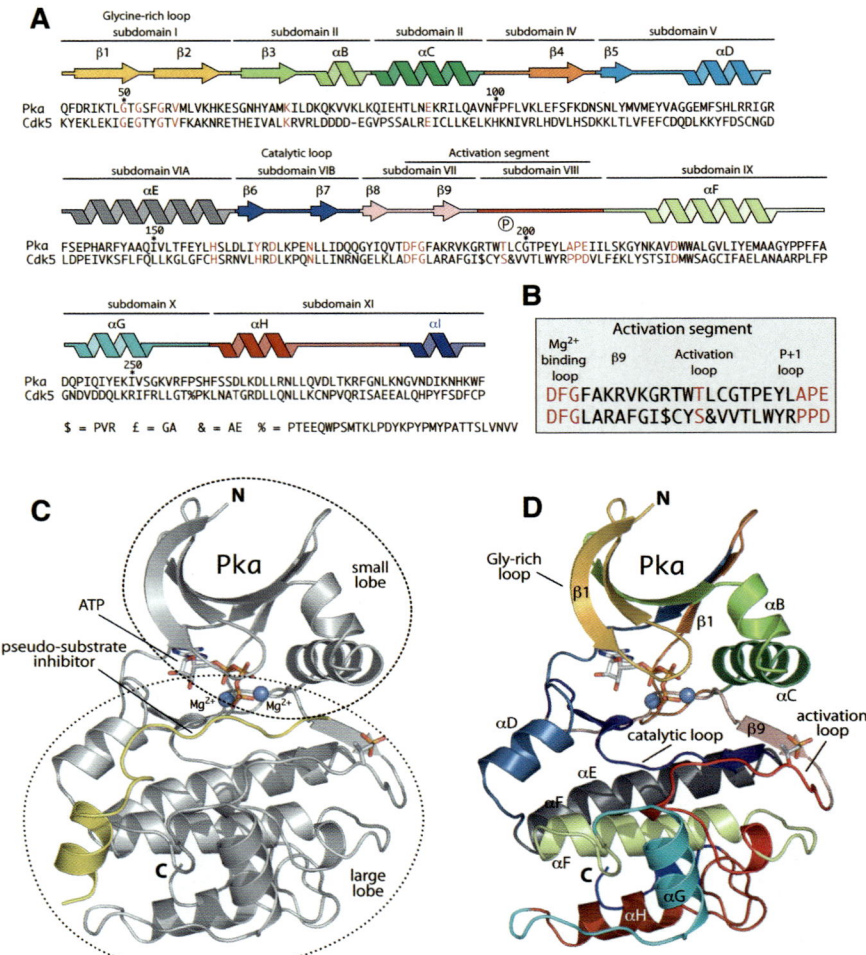

A

Glycine-rich loop

| subdomain I | | subdomain II | | subdomain II | subdomain IV | | subdomain V |
| β1 | β2 | β3 | αB | αC | β4 | β5 | αD |

```
      50                                100
Pka QFDRIKTLGTGSFGRVMLVKHKESGNHYAMKILDKQKVVKLKQIEHTLNEKRILQAVNFPFLVKLEFSFKDNSNLYMVMEYVAGGEMFSHLRRIGR
Cdk5 KYEKLEKIGEGTYGTVFKAKNRETHEIVALKRVRLDDDD-EGVPSSALREICLLKELKHKNIVRLHDVLHSDKKLTLVFEFCDQDLKKYFDSCNGD
```

Catalytic loop   Activation segment

| subdomain VIA | subdomain VIB | subdomain VII | subdomain VIII | subdomain IX |
| αE | β6 | β7 | β8 | β9 | αF |

```
       150                                        ⓟ 200
Pka FSEPHARFYAAQIVLTFEYLHSLDLIYRDLKPENLLIDQQGYIQVTDFGFAKRVKGRTWTLCGTPEYLAPEIILSKGYNKAVDWWALGVLIYEMAAGYPPFFA
Cdk5 LDPEIVKSFLFQLLKGLGFCHSRNVLHRDLKPQNLLINRNGELKLADFGLARAFGI$CYS&VVTLWYRPPDVLF£KLYSTSIDMWSAGCIFAELANAARPLFP
```

| subdomain X | subdomain XI | |
| αG | αH | αI |

```
      250
Pka DQPIQIYEKIVSGKVRFPSHFSSDLKDLLRNLLQVDLTKRFGNLKNGVNDIKNHKWF
Cdk5 GNDVDDQLKRIFRLLGT%PKLNATGRDLLQNLLKCNPVQRISAEEALQHPYFSDFCP
```

$ = PVR   £ = GA   & = AE   % = PTEEQWPSMTKLPDYKPYPMYPATTSLVNVV

B

Activation segment			
Mg²⁺ binding loop	β9	Activation loop	P+1 loop
DFG	FAKRVKGRTW	TLCGTPEYL	APE
DFG	LARAFGI$CYS	&VVTLWYR	PPD

C

D

Chapter 14, Fig. 1 Structural organization of the ePK (**A**) The aligned sequences of the catalytic cores of PKA and CDK5 are presented, together with the secondary structure. The different subdomains are identified by a different color and they are marked. Prominent loops discussed in the text are also marked. Residues that are often referred to in the text are shown in red. The phosphoryl group of T197 of PKA is also shown. CDK5's equivalent residue, Ser159, is not phosphorylated. The &, %, £, and & signs mark insertions in the CDK5 sequence that were not included in the alignment to preserve clarity. The sequence of the insertions is shown at the bottom of the panel. (**B**) Enlargement of the activation segment and definition of regions within it. The activation loop corresponds to subdomain VIII. (**C**) Ribbon model of the PKA catalytic domain (PDB ID code 1 ATP). Structural elements flanking the catalytic domain were removed to improve clarity. The small and large lobes are marked. A pseudo-substrate inhibitor binds near the catalytic cleft, which contains ATP. (**D**) The subdomains of PKA were coloured with the same color scheme used in panel A. The activation segment is stretched. Most secondary structure elements are marked

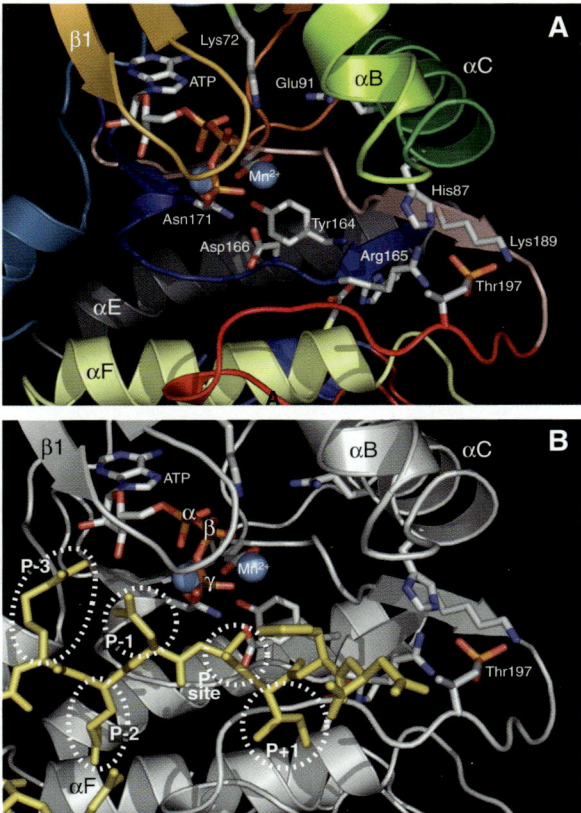

Chapter 14, Fig. 2 Catalytic apparatus of ePKs (**A**) Close-up of the catalytic cleft of PKA (PDB ID code 1 ATP). The coloring scheme is as in Fig. 1D. The side chains of residues that are important for catalysis are highlighted. Thr197 is phosphorylated and acts as an organization center for several positively charged residues. This, in turn, results in the stabilization of the activation segment and poises the active site for catalysis. (**B**) The view is as in panel A. The pseudo-substrate inhibitor, which was omitted in panel A, is now visible in dark yellow, while the kinase is drawn in gray. It occupies the front of the active cleft cavity. In the pseudo-substrate inhibitor peptide, the P-site residue is Ala. It faces the γ-phosphate of ATP, which is held in place by a Mg^{2+} atom (Mn^{2+} in the particular case of this crystal structure) through a bunch of different interactions. Arginine residues at the P-3 and P-2 sites interact with an array of negatively charged side chains that are described in the text but are not shown here. An Ile side chain of the pseudo-substrate inhibitor at the P+1 site binds in a hydrophobic pocket contributed predominantly by the activation loop and by the P+1 loop (*See* Color Insert)

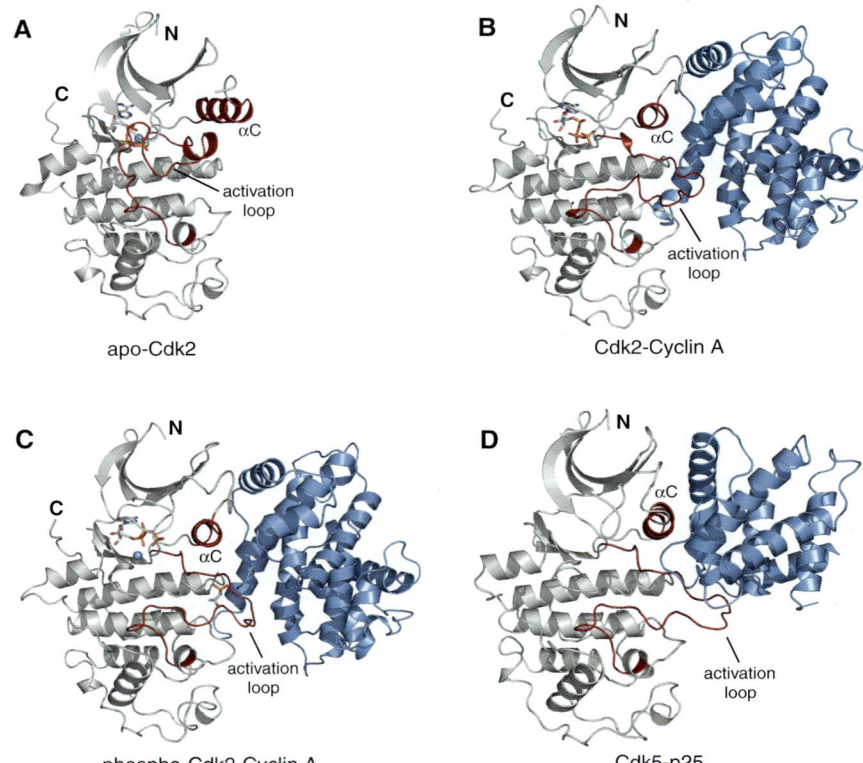

Chapter 14, Fig. 3 The CDK activation cycle (**A**) Ribbon model of the inactive apo-CDK2 structure (PDB ID code 1 HCK). The activation loop and the αC helix are displayed in red. The activation loop meanders in the form of the catalytic cleft and blocks access to it. (**B**) The structure of the partially active form of CDK2, consisting of the CDK2–cylin A complex (PDB ID code 1FIN). The color code is as in panel **A**, and the cyclin is shown in blue. Note that the large lobe is essentially static in panels **A** and **B**. Thus, the cyclin imparts an important conformational change on the small lobe and on the activation loop. (**C**) The structure of the phospho-CDK2–cylin A complex (PDB ID code 1 JST), which is fully active. The activation loop is completely stretched out. (**D**) The CDK5–p25 complex (PDB ID code 1 H4L) displays an activation loop that is essentially indistinguishable from that of the phospho-CDK2–cylin A complex shown in panel **C**

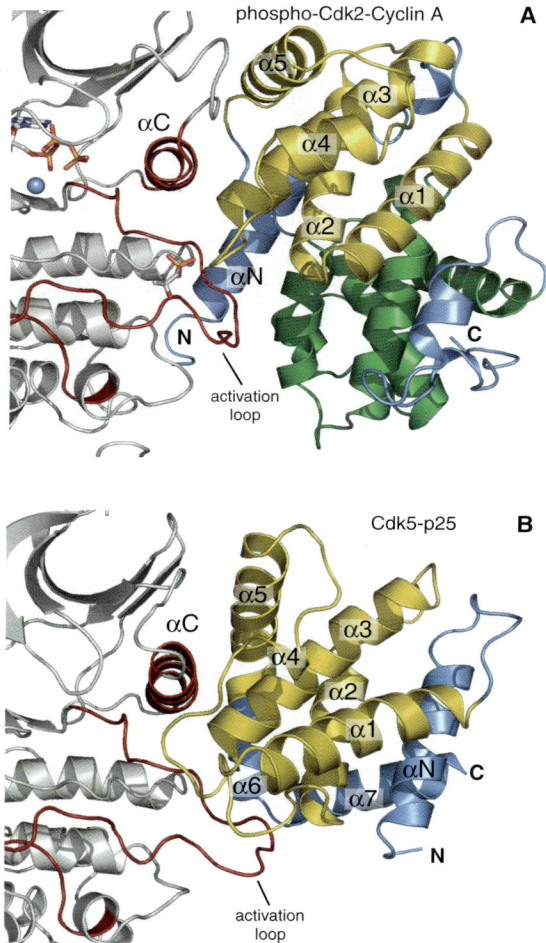

Chapter 14, Fig. 4 The cyclin-box fold (**A**) Close-up of the phospho-CDK2–cylin A structure (se Fig. 3) at the CDK2–cyclin A interface. The cyclin-box fold motifs (*yellow and green*) are 5-helical arrays of about 100 residues. Additional helices and coils complete the fold. (**B**) The p25 activator contains a cyclin-box motif (*yellow*) embedded within a larger globular domain containing three additional helices. There is striking similarity with the CDK2–cyclin A interface, but also conspicuous differences (for details, see Tarricone et al., 2001)

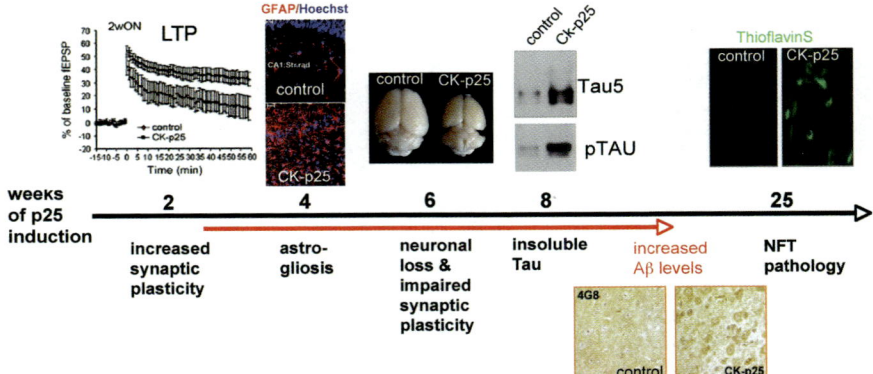

Chapter 17, Fig. 2 The phenotypes of CK-p25 Tg mice. In bi-transgenic CK-p25 mice, inducible GFP-tagged human p25 expression is mediated by the CamK promoter (CK)-regulated tet-off system. As such p25 expression is repressed in the presence of doxycycline. We reported that these mice exhibit increased hippocampal-dependent learning and synaptic plasticity such as facilitated LTP and NMDA conductivity upon 2 weeks of p25 expression. The first pathological hallmark, hippocampal astrogliosis, is detected upon 4 weeks of induction followed by significant neuronal loss after 6 weeks of p25 expression. Neuronal loss is accompanied by severe learning and memory impairments and synaptic plasticity defects. Insoluble tau is first detected after 8 weeks of induction and neurofibrillary tangles (NFT) occur after 25 weeks. Interestingly, an increase in amyloid-beta peptides is detectable from 2 to 9 weeks of p25 induction (Cruz et al., 2003, 2006a; Fischer et al., 2005, 2007).

concentrations to Cdk5–p35. IC50 is about 1 μM and largely inhibited at 20 μM (Yamada et al. 2007).

The phosphorylation of several proteins by Cdk5–p39 was compared with phosphorylation by Cdk5–p35. Histone H1 is the best substrate for Cdk5–p39. Tau is the second best, and MAP2 and the neurofilament H subunit are similarly phosphorylated by Cdk5–p39 or Cdk5–p35 (Yamada et al. 2007). Inhibitor 1, mDab1, p35, and p39 are also phosphorylated by both Cdk5–p39 and Cdk5–p35 (Ohshima et al. 2005; Yamada et al. 2007). Tau and mDab1 are phosphorylated at the same sites by Cdk5–p39 and Cdk5–p35 (Sakaue et al. 2005 and unpublished results). Considering the high level of conservation of possible substrate-docking site in p35 and p39 (Tarricone et al. 2001), it is not surprising that the activator of Cdk5 does not dramatically affect its substrate preferences. However, it is possible that Cdk5–p39 or Cdk5–p35 may exhibit specific activities toward other substrate proteins in more intensive substrate specificity survey.

A major difference in the kinase activities between Cdk5–p39 and Cdk5–p35 is the sensitivity to nonionic detergent; Cdk5–p39 is inactivated by detergent, whereas Cdk5–p35 is activated (Fig. 2C). Inactivation of Cdk5–p39 by detergent is due to dissociation of the complex. When Cdk5–p39his (p39 with a his-tag at the C-terminus) complex bound to Ni-beads or anti-p39 immunocomplex bound to protein G beads is washed with a solution containing nonionic detergent, Cdk5 protein is lost from the complexes (Fig. 2D). Concentrated salt solution also appears to weaken the interaction, although not so markedly as detergent. Therefore, the enzyme fraction purified using Ni-beads contained less Cdk5 protein compared with p39his. This labile property is derived from the C-terminal Cdk5-activation domain, which is relatively well conserved between p35 and p39, as shown by the instability of Cdk5 complexed with the C-terminal Cdk5-binding region of p39.

Effects of the N-terminal p10 Domain on the Kinase Activity of Cdk5

The first active Cdk5 purified from mammalian brains was a complex with p25 (Ishiguro et al. 1992; Lew et al. 1992; Shetty et al. 1993; Hisanaga et al. 1995). This was later shown to be the N-terminal truncated form of p35 (Tsai et al. 1994; Lew et al. 1994; Uchida et al. 1994). This truncated form, p25, is the C-terminal two-thirds of p35, consisting of the Cdk5 activation domain, and generated during purification by cleavage with calpain, a calcium-activated cysteine protease (Kusakawa et al. 2000; Lee et al. 2000). The following reasons may explain why the Cdk5–p25 complex, but not Cdk5–p35, was purified from brains. Bovine or porcine brains were used for large-scale purification, and there is inevitable post-mortem delay until homogenization in the buffers containing protease inhibitors, during which time proteases such as calpains are activated (Taniguchi et al. 2001). The high-speed supernatant of brain homogenates was used as the starting material for purification. Most of the Cdk5–p35 bound to membranes or cytoskeletons was pelleted by centrifugation, and the Cdk5 complex remaining in the supernatant was Cdk5–p25 that had lost the N-terminal membrane-anchoring site. We have tried to purify Cdk5–p35 from rat brain extract treated immediately after sacrifice, but it was still very difficult to completely inhibit proteolytic cleavage of p35, even in the presence of EGTA, a Ca^{2+} chelator. The p35 protein must be vulnerable to calpain, or there may be other proteases that attack p35.

Cdk5–p25 displays a higher kinase activity than Cdk5–p35 both *in vivo* and *in vitro*. This was first demonstrated by phosphorylation of tau co-transfected with Cdk5–p35 or Cdk5–p25 in COS-7 cells (Patrick et al. 1999). MAP1B phosphorylation is another example. MAP1B was phosphorylated by Cdk5–p25 but not by Cdk5–p35 in COS-7 cells (Kawauchi et al. 2005). The stronger activation by p25 is caused by several different mechanisms. First, the cleavage increases the intrinsic kinase activity of Cdk5. When Cdk5–p25 and Cdk5–p35 were prepared from Sf9 cells by the same method and kinase activities for tau were compared, Cdk5–p25 showed higher kinase activity than Cdk5–p35, although the phosphorylation sites were the same. This increased activity might be due to the increased catalytic activity (k_{cat}), but not the affinity (K_m) of Cdk5–p25 for tau (Hashiguchi et al. 2002). Secondly, the cleavage to p25 changes the cellular localization of the Cdk5 complex from membranes to soluble cytoplasm or to the nucleus. Cdk5 complexed with p35 binds to membranes through myristoylation of p35 at Gly2. Thus, p25 loses the membrane-binding site, resulting in its dissociation from membranes. This dissociation may cause easier access to several additional protein substrates whose localization is not restricted to membranes. Thirdly, p35 is a short-lived protein degraded by proteasome in neurons, but p25 has a longer life (Patrick et al. 1998). This indicates that the cleavage to p25 both lengthens the active state duration of Cdk5 and increases the size of the active complex population. Fourthly, dissociation from membranes is another activation mechanism. Binding to membranes inhibits the kinase activity of Cdk5–p35 as described below. Cleavage dissociates Cdk5–p25 from membranes, resulting in liberation from the suppression by membranes. Cleavage to p25 thus results in a considerable, possibly abnormal, increase in Cdk5 kinase activity.

The substrate specificity appears to be affected by the N-terminal region slightly. The N-terminal deletion mutants, Cdk5–p39$^{\Delta N129}$ (the N-terminal 129 amino acid–truncated fragment of p39) and Cdk5–p35$^{\Delta N98}$ (the N-terminal 98 amino acid–truncated fragment of p35 identical to p25), have slightly different phosphorylating activity for several proteins from those of Cdk5 bound to the corresponding full-length activators, although the overall substrate preference does not change. Both Cdk5–p39$^{\Delta N129}$ and Cdk5–p25 show decreased kinase activity for NF-H, and Cdk5–p25 has increased kinase activity for tau when compared with Cdk5–p35 and Cdk5–p39 *in vitro* (Yamada et al. 2007). Cdk5–p35 and Cdk5–p25 show different kinase activities for amyloid precursor protein (APP) in transfected cultured cells. Cdk5–p35 phosphorylates both mature and immature APP, whereas Cdk5–p25 primarily phosphorylates immature APP (Liu et al. 2003). In the brains of p25 transgenic mice, phosphorylation of several pathological substrate proteins including tau, APP, and NF-L is specifically increased (Cruz et al. 2003), indicating that substrate specificity of Cdk5–p25 is different from Cdk5–p35 in brain. These results are interpreted as showing that the N-terminal domain of the activator may affect substrate recognition of Cdk5 in neurons through cellular localization.

A number of papers have shown that the conversion of p35 to p25 occurs in neurons undergoing cell death induced by various insults, including serum depletion, staurosporin, glutamate cytotoxicity, ER stress, O_2 stress, and beta amyloid (see Cruz and Tsai 2004; Shelton and Johnson 2004 for reviews). Is the cleavage a cause or a result of cell death? Involvement in cell death induction is indicated by the following studies. Forced expression of Cdk5–p25 induces cell death in cultured neurons to a greater extent than expression of Cdk5–p35 does (Patrick et al. 1999). Inducible overexpression of p25 in postnatal brains causes neuronal cell death (Cruz et al. 2003), or hyperphosphorylation of FTDP-mutant tau is observed in p25 transgenic mice brains (Noble et al. 2003). Dopaminergic neuron loss induced by 1-methyl-4-phenyl-1,2,3,6-tetrahydropyridine (MPTP) is attenuated by dominant-negative Cdk5 and Cdk5 deficiency (Smith et al. 2003, 2006). Roscovitine or dominant-negative Cdk5 suppresses the neuronal cell death induced by ER stress, which induces the cleavage of p35 to p25 and the translocation of Cdk5–p25 into the nucleus (Saito et al. 2007). These results suggest that the calpain cleavage of p35 to p25 is positively involved in neuronal cell death. The prosurvival transcription factor MEF2 is a nuclear target of Cdk5–p25 (Gong et al. 2003).

Regulation of Cdk5 Activity

Considerations from the Activation Mechanisms of Cdk1–Cyclin B in Proliferating Cells

Because of the extensive investigation of Cdks in proliferating cells (Morgan 1995), a discussion of their activation mechanisms is relevant to understanding Cdk5 mechanisms. There are some variations in the activation mechanism among cell cycle Cdks, and here we consider the well-understood Cdk1–cyclin B as an example (Fig. 3). Binding to cyclin B is necessary for activation of Cdk1. Upon cyclin B binding, Cdk1 undergoes two types of phosphorylation: inhibitory phosphorylation at Thr14 and Tyr15 by Wee1 or Myt1, and activatory phosphorylation at Thr161 in the activation T loop by Cdk-activating kinase (CAK; also known as Cdk7–cyclin H). As a result, the Cdk1–cyclin B complex is kept in the inactive form until mitosis, when Cdk1–cyclin B is coordinately activated by dephosphorylation at Thr14 and Tyr15 by the Cdc25 phosphatase.

Cdk1-cycB in proliferating cell

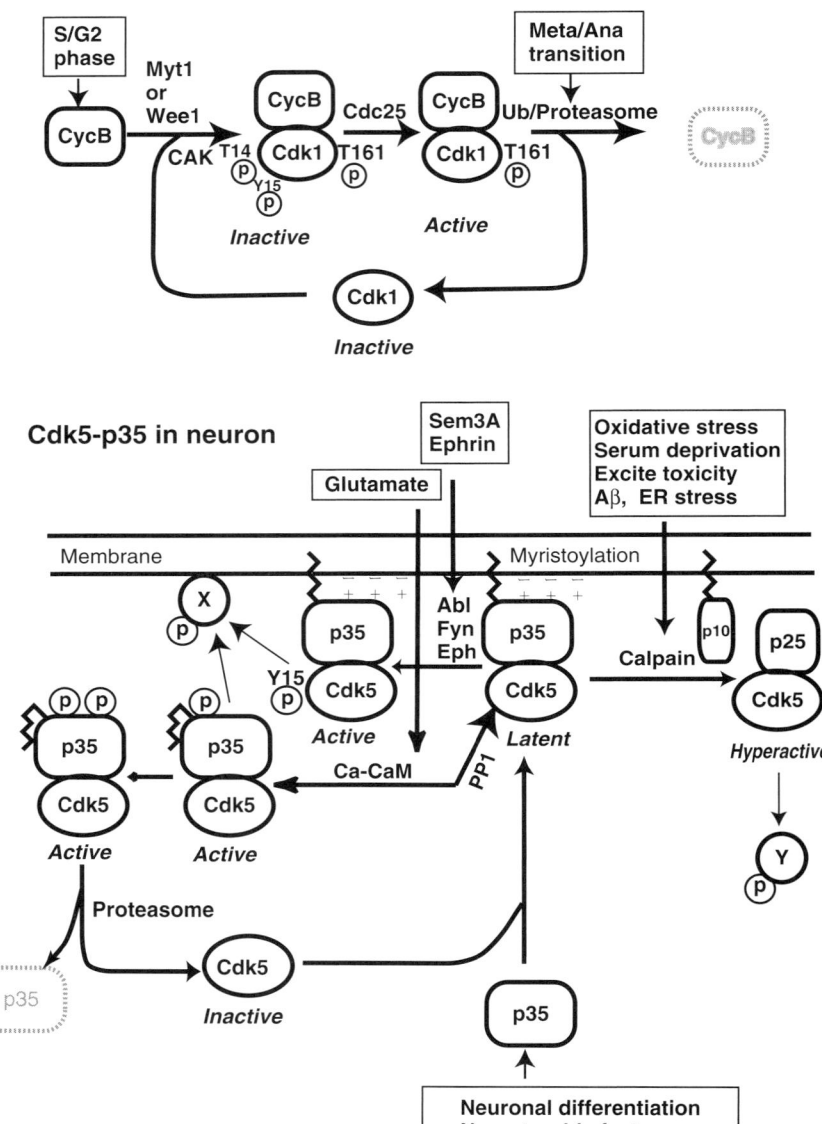

Fig. 3 The regulation of the kinase activity of Cdk1–cyclin B and Cdk5–p35. (**A**) Cdk1–cyclin B in proliferating cells. Cyclin B synthesized during S and G2 phase binds to Cdk1 to form a complex. Upon binding, Cdk1 is phosphorylated at Thr14 and Tyr15 by Wee1 or Myt1 in addition to phosphorylation at Thr161 by CAK, and remains inactive until activation by dephosphorylation at Thr14 and Tyr15 with Cdc25 at mitosis. Cdk1 is inactivated by

After completion of its functions at mitosis, Cdk1–cyclin B is inactivated by degradation of cyclin B by proteasome following ubiquitination by anaphase-promoting complex/cyclosome (APC/C). Inactivation of Cdk1–cyclin B is also an important step in the completion of mitosis.

Cdk5, as well as other Cdks, requires the binding of p35 (p25) or p39 for activation. Cdk5 has a high identity (50–60%) in the primary sequence to other Cdks (Meyerson et al. 1992) and is structurally very similar to Cdk2 (Tarricone et al. 2001). Although p35 shows the limited similarity in amino acid sequence to cell cycle cyclins, it can activate Cdk5 because the Cdk5-binding domain of p35 conforms a ternary structure similar to the Cdk2-binding domain of cyclin A, called the cyclin box fold (CBF) (Tarricone et al. 2001). However, subtle differences in the structure and sequence of p35 (p25) from cell cycle cyclins impose unique activation mechanism on Cdk5–p35. In particular, the regulation by membrane association through N-terminal myristoylation is a characteristic property of Cdk5. Furthermore, expression of p35 or p39 in differentiated neurons without cell cycle couples the activation and inactivation of Cdk5 to neuronal activities unrelated to cell cycle.

Regulation of Cdk5 Activity by Synthesis and Degradation of p35

The kinase activity of Cdk5 is determined primarily by the amount of p35 or p39 protein in neurons. This is indicated by the excess of Cdk5 relative to p35 or p39 in brain tissue. When brain extract was fractionated by gel filtration, a large quantity of free-form Cdk5 was detected in the low molecular weight fractions, corresponding to monomeric molecular mass of about 30 kDa, whereas a small amount of Cdk5 complexed with p35 was eluted at the void volume fractions in association with membrane (Qi et al. 1995; Zhu et al. 2005). When the brain extract was subjected to ultracentrifugation, the soluble Cdk5 found in the supernatant was more abundant than p35-bound Cdk5 in the pellet fraction (Qi et al. 1995; Zhu et al. 2005). These results indicate that the availability of p35 or p39 is the limiting factor for determining Cdk5 activity.

Fig. 3 (continued) degradation of cyclin B with proteasome at the time of anaphase initiation. (**B**) Cdk5–p35 in neurons. Synthesis of p35 is stimulated by extracellular signals such as neurotrophic factors or extracellular matrix. Cdk5 bound to p35 is recruited to membranes through myristoylation of p35, and membrane-bound Cdk5–p35 is low in the kinase activity. The latent Cdk5–p35 is activated by phosphorylation at Tyr15 with Abl, Fyn, Eph, or TrkB tyrosine kinases upon stimulation with a respective ligand. Cdk5–p35 is also activated transiently by excitatory neurotransmitter glutamate probably through Ca-calmodulin to phosphorylate p35 tagging a degradation signal for ubiquitin–proteasome for inactivation. Membrane binding is also regulated by dephosphorylation with protein phosphatase 1 (PP1). Stress signals activate calpain to cleavage p35 to p25, resulting in the release of Cdk5–p25 complex from membranes to induce hyperactivation

The amount of p35 or p39 protein is controlled by the balance between synthesis and degradation. The kinase activity of Cdk5 begins to increase in brain along with neuronal differentiation at about E13 when the synthesis of p35 begins (Tsai et al. 1993; Wu et al. 2000). It was recently reported that p35 and p39 genes are targets for heat shock factor 2 (HSF2) in cortical development (Chang et al. 2006). The synthesis of p35 has been studied with cultured neurons and neuroblastoma cells that differentiate into neuron-like cells *in vitro*. In cultured cortical neurons and medium-sized spiny neurons, the expression of p35 is induced by brain-derived neurotrophic factor (BDNF) (Tokuoka et al. 2000; Bogush et al. 2007). The increase in p35 expression in medium-sized spiny neurons is prevented by phosphatidylinositol 3-kinase (PI3 K) inhibitors such as dominant-negative p85 or LY294002, suggesting the mediation of PI3 K (Bogush et al. 2007). In PC12 cells, expression of p35 is strongly induced by NGF, and the Cdk5 kinase activity is thereby greatly enhanced. Activation of the ERK pathway is required for NGF-induced p35 synthesis, and Egr1 mediates the induction of p35 (Harada et al. 2001). Fas engagement also induces sustained activation of ERK and MEK1-dependent expression of p35 in SH-SY5Y neuroblastoma cells and primary sensory neurons (Desbarats et al. 2003). Retinoic acid treatment markedly increases the protein expression of p35 and Cdk5 through the ERK1/2 pathway in SKNB-2C neuroblastoma cells. In this case, the expression of p35 is dependent on Egr1, whereas expression of Cdk5 is dependent on c-fos and CREB (Lee and Kim 2004). Synthesis of p35 is also enhanced by the presence of extracellular matrix. When cerebellar macroneurons are cultured on laminin, p35 protein level increases dramatically, resulting in increased Cdk5 activity (Paglini et al. 1998). Laminin induces higher expression of p35 mRNA and p35 protein, followed by increased Cdk5 activity as SH-SY5Y cells are differentiated in the presence of retinoic acid (Li et al. 2000). Laminin interaction with integrin $\alpha1\beta1$ triggers Cdk5 kinase activation. Thus, the synthesis of p35 is stimulated by neuronal growth factors or extracellular matrix proteins.

Cdk5 is inactivated by degradation of p35 or p39 by the proteasome (Patrick et al. 1998; Saito et al. 1998; Patzke and Tsai 2002). p35 and p39 are short-lived proteins with a half-life of 20–30 min (Patrick et al. 1998; Patzke and Tsai 2002). Ubiquitination of p35 is detected in C33A and Cos7 cells (Patrick et al. 1998), but not yet in neurons. Both of p35 and p39 do not contain the amino acid sequence corresponding to "destruction box" of cyclin B and A, which is recognized by APC/C. Degradation of p35 is stimulated by okadaic acid, an inhibitor of protein phosphatase 1 and 2A (Patrick et al. 1998; Saito et al. 1998), indicating that p35 degradation is triggered by phosphorylation of p35. In this respect, p35 rather resembles cyclin D or E, whose degradation by SKP1-CUL1-F-box protein (SCF) complex is induced by phosphorylation (Nakayama and Nakayama 2007). The unidentified p35, or p39 E3 enzyme, may be an SCF-type ubiquitin ligase, which uses phosphorylation sites as degradation tags (Orlicky et al. 2003).

Degradation of p35 is induced in cortical neurons by the excitatory neurotransmitter, glutamate (Wei et al. 2005). Glutamate treatment activates

Cdk5–p35 transiently and phosphorylates p35, which would then be recognized by an unidentified p35 E3 ligase, leading to ubiquitination followed by degradation of p35. Thus, the transient activation induces the downregulation of Cdk5 by degradation of p35. Inactivation of Cdk5 would be involved in long-term potentiation induction by reducing the threshold of excitation (Wei et al. 2005).

Regulation of Cdk5 Activity by Phosphorylation

Cdk5 preserves all three phosphorylation sites corresponding to Thr14, Tyr15, and Thr161 in Cdk1, although Thr161 in Cdk1 is changed to Ser159 in Cdk5. It is interesting, however, that the effect of phosphorylation of Cdk5 is completely different from those of cell cycle Cdks. First, although Cdk1 needs T-loop phosphorylation at Thr161 for activation, Cdk5 does not require the phosphorylation at Ser159 for activation. In fact, Cdk5 prepared from *E. coli* becomes active only by binding to p35 *in vitro* (Qi et al. 1995; Tang et al. 1997; Amin et al. 2002). Addition of CAK neither causes phosphorylation of Cdk5 nor enhances the kinase activity (Qi et al. 1995; Lee et al. 1997; Poon et al. 1997). It is clearly shown by the crystal structure of Cdk5–p25 how Cdk5 does not need T-loop phosphorylation for activation. The binding of p25 to Cdk5 forces the activation loop to adopt an extended conformation typical of active proline-directed kinases (Tarricone et al. 2001). Although phosphorylation of Ser159 is not required for activation, it is reported that the site is phosphorylated by CK1 (casein kinase 1), resulting in enhancement of the Cdk5–p25 catalytic activity (Sharma et al. 1999).

    The role of Thr14/Tyr15 phosphorylation is in complete contrast to that of Cdk1–cyclin B. Whereas Cdk1 is inhibited by Thr14/Tyr15 phosphorylation with Wee1 or Myt1 (Fattaey and Booher 1997), the kinase activity of Cdk5 is stimulated by phosphorylation at Tyr15 by nonreceptor tyrosine kinases (Abl and Fyn) or receptor-type tyrosine kinases (Eph4 and TrkB) (Zukerberg et al. 2000; Sasaki et al. 2002; Cheng et al. 2003; Fu et al. 2007; Cheung et al. 2007). Phosphorylation of Cdk5 at Tyr15 by Abl is stimulated by Cables, a Cdk5-binding protein (Zukerberg et al. 2000). The Cables-mediated Tyr15 phosphorylation is involved in axonal growth regulation. Phosphorylation by Fyn is detected when a growth cone encounters Sema3, a repellent cue (Sasaki et al. 2002). It was recently reported that Tyr15 is also phosphorylated when the Eph4 receptor kinase is activated by ephrin A1 (Fu et al. 2007). Further, upon BDNF stimulation, Cdk5 is recruited to TrkB BDNF receptor and phosphorylated by TrkB at Tyr15, thus leading to enhanced Cdk5 activity that promotes the phosphorylation of TrkB at Ser478 (Cheung et al. 2007). There is a report that protein kinase purified from bovine thymus, as the Thr14 kinase of Cdk1 and Cdk2, phosphorylated Cdk5 and inhibited the kinase activity (Matsuura and Wang 1996).

Protein Modulators of Cdk5 Activity

The kinase activity of cell cycle Cdks is regulated by a number of Cdk inhibitors (CKI) known as Cip/Kip (p21, p27, and p57) or INK4 family proteins (p16, p18, and p19). However, the kinase activity of Cdk5–p35 is not affected by these Cdk inhibitors (Harper et al. 1995; Lee et al. 1996b). So far, no inhibitor like the Cdk inhibitors in proliferating cells has been found for Cdk5. However, there are different types of protein factors regulating the Cdk5 activity. CK2 (casein kinase 2) inhibits the Cdk5 activity by inhibiting the complex formation via binding to both subunits of Cdk5 and p35 independently (Lim et al. 2004). C42 protein, isolated as one of p35-binding proteins by yeast two-hybrid screening, shows inhibitory activity to Cdk5–p35 when it is preincubated with p35 before complex formation (Ching et al. 2002). In addition, there are activating factors that stimulate kinase activity by binding to the Cdk5-activator complex. Munc-18, a synaptic protein, is co-purified with the Cdk5 kinase activity through several columns during purification; it stimulates kinase activity several fold (Shetty et al. 1995). Nuclear protein SET with an inhibitor activity to protein phosphatase 2A binds to p35 either monomerically or in a complex with Cdk5. It also stimulates kinase activity (Qu et al. 2002). Although the presence of the kinase-inactive 700 kDa Cdk5–p35 complex in bovine brain extract supports the presence of the kinase–inhibitor complex (Lee et al. 1996a), more studies will be required to determine the biological meaning of these modulator proteins.

Suppression of the Cdk5 Activity by Membranes

Autocatalytic phosphorylation of p35 is one of the methods used to estimate the kinase activity of Cdk5 (Fig. 2A). Phosphorylated p35 can be distinguished easily by the electrophoretic mobility shift of p35 (Kerokoski et al. 2002; Saito et al. 2003; Zhu et al. 2005). The electrophoretic mobility of p35 shifts upward upon phosphorylation when Cdk5–p35 is highly active. Thus, the upward shift of p35 is the hallmark of Cdk5 activity. This is particularly useful for looking at the kinase activity *in vivo* or in cells. The electrophoretic upward mobility of p35 changes gradually to downward during brain development (Saito et al. 2003), indicating that the kinase activity of Cdk5–p35 is high in fetal brain and low in adult brain, as has been recently shown (Sato et al. 2007a).

This developmental change in the kinase activity is partly dependent on the binding of Cdk5–p35 to membranes. When the brain extract is fractionated by a Sepharose CL-4B gel filtration column, most adult Cdk5–p35 is eluted at void volume fractions as a membrane-bound form, whereas a part of fetal Cdk5–p35 is obtained as a soluble complex in the low molecular weight region. Measurement of kinase activity shows that the membrane-bound Cdk5–p35 has low histone H1 phosphorylating activity and no autophosphorylation activity for p35, whereas soluble Cdk5–p35 has high activity in both histone H1 and p35. The membrane-bound and low kinase active Cdk5–p35 is stimulated when assayed in the presence of nonionic detergents (Zhu et al. 2005). A portion of Cdk5–p35 is dissociated from membranes in concentrated salt solutions. The kinase activity of the high-salt dissociated Cdk5–p35 is suppressed by addition of membrane preparations (Fig. 2E). Thus, the membrane association of Cdk5–p35 is regulated by phosphorylation of p35. The soluble, active Cdk5–p35 in fetal brain becomes the membrane-bound and latent complex after dephosphorylation with protein phosphatase 1 (Sato et al. 2007a). This evidence shows that the kinase activity of Cdk5 is regulated through its association with membranes, which in turn is under the control of Cdk5-dependent phosphorylation and protein phosphatase-1-dependent dephosphorylation of p35.

Summary

The kinetic properties of Cdk5–p35 are very similar to those of Cdk1–cyclin B. This is because of the similarity of Cdk catalytic subunits. However, the activation mechanisms of Cdk5–p35 are quite different from those of cell cycle Cdks (Fig. 3). Whereas the activation of cell cycle Cdks is associated with cell cycle events occurring predominantly in the nucleus, the activation of Cdk5 is regulated in neurons by extracellular signals. Neurotrophic factors and extracellular matrix activate Cdk5 by increasing synthesis of p35 and by phosphorylation on Tyr15 of Cdk5. The excitatory neurotransmitter glutamate inactivates Cdk5 by degradation of p35. Stress signals deregulate the Cdk5 activity through the cleavage of p35 to p25. Considering that the neuronal functions are based on cell–cell communication, Cdk5 may have evolved with an activation mechanism suitable for neurons. For this adaptation, Cdk5 may change the cellular localization from the nucleus to the vicinity of cytoplasmic membranes. This is achieved by myristoylation of p35. The membrane association may confer upon neurons the additional characteristic feature, the permanent inactivation of cell cycle machinery. The myristoylation of p35 keeps the active Cdk5 out of the nucleus. The cleavage of p35 to p25 liberates the active Cdk5 complex from membranes and allows it to enter into the nucleus, leading to cell death of neurons by ectopic activation of cell cycle. The membrane association of the active Cdk5 may be crucial for neurons to live a long life.

References

Amin ND, Albers W, Pant HC (2002) Cyclin-dependent kinase 5 (cdk5) activation requires interaction with three domains of p35. J Neurosci Res 67:354–362
Beaudette KN, Lew J, Wang JH (1993) Substrate specificity characterization of a cdc2-like protein kinase purified from bovine brain. J Biol Chem 268:20825–20830

Bogush A, Pedrini S, Pelta-Heller J, Chan T, Yang Q, Mao Z, Sluzas E, Gieringer T, Ehrlich ME (2007) AKT and CDK5/p35 mediate brain-derived neurotrophic factor induction of DARPP-32 in medium size spiny neurons in vitro. J Biol Chem 282:7352–7359

Chae T, Kwon YT, Bronson R, Dikkes P, Li E, Tsai LH (1997) Mice lacking p35, a neuronal specific activator of Cdk5, display cortical lamination defects, seizures, and adult lethality. Neuron 18:29–42

Chang Y, Ostling P, Akerfelt M, Trouillet D, Rallu M, Gitton Y, El Fatimy R, Fardeau V, Le Crom S, Morange M, Sistonen L, Mezger V (2006) Role of heat-shock factor 2 in cerebral cortex formation and as a regulator of p35 expression. Genes Dev 20:836–847

Cheng Q, Sasaki Y, Shoji M, Sugiyama Y, Tanaka H, Nakayama T, Mizuki N, Nakamura F, Takei K, Goshima Y (2003) Cdk5/p35 and Rho-kinase mediate ephrin-A5-induced signaling in retinal ganglion cells. Mol Cell Neurosci 24:632–645

Cheung ZH, Chin WH, Chen Y, Ng YP, Ip NY (2007) Cdk5 Is Involved in BDNF-Stimulated Dendritic Growth in Hippocampal Neurons. PLoS Biol 5:e63

Ching YP, Pang AS, Lam WH, Qi RZ, Wang JH (2002) Identification of a neuronal Cdk5 activator-binding protein as Cdk5 inhibitor. J Biol Chem 277:15237–15240

Cruz JC, Tsai LH (2004) Cdk5 deregulation in the pathogenesis of Alzheimer's disease. Trends Mol Med 10:452–458.

Cruz JC, Tseng HC, Goldman JA, Shih H, Tsai LH (2003) Aberrant Cdk5 activation by p25 triggers pathological events leading to neurodegeneration and neurofibrillary tangles. Neuron 40:471–483

Desbarats J, Birge RB, Mimouni-Rongy M, Weinstein DE, Palerme JS, Newell MK (2003) Fas engagement induces neurite growth through ERK activation and p35 upregulation. Nat Cell Biol 5:118–125

Dhavan R, Tsai LH (2001) A decade of CDK5. Nat Rev Mol Cell Biol 2:749–759

Fattaey A, Booher RN (1997) Myt1: a Wee1-type kinase that phosphorylates Cdc2 on residue Thr14. Prog Cell Cycle Res 3:233–240

Fu WY, Chen Y, Sahin M, Zhao XS, Shi L, Bikoff JB, Lai KO, Yung WH, Fu AK, Greenberg ME, Ip NY (2007) Cdk5 regulates EphA4-mediated dendritic spine retraction through an ephexin1-dependent mechanism. Nat Neurosci 10:67–76

Gray N, Dtivaud L, Doerig C, Meijer L (1999) ATP-site directed inhibitos of cyclin-dependent kinases. Curr Med Chem 6:859–875

Gong X, Tang X, Wiedmann M, Wang X, Peng J, Zheng D, Blair LAC, Marshall J, Mao Z (2003) Cdk5-mediated inhibition of the protective effects of transcription factor MEF2 in neurotoxicity-induced apoptosis. Neuron 38:33–46

Harada T, Morooka T, Ogawa S, Nishida E (2001) ERK induces p35, a neuron-specific activator of Cdk5, through induction of Egr1. Nat Cell Biol 3:453–459

Harper JW, Elledge SJ, Keyomarsi K, Dynlacht B, Tsai LH, Zhang P, Dobrowolski S, Bai C, Connell-Crowley L, Swindell E,, Fox MP, Wei N (1995) Inhibition of cyclin-dependent kinases by p21. Mol Biol Cell 6:387–400

Hashiguchi M, Saito T, Hisanaga S, Hashiguchi T (2002) Truncation of CDK5 activator p35 induces intensive phosphorylation of Ser202/Thr205 of human tau. J Biol Chem 277:44525–44530

Hisanaga S, Ishiguro K, Uchida T, Okumura E, Okano T, Kishimoto T (1993) Tau protein kinase II has a similar characteristic to cdc2 kinase for phosphorylating neurofilament proteins. J Biol Chem 268:15056–15060

Hisanaga S, Saito T (2003) The regulation of cyclin-dependent kinase 5 activity through the metabolism of p35 or p39 Cdk5 activator. Neurosignals 12:221–229

Hisanaga S, Uchiyama M, Hosoi T, Yamada K, Honma N, Ishiguro K, Uchida T, Dahl D, Ohsumi K, Kishimoto T (1995) Porcine brain neurofilament- H tail domain kinase: its identification as cdk5/p26 complex and comparison with cdc2/cyclin B kinase. Cell Motil Cytoskeleton 31:283–297

Humbert S, Dhavan R, Tsai L (2000) p39 activates cdk5 in neurons, and is associated with the actin cytoskeleton. J Cell Sci 113:975–983

Ishiguro K, Omori A, Sato K, Tomizawa K, Imahori K, Uchida T (1991) A serine/threonine proline kinase activity is included in the tau protein kinase fraction forming a paired helical filament epitope. Neurosci Lett 128:195–198

Ishiguro K, Takamatsu M, Tomizawa K, Omori A, Takahashi M, Arioka M, Uchida T, Imahori K (1992) Tau protein kinase I converts normal tau protein into A68-like component of paired helical filaments. J Biol Chem 267:10897–10901

Jeong YG, Rosales JL, Marzban H, Sillitoe RV, Park DG, Hawkes R, Lee KY (2003) The cyclin-dependent kinase 5 activator, p39, is expressed in stripes in the mouse cerebellum. Neuroscience 118:323–334

Kawauchi T, Chihama K, Nishimura YV, Nabeshima Y, Hoshino (2005) MAP1B phosphorylation is differentially regulated by Cdk5/p35, Cdk5/p25, and JNK. Biochem Biophys Res Commun 331:50–55

Kerokoski P, Suuronen T, Salminen A, Soininen H, Pirttila T (2002) Influence of phosphorylation of p35, an activator of cyclin-dependent kinase 5 (cdk5), on the proteolysis of p35. Brain Res Mol Brain Res 106:50–56

Ko J, Humbert S, Bronson RT, Takahashi S, Kulkarni AB, Li E, Tsai LH (2001) p35 and p39 are essential for cyclin-dependent kinase 5 function during neurodevelopment. J Neurosci 21:6758–6871

Kusakawa G, Saito T, Onuki R, Ishiguro K, Kishimoto T, Hisanaga S (2000) Calpain-dependent proteolytic cleavage of the p35 cyclin-dependent kinase 5 activator to p25. J Biol Chem 275:17166–17172

Lee KY, Helbing CC, Choi, KS, Johnston RN, Wang JH (1997) Neuronal Cdc2-like kinase (Nclk) binds and phosphorylates the retinoblastoma protein. J Biol Chem 272:5622–5626

Lee JH, Kim KT (2004) Induction of cyclin-dependent kinase 5 and its activator p35 through the extracellular-signal-regulated kinase and protein kinase A pathways during retinoic-acid mediated neuronal differentiation in human neuroblastoma SK-N-BE(2)C cells. J Neurochem 91:634–647

Lee MS, Kwon YT, Li M, Peng J, Friedlander RM, Tsai LH (2000) Neurotoxicity induces cleavage of p35 to p25 by calpain. Nature 405:360–364

Lee MH, Nikolic M, Baptista CA, Lai E, Tsai LH, Massague J (1996b) The brain-specific activator p35 allows Cdk5 to escape inhibition by p27[Kip1] in neurons. Proc Natl Acad Sci USA 93:3259–3263

Lee KY, Rosales JL, Tang D, Wang JH (1996a) Interaction of cyclin-dependent kinase 5 (Cdk5) and neuronal Cdk5 activator in bovine brain. J Biol Chem 271:1538–1543

Ledee DR, Gao CY, Seth R, Fariss RN, Tripathi BK, Zelenka PS (2005) A specific interaction between muskelin and the cyclin-dependent kinase 5 activator p39 promotes peripheral localization of muskelin. J Biol Chem 280:21376–21383

Lew J, Huang QQ, Qi Z, Winkfein RJ, Aebersold R, Hunt T, Wang JH (1994) A brain-specific activator of cyclin-dependent kinase 5. Nature 37:423–426

Lew J, Winkfein RJ, Paudel HK, Wang JH (1992) Brain proline-directed protein kinase is a neurofilament kinase which displays high sequence homology to p34[cdc2]. J Biol Chem 267:25922–25926

Li BS, Zhang L, Gu J, Amin ND, Pant HC (2000) Integrin $\alpha1\beta1$-mediated activation of cyclin-dependent kinase 5 activity is involved in neurite outgrowth and human neurofilament protein H Lys-Ser-Pro tail domain phosphorylation. J Neurosci 20:6055–6062

Lilja L, Johansson JU, Gromada J, Mandic SA, Fried G, Berggren PO, Bark C (2004) Cyclin-dependent kinase 5 associated with p39 promotes Munc18-1 phosphorylation and Ca^{2+}-dependent exocytosis. J Biol Chem 279:29534–29541

Lim AC, Hou Z, Goh CP, Qi RZ (2004) Protein kinase CK2 is an inhibitor of the neuronal Cdk5 kinase. J Biol Chem 279:46668–46673

Liu F, Su Y, Li B, Zhou Y, Ryder J, Gonzalez-DeWhitt P, May PC, Ni B (2003) Regulation of amyloid precursor protein (APP) phosphorylation and processing by p35/Cdk5 and p25/Cdk5. FEBS Lett 547:193–196

Matsuura I, Wang JH (1996) Demonstration of cyclin-dependent kinase inhibitory serine/threonine kinase in bovine thymus. J Biol Chem 271:5443–5450

Meijer L, Borgne A, Mulner O, Chong JP, Blow JJ, Inagaki N, Inagaki M, Delcros JG, Moulinoux JP (1997) Biochemical and cellular effects of roscovitine, a potent and selective inhibitor of the cyclin-dependent kinases cdc2, cdk2 and cdk5. Eur J Biochem 243:527–536

Meyerson M, Enders GH, Wu C-L, Su L-K, Gorka C, Nelson C, Harlow E, Tsai L-H (1992) A family of human cdc2-related protein kinases. EMBO J 11:2909–2917

Morgan DO (1995) Principles of CDK regulation. Nature 374:131–134

Nakayama KI, Nakayama K (2006) Ubiquitin ligases: cell-cycle control and cancer. Nat Rev Cancer 6:369–381

Ohshima T, Ogura H, Tomizawa K, Hayashi K, Suzuki H, Saito T, Kamei H, Nishi A, Bibb JA, Hisanaga S, Matsui H, Mikoshiba K (2005) Impairment of hippocampal long-term depression and defective spatial learning and memory in p35 mice. J Neurochem 94:917–925

Ohshima T, Ward JM, Huh CG, Longenecker G, Veeranna, Pant HC, Brady RO, Martin LJ, Kulkarni AB (1996) Targeted disruption of the cyclin-dependent kinase 5 gene results in abnormal corticogenesis, neuronal pathology and perinatal death. Proc Natl Acad Sci USA 93:11173–11178

Orlicky S, Tang X, Willems A, Tyers M, Sicheri F (2003) Structural basis for phosphodependent substrate selection and orientation by the SCFCdc4 ubiquitin ligase. Cell 112:243–256

Paglini G, Pigino G, Kunda P, Morfini G, Maccioni R, Quiroga S, Ferreira A, Caceres A (1998) Evidence for the participation of the neuron-specific CDK5 activator p35 during laminin-enhanced axonal growth. J Neurosci 18:9858–9869

Patrick GN, Zhou P, Kwon YT, Howley PM, Tsai LH (1998) p35, the neuronal-specific activator of cyclin-dependent kinase 5 (Cdk5) is degraded by the ubiquitin-proteasome pathway. J Biol Chem 273:24057–24064

Patrick GN, Zukerberg L, Nikolic M, de la Monte S, Dikkes P, Tsai LH (1999) Conversion of p35 to p25 deregulates Cdk5 activity and promotes neurodegeneration. Nature 402:615–622

Patzke H, Tsai LH (2002) Calpain-mediated cleavage of the cyclin-dependent kinase-5 activator p39 to p29. J Biol Chem 277:8054–8060

Poon RY, Lew J, Hunter T (1997) Identification of functional domains in the neuronal Cdk5 activator protein. J Biol Chem 272:5703–5708

Qi Z, Huang QQ, Lee KY, Lew J, Wang JH (1995) Reconstitution of neuronal Cdc2-like kinase from bacteria-expressed Cdk5 and an active fragment of the brain-specific activator. J Biol Chem 270:10847–10854

Qu D, Li Q, Lim HY, Cheung NS, Li R, Wang JH, Qi RZ (2002) The protein SET binds the neuronal Cdk5 activator p35[nck5a] and modulates Cdk5/p35[nck5a] activity. J Biol Chem 277:7324–7332

Saito T, Ishiguro K, Onuki R, Nagai Y, Kishimoto T, Hisanaga S (1998) Okadaic acid-stimulated degradation of p35, an activator of CDK5, by proteasome in cultured neurons. Biochem Biophys Res Commun 252:775–778

Saito T, Konno T, Hosokawa T, Asada A, Ishiguro K, Hisanaga S (2007) p25/Cyclin-dependent kinase 5 promotes the progression of cell death in nucleus of endoplasmic reticulum-stressed neurons. J Neurochem 102:133–140

Saito T, Onuki R, Fujita Y, Kusakawa G, Ishiguro K, Bibb JA, Kishimoto T, Hisanaga S (2003) Developmental regulation of the proteolysis of the p35 cyclin-dependent kinase 5 activator by phosphorylation. J Neurosci 23:1189–1197

Sakaue F, Saito T, Sato Y, Asada A, Ishiguro K, Hasegawa M, Hisanaga S (2005) Phosphorylation of FTDP-17 mutant tau by cyclin-dependent kinase 5 complexed with p35, p25, or p39. J Biol Chem 280:31522–31529

Sasaki Y, Cheng C, Uchida Y, Nakajima O, Ohshima T, Yagi T, Taniguchi M, Nakayama T, Kishida R, Kudo Y, Ohno S, Nakamura F, Goshima Y (2002) Fyn and Cdk5 mediate semaphorin-3A signaling, which is involved in regulation of dendrite orientation in cerebral cortex. Neuron 35:907–920

Sato Y, Taoka M, Sugiyama N, Kubo K, Fuchigami T, Asada A, Saito T, Nakajima K, Isobe T, Hisanaga S (2007b) Regulation of the interaction of disabled-1 with CIN85 by phosphorylation with cyclin-dependent kinase 5. Genes Cells 12:1315–1327.

Sato K, Zhu YS, Saito T, Yotsumoto K, Asada A, Hasegawa M, Hisanaga S (2007a) Regulation of the membrane association and kinase activity of Cdk5-p35 by phosphorylation of p35. J Neurosci Res 85:3071–3078

Sharma P, Sharma M, Amin ND, Albers RW, Pant HC (1999) Regulation of cyclin-dependent kinase 5 catalytic activity by phosphorylation. Proc Natl Acad Sci USA 96:11156–11160

Shelton SB, Johnson GV (2004) Cyclin-dependent kinase-5 in neurodegeneration. J Neurochem 88:1313–1326.Erratum in: J Neurochem 89:528 (2004)

Shetty KT, Kaech S, Link WT, Jaffe H, Flores CM, Wray S, Pant HC, Beushausen S (1995) Molecular characterization of a neuronal-specific protein that stimulates the activity of Cdk5. J Neurochem 64:1988–1995

Shetty KT, Link WT, Pant HC (1993) cdc2-like kinase from rat spinal cord specifically phosphorylates KSPXK motifs in neurofilament proteins: isolation and characterization. Proc Natl Acad Sci USA 90:6844–6848

Smith PD, Crocker SJ, Jackson-Lewis V, Jordan-Sciutto KL, Hayley S, Mount MP, O'Hare MJ, Callaghan S, Slack RS, Przedborski S, Anisman H, Park DS (2003) Cyclin-dependent kinase 5 is a mediator of dopaminergic neuron loss in a mouse model of Parkinson's disease. Proc Natl Acad Sci USA 100:13650–13655

Smith PD, Mount MP, Shree R, Callaghan S, Slack RS, Anisman H, Vincent I, Wang X, Mao Z, Park DS (2006) Calpain-regulated p35/cdk5 plays a central role in dopaminergic neuron death through modulation of the transcription factor myocyte enhancer factor 2. J Neurosci 26:440–447

Takahashi S, Saito T, Hisanaga S, Pant HC, Kulkarni AB (2003) Tau phosphorylation by cyclin-dependent kinase 5/p39 during brain development reduces its affinity for microtubules. J Biol Chem 278:10506–10515

Tang D, Chun AC, Zhang M, Wang JH (1997) Cyclin-dependent kinase 5 (Cdk5) activation domain of neuronal Cdk5 activator. J Biol Chem 272:2318–2327

Tang D, Wang JH (1996) Cyclin-dependent kinase 5 (Cdk5) and neuron-specific Cdk5 activators. Prog Cell Cycle Res 2:205–216

Tang D, Yeng J, Lee KY, Matsushita M, Matsui H, Tomizawa K, Hatase O, Wang JH (1995) An isoform of the neuronal cyclin-dependent kinase 5 (Cdk5) activator. J Biol Chem 270:26897–26903

Taniguchi S, Fujita Y, Hayashi S, Kakita A, Takahashi H, Murayama S, Saido TC, Hisanaga S, Iwatsubo T, Hasegawa M (2001) Calpain-mediated degradation of p35 to p25 in postmortem human and rat brains. FEBS Lett 489:46–50

Tarricone C, Dhavan R, Peng J, Areces LB, Tsai LH, Musacchio A (2001) Structure and regulation of the CDK5-p25[nck5a] complex. Mol Cell 8:657–669

Tokuoka H, Saito T, Yorifuji H, Wei F, Kishimoto T, Hisanaga S (2000) Brain-derived neurotrophic factor-induced phosphorylation of neurofilament-H subunit in primary cultures of embryo rat cortical neurons. J Cell Sci 113:1059–1068

Tsai LH, Delalle I, Caniness VS, Chae T, Harlow E (1994) p35 is a neural-specific regulatory subunit of cyclin-dependent kinase 5. Nature 371:419–423

Tsai LH, Takahashi T, Caviness Jr VS, Harlow E (1993) Activity and expression pattern of cyclin-dependent kinase 5 in the embryonic mouse nervous system. Devlopment 119:1029–1040

Uchida T, Ishiguro K, Ohnuma J, Takamatsu M, Yonekura S, Imahori K (1994) Precursor of cdk5 activator, the 23 kDa subunit of tau protein kinase II: its sequence and developmental change in brain. FEBS Lett 355:35–40

Wada Y, Ishiguro K, Itoh TJ, Uchida T, Hotani H, Saito T, Kishimoto T, Hisanaga S (1998) Microtubule-stimulated phosphorylation of tau at Ser202 and Thr205 by cdk5 decreases its microtubule nucleation activity. J Biochem (Tokyo) 124:738–746

Wei FY, Tomizawa K, Ohshima T, Asada A, Saito T, Nguyen C, Bibb JA, Ishiguro K, Kulkarni AB, Pant HC, Mikoshiba K, Matsui H, Hisanaga S (2005) Control of cyclin-dependent kinase 5 (Cdk5) activity by glutamatergic regulation of p35 stability. J Neurochem 93:502–512

Wu DC, Yu YP, Lee NT, Yu AC, Wang JH, Han YF (2000) The expression of Cdk5, p35, p39, and Cdk5 kinase activity in developing, adult and aged rat brains. Neurochem Res 25:923–929

Yamada M, Saito T, Sato Y, Kawai Y, Sekigawa A, Hamazumi Y, Asada A, Wada M, Doi H, Hisanaga S (2007) Cdk5-p39 is a labile complex with the similar substrate specificity to Cdk5-p35. J Neurochem 102:1477–1487

Zheng Y-L, Kesavapany S, Gravell M, Hamilton RS, Schebert M, Amin N, Albers W, Grant P, Pant HC (2005) A Cdk5 inhibitory peptide reduces tau hyperphosphorylation and apoptosis in neurons. EMBO J 24:209–220

Zheng M, Leung CL, Liem RKH (1998) Region-specific expression of cyclin-dependent kinase 5 (cdk5) and its activators, p35 and p39, in the developing and adult rat central nervous system. J Neurobiol 35:141–159

Zukerberg LR, Patrick GN, Nikolic M, Humbert S, Wu CL, Lanier LM, Gertler FB, Vidal M, Van Etten RA, Tsai LH (2000) Cables links Cdk5 and c-Abl and facilitates Cdk5 tyrosine phosphorylation, kinase upregulation, and neurite outgrowth. Neuron 26:633–646

Zhu YS, Saito T, Asada A, Maekawa S, Hisanaga S (2005) Activation of latent cyclin-dependent kinase 5 (Cdk5)-p35 complexes by membrane dissociation. J Neurochem. 94:1535–1545

The Structural Bases of CDK5 Activity

Andrea Musacchio

Abstract In the last 15 years, a wealth of structural investigations on protein kinases has been reported. These studies have revealed that the active states of protein kinases are usually structurally alike, a requirement imposed by the necessity to maintain the basic geometry of a highly conserved machinery required for good catalytic output. Conversely, the structures of the inactive states of kinase-family members can vary widely from each other, a principle that can be exploited to improve the specificity of kinase inhibitors. In this chapter, we discuss the activation mechanism of the CDK5 kinase within the general frame of reference of kinase activation mechanisms, and in comparison to other members of the CDK family. We explain how CDK5, not unlike other kinases, has made its own capricious decisions to design an original activation mechanism and distinguish itself from CDK-family relatives.

Introduction

This chapter aims to deciphering the structural bases of cyclin-dependent kinase 5 (CDK5) function, using the crystal structure of the CDK5/p25 complex as the main reference (Tarricone et al., 2001). We will begin our analysis by constructing a general framework to understand the structural bases of regulation of the catalytic domain of eukaryotic protein kinases (ePKs), the protein family to which CDK5 belongs. Specifically, we will discuss a certain number of paradigmatic observations on the mechanisms whereby the activity of protein kinases can be turned on and off, which often rely on interactions with regulatory subunits and on post-translational modifications. Such an analysis can rely on a wealth of examples that have been made available through structural analysis of many protein kinases. With such a scheme of reference in hand, we will discuss in more detail the specific features of CDK5 that distinguish it from

A. Musacchio
Department of Experimental Oncology, European Institute of Oncology, Via
Adamello 16, I-20139 Milan, Italy
e-mail: andrea.musacchio@ifom-ieo-campus.it

N.Y. Ip, L.-H. Tsai (eds.), *Cyclin Dependent Kinase 5 (Cdk5)*,   191
DOI: 10.1007/978-0-387-78887-6_14, © Springer Science+Business Media, LLC 2008

other kinases, and in particular from the other members of the CDK family. We will also briefly discuss the mechanisms of specific substrate recognition by CDK5 and other kinases, an argument that remains largely unexplored.

Regulation of Protein Kinases: General Remarks

The ePK catalytic domain contains 518 recognizable members in humans and is therefore one of the most numerous domains in our genome (Cheek et al., 2005; Johnson and Hunter, 2005; Kostich et al., 2002; Manning et al., 2002). With almost 2% of the genome being made of kinases, and with the abundance of potential target sites, it is hardly surprising if every single physiological process in our cells is directly or indirectly regulated by one or more protein kinases (Ubersax and Ferrell, 2007).

In most cases, the ability of kinases to regulate a process is exerted directly through the ePK's catalytic activity, which consists in the reversible addition of a phosphate group (phospho-transfer) to defined residues (serine/threonine or tyrosine) within a specific sequence motif of a target protein. A long string of cases is available to illustrate the variety of effects that phosphorylation imparts on target proteins (Jimenez et al., 2007; Johnson and Barford, 1993). For instance, the phosphorylation of a target residue may result in a conformational change on the target protein that modifies its activity (Barford et al., 1991), or it may introduce a steric blockade to an enzyme's active site that precludes substrate binding (Hurley et al., 1990). Phosphorylation sites can also act as sites of recognition for specific phospho-residue-binding domains, several of which have now been identified (Bhattacharyya et al., 2006; Seet et al., 2006; Yaffe and Smerdon, 2004), or for specific enzymes (Lu et al., 2003). More recently, the modification of a protein's bulk electrostatics has emerged as a new mechanism whereby phosphorylation can influence a protein's activity, localization, or stability (Cohen, 2000; Serber and Ferrell, 2007).

Besides exerting a direct effect on substrates through phosphorylation, the catalytic domains of ePKs are often incorporated in larger protein assemblies, within which they are endowed with additional scaffolding functions that frequently appear to be independent of the catalytic activity (Pellicena and Kuriyan, 2006). The protein–protein interaction potential of protein kinases is also exploited as a means to regulate their activity, cellular localization, half-life, and substrate docking (Bhattacharyya et al., 2006; Biondi and Nebreda, 2003; Pellicena and Kuriyan, 2006; Remenyi et al., 2006; Ubersax and Ferrell, 2007). Similar regulatory functions can be also played *in cis* by other flanking functional domains, a census of which has been recently examined (Bhattacharyya et al., 2006; Manning et al., 2002; Pawson and Nash, 2003).

The split personality of Jekyll and Hide has served as a metaphor to describe the occasional transformation of kinases from competent controllers of cellular mechanisms to scourges capable of causing deadly human diseases (Cruz and

Tsai, 2004). Mutations in kinases or changes in their expression levels have been often correlated with disease, and have been shown in several cases to be their causative pathogenic agents (for instance, see Altomare and Testa, 2005; Baselga, 2006; Dhomen and Marais, 2007; Reindl and Spiekermann, 2006; Tan and Skipper, 2007; Wood-Kaczmar et al., 2006). Given the breadth of the implications of kinase deregulation for human disease, the identification of small-molecule inhibitors of protein kinases is an effort that extends horizontally across the entire pharmaceutical industry. The goal of identifying selective kinase inhibitors is fiercely opposed by the remarkable similarity of the protein kinase scaffold across different kinase groups, which makes the probability of side effects due to the inhibition of innocent kinase bystanders particularly high (Knight and Shokat, 2005).

Structure of the ePK Catalytic Domain

The determination of the crystal structure of the catalytic domain of cAMP-dependent protein kinase, or protein kinase A (PKA, PDB ID code 1 ATP) by Taylor, Sowadski, and colleagues provided the first snapshot of the general topology of the ePK family (Knighton et al., 1991a, b). The structure revealed a bilobar organization, with a smaller N-terminal lobe consisting predominantly of an anti-parallel β-sheet structure and containing a previously uncharacterized nucleotide-binding motif, and a larger C-terminal lobe of predominantly helical content (Fig. 1). The small lobe is primarily involved in anchoring the nucleotide to orient it for phospho-transfer. The large lobe contributes to substrate recognition and provides several conserved residues that are essential for phospho-transfer (Fig. 1C). The large, deep cleft between the two lobes is the site of catalysis. In their capacity as ATP analogues, most kinase inhibitors bind to this predominantly hydrophobic cleft, exploiting in addition the hydrogen-bonding potential of the main chain carbonyl and amide groups of two or more residues lining the active site in subdomain V (defined below) (Knight and Shokat, 2005). The importance of the PKA structure was further increased by the fact that it depicted the complex of PKA with a co-crystallized pseudo-substrate inhibitor (Fig. 1C) (Knighton et al., 1991b). The kinase–pseudo-substrate complex revealed the molecular bases of substrate recognition, identifying the site of phospho-transfer (the so-called P-site) and the critical interactions that allow the docking of the peptide substrate upstream and downstream from the P-site (see below).

There is a group of extremely well-conserved residues scattered along the catalytic domain of ePKs (Fig. 1A). The function of most of the conserved residues has been elucidated. Not surprisingly, it can be related to the phospho-transfer reaction or to substrate recognition (Fig. 2) (Johnson et al., 1996). Specific regions of the kinase domain whose role is closely associated with catalysis are usually referred to with specific names (Fig. 1A, B). In particular,

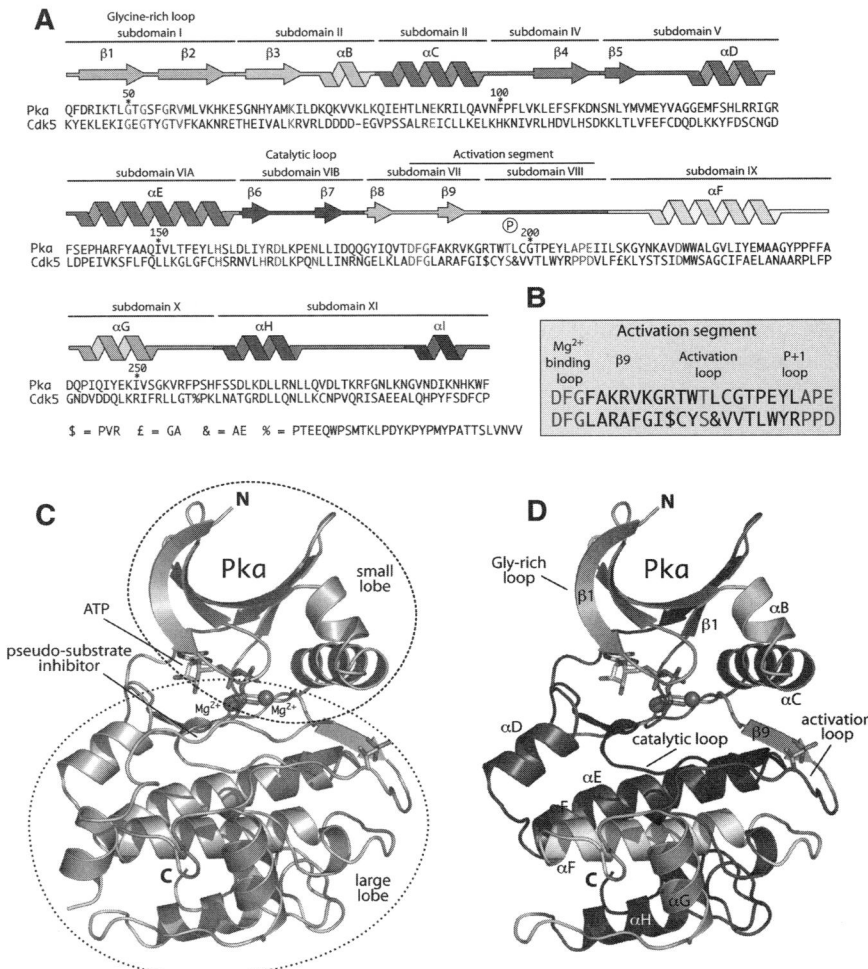

Fig. 1 Structural organization of the ePK (**A**) The aligned sequences of the catalytic cores of PKA and CDK5 are presented, together with the secondary structure. The different subdomains are identified by a different color and they are marked. Prominent loops discussed in the text are also marked. Residues that are often referred to in the text are shown in red. The phosphoryl group of T197 of PKA is also shown. CDK5's equivalent residue, Ser159, is not phosphorylated. The &, %, £, and & signs mark insertions in the CDK5 sequence that were not included in the alignment to preserve clarity. The sequence of the insertions is shown at the bottom of the panel. (**B**) Enlargement of the activation segment and definition of regions within it. The activation loop corresponds to subdomain VIII. (**C**) Ribbon model of the PKA catalytic domain (PDB ID code 1ATP). Structural elements flanking the catalytic domain were removed to improve clarity. The small and large lobes are marked. A pseudo-substrate inhibitor binds near the catalytic cleft, which contains ATP. (**D**) The subdomains of PKA were coloured with the same color scheme used in panel A. The activation segment is stretched. Most secondary structure elements are marked (*See* Color Insert)

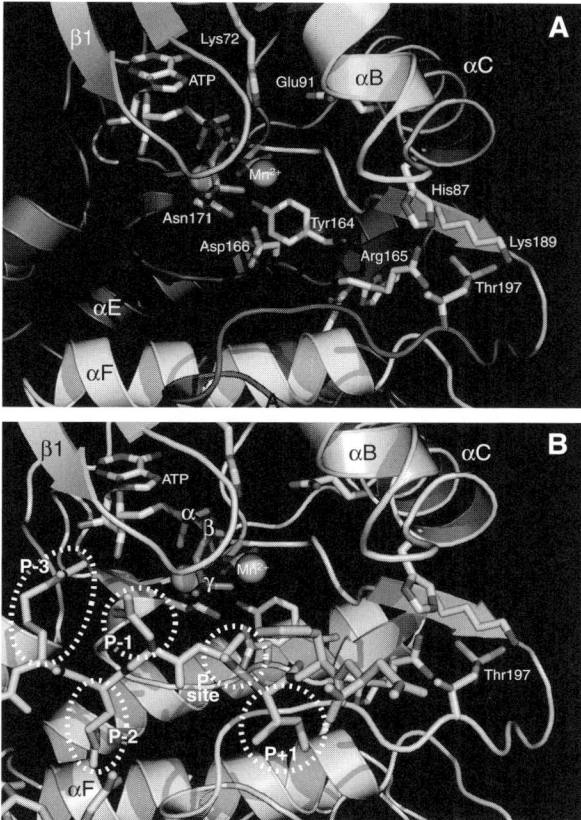

Fig. 2 Catalytic apparatus of ePKs (**A**) Close-up of the catalytic cleft of PKA (PDB ID code 1 ATP). The coloring scheme is as in Fig. 1D. The side chains of residues that are important for catalysis are highlighted. Thr197 is phosphorylated and acts as an organization center for several positively charged residues. This, in turn, results in the stabilization of the activation segment and poises the active site for catalysis. (**B**) The view is as in panel A. The pseudo-substrate inhibitor, which was omitted in panel A, is now visible in dark yellow, while the kinase is drawn in gray. It occupies the front of the active cleft cavity. In the pseudo-substrate inhibitor peptide, the P-site residue is Ala. It faces the γ-phosphate of ATP, which is held in place by a Mg^{2+} atom (Mn^{2+} in the particular case of this crystal structure) through a bunch of different interactions. Arginine residues at the P-3 and P-2 sites interact with an array of negatively charged side chains that are described in the text but are not shown here. An Ile side chain of the pseudo-substrate inhibitor at the P+1 site binds in a hydrophobic pocket contributed predominantly by the activation loop and by the P+1 loop (*See* Color Insert)

we distinguish (1) a "glycine-rich loop," which contains the conserved Gly-X-Gly-X-X-Gly-X-Val motif, which starts at Gly50 (using PKA sequence numbering; when necessary, we will add a superscript to the residue name to indicate the specific kinase sequence to which the residue belongs); (2) a "catalytic loop," which contains Tyr164 (usually a His in other kinases), Asp166, and Asn171; and (3) an "activation segment," which begins at the conserved residue Asp184 in the Asp-Phe-Gly (DFG) motif and ends in the conserved Ala-Pro-Glu (APE) motif, and which is further subdivided into a Mg^{2+}-binding loop (the DFG motif), the β9 strand, the activation loop, and the P+1 loop (P+1 indicates the substrate position immediately C-terminal to the phosphate acceptor, and the name P+1 loop reflects the involvement of this loop in substrate recognition at the P+1 site. See Fig. 2B for details) (Nolen et al., 2004).

Sequence conservation among protein ePKs within the three loops is easily explained in relationship to the kinase scaffold. The glycine-rich loop forms a flexible flap that surrounds the non-transferable phosphates of ATP (Hanks and Hunter, 1995; Knighton et al., 1991a, b). Within the catalytic loop, Asp166 acts as a catalytic base that accepts the proton from the hydroxyl group of the attacking nucleophile substrate; Asn171, on the other hand, stabilizes the main chain of Asp166 through a hydrogen bond, and it also contributes to chelating the secondary Mg^{2+} ion bridging between the α- and γ-phosphates of ATP. The activation segments contribute to stabilizing the ATP. Asp184, in the first part of the segment, chelates the primary Mg^{2+} ion bridging between the β- and γ-phosphates of ATP (Fig. 2). The activation segments also act as a critical component of the substrate-binding interface, as explained more thoroughly below; the conformation of the activation segment is endowed with remarkable conformational plasticity; and several distinct control mechanisms have evolved to regulate kinase activity and substrate access by modulating the conformation of this loop (Huse and Kuriyan, 2002; Nolen et al., 2004).

Additional conserved residues in the ePK family are Lys72 in the β3 loop and Glu91 in the αC helix (Fig. 2). These residues form a buried salt bridge that is an almost invariant feature of protein kinases. Other conserved residues, including His158 (in the αE helix) and Asp220 (in the αF helix), are involved in a network of hydrogen bonds whose ultimate goal appears to be the stabilization of the catalytic apparatus of the ePK catalytic domain (Kornev et al., 2006) (Fig. 2).

Besides the "classical" structural description based on the succession of secondary structure elements and "special" functional loops, the ePK domain has often been described based on a sub-division into 12 subdomains, that are usually indicated by roman numerals (Hanks and Hunter, 1995; Hanks et al., 1988). Such subdomains have been defined as "regions never interrupted by large amino acid insertions and containing characteristic patterns of conserved residues" (Hanks and Hunter, 1995; Hanks et al., 1988). The relationship between secondary structure elements, functional loops, and subdomains is illustrated in Fig. 1A.

The availability of metagenomic data has led to the recent exciting realization that the ePK catalytic domain represents only a portion of a much larger and diverse superfamily of enzymes characterized for the presence of a common protein kinase–like (PLK) fold, and that includes a variety of microbial enzymes (Kannan and Neuwald, 2005; Kannan et al., 2007; Scheeff and

Bourne, 2005). The residues described above are conserved in the majority of the sequences in the PLK fold, and therefore represent its hallmark.

Conformational Regulation of Kinase Activity

The bilobar scaffold unveiled by the structure of PKA is largely conserved in other ePK family members. The structures of the apo form of CDK2 (i.e., non-bound to a cyclin subunit: PDB ID code 1HCK) and of Erk2 (PDB ID code 1ERK), reported shortly after the structure of PKA (De Bondt et al., 1993; Zhang et al., 1994), were indeed very similar to PKA (Fig. 3A). But in addition, they revealed a set of structural differences that were progressively interpreted as the manifestation of a state of inactivity of the kinase (Hanks and Hunter, 1995; Huse and Kuriyan, 2002; Nolen et al., 2004). Specifically, the structural differences with PKA in the structures of apo-CDK2 and Erk2 concentrated in the small lobe and around the activation loop. In the PKA-pseudo-substrate structure, the activation loop (phosphorylated on Thr197) was fully extended and allowed the pseudo-substrate peptide to dock neatly onto the kinase surface. By contrast, the structures of CDK2 and Erk2, both of which were obtained with non-phosphorylated proteins (De Bondt et al., 1993; Zhang et al., 1994), revealed a conformation of the unphosphorylated activation loop that, by creating a blockade to the active site, was incompatible with the docking of substrates for phospho-transfer (Fig. 2 shows the CDK2 structure). Furthermore, in both cases, the αC helix was displaced outwards, in a way that prevented the formation of the buried salt bridge between Lys72 and Glu91 (Lys31^{CDK2}–Glu51^{CDK2} and Lys52^{ERK2}–Glu69^{ERK2}).

Thanks to subsequent work from several laboratories, we can now enjoy a refined view of the structural requirements for activation of Erk2 and CDK2 (Brown et al., 1999a,b; Canagarajah et al., 1997; Jeffrey et al., 1995; Pavletich, 1999; Russo et al., 1996b; Zhang et al., 2003). In both cases, the ultimate requirement for activation are (1) the extension of the activation loop in a stretched conformation similar to that originally observed in PKA to allow a productive interaction of the substrates with the catalytic cleft and the re-orientation of certain residues involved in catalysis, and (2) the concomitant ingression of the αC helix in the active site.

The details in which Erk2 and CDK2 achieve the active state vary. In both circumstances, the phosphorylation of the activation loop at conserved residues (Thr160^{CDK2}, Thr183^{ERK2}, and Tyr185^{ERK2}) is necessary for the conformational reorganization of the activation loop (Brown et al., 1999a,b; Canagarajah et al., 1997; Jeffrey et al., 1995; Pavletich, 1999; Russo et al., 1996b; Zhang et al., 2003). By changing the energy landscape around the target residues (Barrett and Noble, 2005; Groban et al., 2006; Serber and Ferrell, 2007), phosphate groups added onto the activation loop act as organization centers by establishing interactions with a cluster of Arg residues (e.g. Arg50, Arg126, and Arg150 in CDK2, corresponding to His87, Arg165, and Lys189 of PKA,

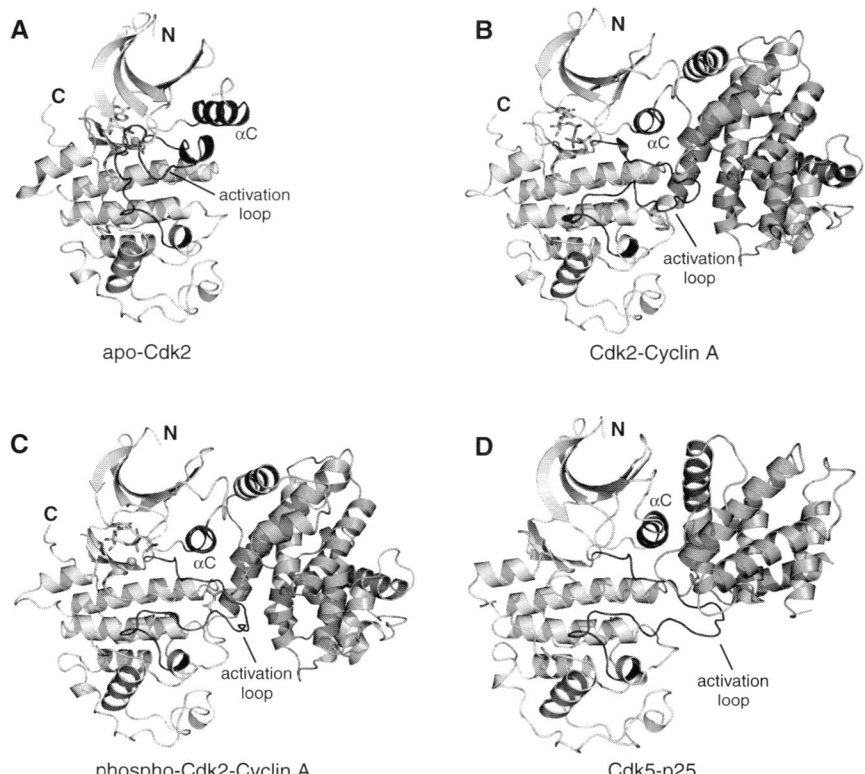

Fig. 3 The CDK activation cycle (**A**) Ribbon model of the inactive apo-CDK2 structure (PDB ID code 1 HCK). The activation loop and the αC helix are displayed in red. The activation loop meanders in the form of the catalytic cleft and blocks access to it. (**B**) The structure of the partially active form of CDK2, consisting of the CDK2–cylin A complex (PDB ID code 1FIN). The color code is as in panel **A**, and the cyclin is shown in blue. Note that the large lobe is essentially static in panels **A** and **B**. Thus, the cyclin imparts an important conformational change on the small lobe and on the activation loop. (**C**) The structure of the phospho-CDK2–cylin A complex (PDB ID code 1 JST), which is fully active. The activation loop is completely stretched out. (**D**) The CDK5–p25 complex (PDB ID code 1 H4L) displays an activation loop that is essentially indistinguishable from that of the phospho-CDK2–cylin A complex shown in panel **C** (*See* Color Insert)

Fig. 2A). In turn, this results in the stabilization of residues directly implicated in catalysis (Kornev et al., 2006). In the case of Erk2, the mere phosphorylation of the activation loop appears to be sufficient to achieve the active state (PDB ID code 2ERK, not shown). But for CDK2, full activation is only achieved through the concomitant binding of activating subunits such as cyclin A, B, and E (Brown et al., 1999a, b; Honda et al., 2005; Jeffrey et al., 1995; Pavletich, 1999; Russo et al., 1996b). The "ternary mode" of CDK2 activation is displayed in Fig. 3A–C. A similar relationship exists between two related kinases, PKA and the Aurora family kinases. In the first case, as we have already observed, the phosphorylation of the activation loop acts as a sufficient stimulus to achieve the fully active conformation (Knighton et al., 1991a, b). In the case of the Aurora family members, on the other hand, activation loop phosphorylation must be supported by the binding of an activating subunit for full activation to be achieved (Bayliss et al., 2003; Sessa et al., 2005).

CDK5: Structural Considerations

Besides the Erk2 and CDK2 cases described above, several structures of active–inactive pairs of different members of the ePK family are now available (Huse and Kuriyan, 2002; Nolen et al., 2004). A remarkable conclusion from the structural studies is that the active states of otherwise diverse kinases tend to be very similar to each other, regardless of the specific mechanism utilized to achieve full kinase activation. Even if there is limited structural information on the interaction of active kinases with their substrates, the similarity of the active states likely reflects restrains imposed by the docking of substrates and ATP (see below). By contrast, the inactive states of the ePK catalytic domains seem to be intrinsically more different from each other (Hanks and Hunter, 1995; Huse and Kuriyan, 2002; Nolen et al., 2004). This observation has important implications for the clinics, as it may provide an opportunity to isolate more selective inhibitors targeting the larger structural variety offered by the inactive forms of protein kinases. Gleevec (Imatinib), a drug used in chronic myeloid leukemia and other illnesses, confirms this idea, as it binds preferentially to the inactive, non-phosphorylated form of its target, the Abl tyrosine kinase (Schindler et al., 2000).

The structural information on CDK5 is currently limited to its active form (PDB ID code 1 H4L), and to the complex of its active form with several small-molecule inhibitors (PDB ID codes 1UNG, 1UNL, 1UNH) (Mapelli et al., 2005; Mapelli and Musacchio, 2003; Meijer et al., 2003; Polychronopoulos et al., 2004; Tarricone et al., 2001). With the caveat that the structure of apo-CDK5 is unknown, the active form of CDK5 seems to arise from the binding of CDK5 to homologous C-terminal domains of the two bonafide CDK5 activators, p35 and p39 (Dhavan and Tsai, 2001). These domains are also retained in p25 and p29, the proteolytic products of p35 and p39 whose accumulation has been associated with several neurodegenerative diseases (Cruz and Tsai, 2004). A remarkable aspect of the mechanism of CDK5 activation revealed by the crystal structure is that full stretching of the activation loop takes place in the absence of activation loop phosphorylation (Tarricone et al., 2001). Lack of phosphorylation is in apparent contrast with the presence of a phosphorylatable residue on the CDK5 activation loop (Ser159^{CDK5}, equivalent to Thr160^{CDK2}) (Figs. 1A and 3D). Despite the report of an activity capable of phosphorylating and activating CDK5 through Ser159^{CDK5} (Sharma et al.,

1999), there is strong evidence that this residue is not phosphorylated in cells (Nishizawa et al., 1999; Poon et al., 1997; Qi et al., 1995; Tarricone et al., 2001). The structural explanation for lack of phosphorylation is that Ser159 forms favourable contacts with the p25 CBF domain that are expected to contribute to the stability of the interaction (Otyepka et al., 2006; Tarricone et al., 2001). While this argument might be weakened by the observation that the substitution of Ser159^{CDK5} with Ala has no visible effect on kinase activity and on the apparent stability of the CDK5–p25 interaction (Tarricone et al., 2001), the substitution of Ser159^{CDK5} with a bulkier residue, including Thr, prevents the interaction of the mutated CDK5 with the p25 activator. Thus, even minimal, conservative changes to larger residues are deleterious for the CDK5-p25 interaction, suggesting that if phosphorylation took place, it likely would act to counteract the interaction of CDK5 with its activating subunits (Tarricone et al., 2001). The liberation of CDK5 from a mechanism of activation based on phosphorylation might have been selected to permit the activation of CDK5 in post-mitotic cells, that is, under conditions where the CDK-activating kinase (CAK) that phosphorylates the activation loop of other CDKs may be scarce.

The structural analysis of the CDK5–p25 complex finally confirmed the prediction, based on very limited sequence similarity, modeling, and threading methods (Brown et al., 1995; Chou et al., 1999; Tang et al., 1995) that p25 is structurally related to the cyclins, the activating subunits of other CDK family members such as CDK1, CDK2, CDK4, and CDK6 (Fig. 4). The structural domain that characterizes the cyclins is the so-called cyclin-box fold (CBF), a 5-helix construction that is found in a variety of proteins (Noble et al., 1997) (for more information, see http://smart.embl-heidelberg.de/). Cyclins often contain tandem repeats of the CBF, in which the first CBF holds the predominant set of interactions with the kinase domain (Fig. 4A). There is a single CBF domain in the CDK5 activators, and this is additionally flanked by an N-terminal helix (αN) and two C-terminal helices (α6 and α7, Fig. 4B) (Tarricone et al., 2001). The α6 helix is directly implicated in the interaction of p25 with CDK5 (Tarricone et al., 2001). Conversely, the function of α7 is primarily in the stabilization of the hydrophobic core of the p35 globular domain through the contributions of residues Val285 and Leu289. A p35 segment starting with αN helix and ending immediately before the α7 helix (residues 154–279) has been reported to be, slightly surprisingly, a more potent CDK5 ligand than p25 (Amin et al., 2002; Zheng et al., 2005).

The globular region containing the CBF and the flanking helices span 140 residues that normally occupy the C-terminal regions of p35 and p39. It is separated from the C-terminus by a 15-residue C-terminal segment that might act as a docking sequence for interacting proteins (Ledee et al., 2007). N-terminal to the CDK5-interacting globular domains of p35 and p39, on the other hand, there are long stretches of low-complexity sequence and, with the exception of short segments, a conspicuous absence of hydrophobic residues. Overall, the sequence of the N-terminal region of p35 and p39 suggests that these regions must be largely unstructured (Sim and Creamer, 2002). As such, they are likely

Fig. 4 The cyclin-box fold (**A**) Close-up of the phospho-CDK2–cylin A structure (se Fig. 3) at the CDK2–cyclin A interface. The cyclin-box fold motifs (*yellow and green*) are 5-helical arrays of about 100 residues. Additional helices and coils complete the fold. (**B**) The p25 activator contains a cyclin-box motif (*yellow*) embedded within a larger globular domain containing three additional helices. There is striking similarity with the CDK2–cyclin A interface, but also conspicuous differences (for details, see Tarricone et al., 2001) (*See* Color Insert)

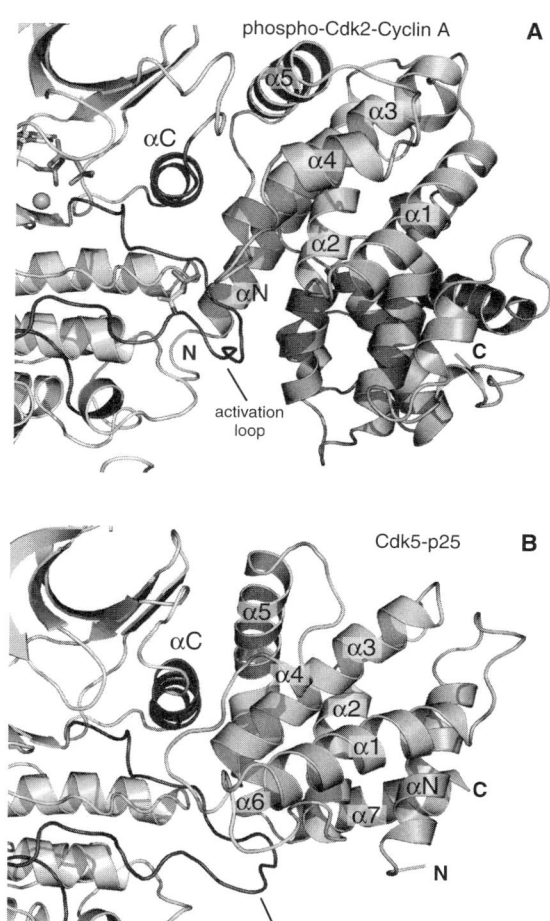

to act as docking sites for interacting partners of p35 and p35 (for instance, see Lim et al., 2003; Sim and Creamer, 2002). The N-terminal region of p35 and p39 also contain a wealth of potential sites of phosphorylation (Kamei et al., 2007).

The CBFs of cyclins and of p25 bind around the αC helix of their cognate CDKs (defined as PSTAIRE and PSAALRE in CDK2 and CDK5, respectively), causing its integration into the kinase active site (Figs. 3 and 4). As we already discussed above, if the binding of cyclin A to CDK2 leads to a partial activation of its kinase activity, it is insufficient to achieve full activation. For full activation, it is required that the activation loop be phosphorylated on $Thr160^{CDK2}$, resulting in a final step of conformational rearrangement required for substrate recognition (Fig. 3A–C). Not only there is an approximately 50-fold increase in substrate binding affinity upon phosphorylation of Thr160 in

CDK2, reflecting a participation of the phoshporylated residue in substrate recognition, but the overall catalytic rate by CDK2–cyclin A increases approximately 3000-fold upon phosphorylation, suggesting that the phosphorylated activation loop participates in the stabilization of the transition state during phospho-transfer (Brown et al., 1999a; Hagopian et al., 2001; Holmes and Solomon, 2001; Lew, 2003; Russo et al., 1996b). A "spine" model describing how the phosphorylation of the activation loop may lead to a global stabilization of a kinase's active conformation has been recently proposed (Kornev et al., 2006). Evidently, in CDK5 and the other large number of kinases whose activation does not require activation loop phosphorylation, alternative mechanisms must be in place to achieve the same goal (Nolen et al., 2004).

In summary, the CBFs of p35 and p39 are "super-CBF domains" that activate CDK5 in a single step, relieving the requirements for activation loop phosphorylation. The precise contacts that allow the p35/p39 CBF to achieve this effect remain unclear, but it seems plausible that the extensive interactions of the p35 CBF with the CDK activation loop must be important (Otyepka et al., 2006; Tarricone et al., 2001). A virus-encoded cyclin (V-cyclin) is also able to activate the CDK6 kinase in the absence of activation loop phosphorylation, providing an additional example of a "super-CBF domain" (Schulze-Gahmen and Kim, 2002).

While other CBFs have been reported to be able to bind CDK5, no catalytic activity was observed, indicating that the interface between CDK5 and p35/p39 contains features that cannot be reproduced with other cyclins (Miyajima et al., 1995; Xiong et al., 1992; Zukerberg et al., 2000). By contrast, CDK5 activation *in vitro* has been observed with a novel family of CDK activators, unrelated to the cyclins, and known as RINGO/Speedy (Dinarina et al., 2005; Nebreda, 2006). The significance of this observation is currently unclear (Nebreda, 2006).

The Inhibition of CDK5

The CDK5–p35 complex (and presumably also the related CDK5–p39 and CDK5–p25 complexes) is insensitive to the inhibition by the CDK inhibitors (CKIs) p21 and p27 (Lacy et al., 2005; Lee et al., 1996). The structural bases for the ability of p27 to inhibit the CDK-cyclin complexes are well known (Russo et al., 1996a). Despite the similarity of helical topology of the CBF motifs in cyclin A and p35, the latter lacks the MRAIL motif that mediates the interaction of cell cycle cyclins with domain 1 of p21 and p27 (Lacy et al., 2005; Lee et al., 1996; Russo et al., 1996a; Tarricone et al., 2001).

CDKs are regulated by inhibitory phosphorylation of conserved residues near the active site (reviewed in Solomon and Kaldis, 1998). Entry into mitosis in most eukaryotes coincides with the dephosphorylation of the main mitotic kinase, CDK1, at Thr14 and Tyr15 (Solomon and Kaldis, 1998; Takizawa and Morgan, 2000). Phosphorylation of CDK1 on Tyr15 is catalyzed by members of the Wee1/Myk1 kinase family, while dephosphorylation requires the members of the Cdc25 family of protein phosphatases (Solomon and Kaldis, 1998). The structure of CDK2–cyclin A containing the phosphorylated form of Tyr15 (PDB ID code 2 CJM; Welburn et al., 2007) shows a "swung-out" conformation in which the hydroxyl group of the tyrosine side chain, rather than being buried into the active site as in the active structure of the CDK2–cyclin A–ATP complex (PDB ID code 1 JST; Russo et al., 1996b), is exposed to solvent with its attached phosphate group. The kinetic and structural investigation of phospho-Tyr15–CDK2–cyclin A reveals that the inhibitory mechanism involves a steric blockade of peptide substrate binding and the creation of a chemical environment that stabilizes a non-productive conformation of the terminal group of ATP (Welburn et al., 2007).

In contrast to the negative effects of Tyr15 phosphorylation on the activity of CDK1, CDK2, and possibly other CDK family members, the phosphorylation of Tyr15 of CDK5 has been reported to bring about CDK5 activation (Zukerberg et al., 2000). *In vitro*, Tyr15 phosphorylation by the Abl tyrosine kinase results in a modest activation of the CDK5–p25 complex toward histone H1 as a substrate (Mapelli et al., 2005). Given the remarkable similarity between CDK5 and other CDK family members, the different consequences on kinase activity upon Tyr15 phosphorylation do not have an immediate explanation and require further structural and functional investigations.

As explained elsewhere in this book, CDK5 deregulation is implicated in several neurodegenerative diseases. CDK5 complexes whose upregulated activity is implicated as a possible cause of disease, such as the CDK5–p25 complex, can be viewed as targets for pharmacological therapy. A great wealth of functional and structural information is available on the mechanisms of small-molecule inhibition of the CDKs, and in particular of those CDKs that being implicated in cell cycle control have been identified as potential targets for the therapy of cancer and other ailments (for recent reviews, see Collins and Garrett, 2005; Dai and Grant, 2003). As expected based on the fact that CDK1, CDK2, and CDK5 share high sequence and structure similarity, most small-molecule inhibitors bind to these kinases with modest selectivity (Sridhar et al., 2006). The availability of crystals for high-resolution structural analysis of small-molecule inhibitors bound to the active site of the CDK5-p25 complex may be key to develop selective CDK5 inhibitors (Mapelli et al., 2005).

The Substrate Specificity of CDK5

Like other CDKs, CDK5 belongs to the CMGC group of ePKs (Hanks and Hunter, 1995; Kannan and Neuwald, 2004; Manning et al., 2002). The CMGC group, whose acronymic name is derived from CDK, MAP kinase, GSK3, and CDK-like kinase (Clk), contains for the most part proline-directed enzymes, that is, kinases that phosphorylate substrates whose Ser or Thr (S/T) target is in the proximity of a Pro residue (P). In the CDKs and in Erk (MAP) kinase family, this requirement takes the form of a phosphate acceptor lying immediately N-terminal to a Pro. In GSK3, the requirement for proline co-exists with an additional phosphate-priming mechanism, whose structural bases have been elegantly elucidated (Dajani et al., 2001; ter Haar et al., 2001).

Other kinase groups are characterized by different substrate specificity. For instance, the kinases of the AGC group (Hanks and Hunter, 1995; Hanks et al., 1988; Manning et al., 2002), which take their name from PKA, PKG, and PKC, tend to be directed toward basic residues, that is, they phosphorylate residues embedded in sequences containing Arg or Lys in the proximity of the P-site. For instance, PKA and Aurora family substrates have basic residues at positions P-3 and P-2. The PKA-pseudo-substrate inhibitor complex provides a valuable frame of reference to understand such preference (Knighton et al., 1991b). The arginine residues at positions P-3 and P-2 of the pseudo-substrate inhibitor interact with acidic residues of PKA, such as Glu127, Glu170, Glu203, and Glu230. Most of these residues are conserved in Aurora family kinases, explaining why its substrate specificity is similar to PKA's (Bayliss et al., 2003; Sessa et al., 2005).

In PKA, residues in the second half of the activation loop and in the P+1 loop are in contact with the P+1 residue of the substrate, and confer specificity toward a hydrophobic residue (Knighton et al., 1991b). Thus, the overall consensus site for this kinase is R-R-X-S/T-Φ (where X is any residue and Φ is hydrophobic). Also for proline-directed kinase, the mechanistic bases for the specificity toward the P+1 proline are well understood (Brown et al., 1999a). Due to the specific conformation of their activation loop, these kinases leave unsatisfied a potential hydrogen bond from the main chain amide nitrogen of the substrate. Proline is unable to form such as hydrogen bond, and can be hosted at the P+1 site without suffering the energy loss from the unsatisfied hydrogen bond (Brown et al., 1999a). Positive residues downstream from the P+1 position, and in particular residues in the P+3 position, further contribute to specificity by binding to the phospho-Thr160 in the CDK2 activation loop (Brown et al., 1999a; Holmes and Solomon, 2001). Thus, the consensus sequence for CDKs is S/T-P-X-R.

The examples above illustrate how the local sequence around the catalytic cleft contributes to the selection of substrates. Kinases, however, have devised several additional strategies to recognize their substrates specifically. Two excellent recent reviews provide the necessary insight into this complex problem, whose discussion goes beyond the scope of the present account (Remenyi et al., 2006; Ubersax and Ferrell, 2007). Here, it suffices to say that it has long been known that additional docking sequences, besides those in the proximity of the catalytic cleft, can increase the specificity of kinases toward their substrates. This is especially well characterized for the CDKs: an RXL motif that is present in many of their substrates binds to a hydrophobic patch on the cyclin subunit, increasing the affinity of the substrate for the kinase (Schulman et al., 1998).

It is well established that CDK5, like other CDKs, is a proline-directed kinase that, like CDK2, prefers substrates with consensus S/T-P-X-R (Dhavan and Tsai, 2001; Tarricone et al., 2001). However, there is a lack of understanding regarding how CDK5 recognizes its substrates specifically. The local binding site around the catalytic cleft is clearly distinct from that of CDK2. For instance, the usually well-conserved "APE" motif in subdomain VIII is mutated in "PPD" in CDK5. Furthermore, CDK5 is not phosphorylated on the activation loop, and we have already observed that the phosphorylated Thr160 of CDK2 is directly engaged in substrate recognition (Brown et al., 1999a; Holmes and Solomon, 2001). As the CDK5–p25 complex retains a preference for an Arg at the P+3 position of the substrate, it has been proposed—based on its position—that a p25 residue, Glu240, might substitute the phosphate group in substrate recognition (Tarricone et al., 2001). Distally from the catalytic cleft, the CDK5 activators lack the hydrophobic patch that binds the RXL motif (the same patch is also involved in the binding of p21 and p27, as discussed previously). In the future, it will be essential to define the rules and mechanisms that sustain specific substrate recognition by the CDK5 kinase.

Web Resources

The "SUGEN" dataset (http://kinase.com/)
Protein kinase resource (http://www.kinasenet.org/pkr/Welcome.do)
Kinase sequence database (http://sequoia.ucsf.edu/ksd/)

References

Altomare, D. A., and Testa, J. R. (2005). Perturbations of the AKT signaling pathway in human cancer. Oncogene *24*, 7455–7464.

Amin, N. D., Albers, W., and Pant, H. C. (2002). Cyclin-dependent kinase 5 (cdk5) activation requires interaction with three domains of p35. J Neurosci Res *67*, 354–362.

Barford, D., Hu, S. H., and Johnson, L. N. (1991). Structural mechanism for glycogen phosphorylase control by phosphorylation and AMP. J Mol Biol *218*, 233–260.

Barrett, C. P., and Noble, M. E. (2005). Molecular motions of human cyclin-dependent kinase 2. J Biol Chem *280*, 13993–14005.

Baselga, J. (2006). Targeting tyrosine kinases in cancer: the second wave. Science *312*, 1175–1178.

Bayliss, R., Sardon, T., Vernos, I., and Conti, E. (2003). Structural basis of Aurora-A activation by TPX2 at the mitotic spindle. Mol Cell *12*, 851–862.

Bhattacharyya, R. P., Remenyi, A., Yeh, B. J., and Lim, W. A. (2006). Domains, motifs, and scaffolds: the role of modular interactions in the evolution and wiring of cell signaling circuits. Annu Rev Biochem *75*, 655–680.

Biondi, R. M., and Nebreda, A. R. (2003). Signalling specificity of Ser/Thr protein kinases through docking-site-mediated interactions. Biochem J *372*, 1–13.

Brown, N. R., Noble, M. E., Endicott, J. A., Garman, E. F., Wakatsuki, S., Mitchell, E., Rasmussen, B., Hunt, T., and Johnson, L. N. (1995). The crystal structure of cyclin A. Structure 3, 1235–1247.

Brown, N. R., Noble, M. E., Endicott, J. A., and Johnson, L. N. (1999a). The structural basis for specificity of substrate and recruitment peptides for cyclin-dependent kinases. Nat Cell Biol 1, 438–443.

Brown, N. R., Noble, M. E., Lawrie, A. M., Morris, M. C., Tunnah, P., Divita, G., Johnson, L. N., and Endicott, J. A. (1999b). Effects of phosphorylation of threonine 160 on cyclin-dependent kinase 2 structure and activity. J Biol Chem 274, 8746–8756.

Canagarajah, B. J., Khokhlatchev, A., Cobb, M. H., and Goldsmith, E. J. (1997). Activation mechanism of the MAP kinase ERK2 by dual phosphorylation. Cell 90, 859–869.

Cheek, S., Ginalski, K., Zhang, H., and Grishin, N. V. (2005). A comprehensive update of the sequence and structure classification of kinases. BMC Struct Biol 5, 6.

Chou, K. C., Watenpaugh, K. D., and Heinrikson, R. L. (1999). A model of the complex between cyclin-dependent kinase 5 and the activation domain of neuronal Cdk5 activator. Biochem Biophys Res Commun 259, 420–428.

Cohen, P. (2000). The regulation of protein function by multisite phosphorylation – a 25 year update. Trends Biochem Sci 25, 596–601.

Collins, I., and Garrett, M. D. (2005). Targeting the cell division cycle in cancer: CDK and cell cycle checkpoint kinase inhibitors. Curr Opin Pharmacol 5, 366–373.

Cruz, J. C., and Tsai, L. H. (2004). A Jekyll and Hyde kinase: roles for Cdk5 in brain development and disease. Curr Opin Neurobiol 14, 390–394.

Dai, Y., and Grant, S. (2003). Cyclin-dependent kinase inhibitors. Curr Opin Pharmacol 3, 362–370.

Dajani, R., Fraser, E., Roe, S. M., Young, N., Good, V., Dale, T. C., and Pearl, L. H. (2001). Crystal structure of glycogen synthase kinase 3 beta: structural basis for phosphate-primed substrate specificity and autoinhibition. Cell 105, 721–732.

De Bondt, H. L., Rosenblatt, J., Jancarik, J., Jones, H. D., Morgan, D. O., and Kim, S. H. (1993). Crystal structure of cyclin-dependent kinase 2. Nature 363, 595–602.

Dhavan, R., and Tsai, L. H. (2001). A decade of CDK5. Nat Rev Mol Cell Biol 2, 749–759.

Dhomen, N., and Marais, R. (2007). New insight into BRAF mutations in cancer. Curr Opin Genet Dev 17, 31–39.

Dinarina, A., Perez, L. H., Davila, A., Schwab, M., Hunt, T., and Nebreda, A. R. (2005). Characterization of a new family of cyclin-dependent kinase activators. Biochem J 386, 349–355.

Groban, E. S., Narayanan, A., and Jacobson, M. P. (2006). Conformational changes in protein loops and helices induced by post-translational phosphorylation. PLoS Comput Biol 2, e32.

Hagopian, J. C., Kirtley, M. P., Stevenson, L. M., Gergis, R. M., Russo, A. A., Pavletich, N. P., Parsons, S. M., and Lew, J. (2001). Kinetic basis for activation of CDK2/cyclin A by phosphorylation. J Biol Chem 276, 275–280.

Hanks, S. K., and Hunter, T. (1995). Protein kinases 6. The eukaryotic protein kinase superfamily: kinase (catalytic) domain structure and classification. Faseb J 9, 576–596.

Hanks, S. K., Quinn, A. M., and Hunter, T. (1988). The protein kinase family: conserved features and deduced phylogeny of the catalytic domains. Science 241, 42–52.

Holmes, J. K., and Solomon, M. J. (2001). The role of Thr160 phosphorylation of Cdk2 in substrate recognition. Eur J Biochem 268, 4647–4652.

Honda, R., Lowe, E. D., Dubinina, E., Skamnaki, V., Cook, A., Brown, N. R., and Johnson, L. N. (2005). The structure of cyclin E1/CDK2: implications for CDK2 activation and CDK2-independent roles. Embo J 24, 452–463.

Hurley, J. H., Dean, A. M., Sohl, J. L., Koshland, Jr., D. E., and Stroud, R. M. (1990). Regulation of an enzyme by phosphorylation at the active site. Science 249, 1012–1016.

Huse, M., and Kuriyan, J. (2002). The conformational plasticity of protein kinases. Cell *109*, 275–282.

Jeffrey, P. D., Russo, A. A., Polyak, K., Gibbs, E., Hurwitz, J., Massague, J., and Pavletich, N. P. (1995). Mechanism of CDK activation revealed by the structure of a cyclinA-CDK2 complex. Nature *376*, 313–320.

Jimenez, J. L., Hegemann, B., Hutchins, J. R., Peters, J. M., and Durbin, R. (2007). A systematic comparative and structural analysis of protein phosphorylation sites based on the mtcPTM database. Genome Biol *8*, R90.

Johnson, L. N., and Barford, D. (1993). The effects of phosphorylation on the structure and function of proteins. Annu Rev Biophys Biomol Struct *22*, 199–232.

Johnson, L. N., Noble, M. E., and Owen, D. J. (1996). Active and inactive protein kinases: structural basis for regulation. Cell *85*, 149–158.

Johnson, S. A., and Hunter, T. (2005). Kinomics: methods for deciphering the kinome. Nat Methods *2*, 17–25.

Kamei, H., Saito, T., Ozawa, M., Fujita, Y., Asada, A., Bibb, J. A., Saido, T. C., Sorimachi, H., and Hisanaga, S. (2007). Suppression of calpain-dependent cleavage of the CDK5 activator p35 to p25 by site-specific phosphorylation. J Biol Chem *282*, 1687–1694.

Kannan, N., and Neuwald, A. F. (2004). Evolutionary constraints associated with functional specificity of the CMGC protein kinases MAPK, CDK, GSK, SRPK, DYRK, and CK2alpha. Protein Sci *13*, 2059–2077.

Kannan, N., and Neuwald, A. F. (2005). Did protein kinase regulatory mechanisms evolve through elaboration of a simple structural component? J Mol Biol *351*, 956–972.

Kannan, N., Taylor, S. S., Zhai, Y., Venter, J. C., and Manning, G. (2007). Structural and functional diversity of the microbial kinome. PLoS Biol *5*, e17.

Knight, Z. A., and Shokat, K. M. (2005). Features of selective kinase inhibitors. Chem Biol *12*, 621–637.

Knighton, D. R., Zheng, J. H., Ten Eyck, L. F., Ashford, V. A., Xuong, N. H., Taylor, S. S., and Sowadski, J. M. (1991a). Crystal structure of the catalytic subunit of cyclic adenosine monophosphate-dependent protein kinase. Science *253*, 407–414.

Knighton, D. R., Zheng, J. H., Ten Eyck, L. F., Xuong, N. H., Taylor, S. S., and Sowadski, J. M. (1991b). Structure of a peptide inhibitor bound to the catalytic subunit of cyclic adenosine monophosphate-dependent protein kinase. Science *253*, 414–420.

Kornev, A. P., Haste, N. M., Taylor, S. S., and Eyck, L. F. (2006). Surface comparison of active and inactive protein kinases identifies a conserved activation mechanism. Proc Natl Acad Sci USA *103*, 17783–17788.

Kostich, M., English, J., Madison, V., Gheyas, F., Wang, L., Qiu, P., Greene, J., and Laz, T. M. (2002). Human members of the eukaryotic protein kinase family. Genome Biol *3*, Research0043.

Lacy, E. R., Wang, Y., Post, J., Nourse, A., Webb, W., Mapelli, M., Musacchio, A., Siuzdak, G., and Kriwacki, R. W. (2005). Molecular basis for the specificity of p27 toward cyclin-dependent kinases that regulate cell division. J Mol Biol *349*, 764–773.

Ledee, D. R., Tripathi, B. K., and Zelenka, P. S. (2007). The CDK5 activator, p39, binds specifically to myosin essential light chain. Biochem Biophys Res Commun *354*, 1034–1039.

Lee, M. H., Nikolic, M., Baptista, C. A., Lai, E., Tsai, L. H., and Massague, J. (1996). The brain-specific activator p35 allows Cdk5 to escape inhibition by p27Kip1 in neurons. Proc Natl Acad Sci USA *93*, 3259–3263.

Lew, J. (2003). MAP kinases and CDKs: kinetic basis for catalytic activation. Biochemistry *42*, 849–856.

Lim, A. C., Qu, D., and Qi, R. Z. (2003). Protein-protein interactions in Cdk5 regulation and function. Neurosignals *12*, 230–238.

Lu, K. P., Liou, Y. C., and Vincent, I. (2003). Proline-directed phosphorylation and isomerization in mitotic regulation and in Alzheimer's disease. Bioessays *25*, 174–181.

Manning, G., Whyte, D. B., Martinez, R., Hunter, T., and Sudarsanam, S. (2002). The protein kinase complement of the human genome. Science 298, 1912–1934.

Mapelli, M., Massimiliano, L., Crovace, C., Seeliger, M. A., Tsai, L. H., Meijer, L., and Musacchio, A. (2005). Mechanism of CDK5/p25 binding by CDK inhibitors. J Med Chem 48, 671–679.

Mapelli, M., and Musacchio, A. (2003). The structural perspective on CDK5. Neurosignals 12, 164–172.

Meijer, L., Skaltsounis, A. L., Magiatis, P., Polychronopoulos, P., Knockaert, M., Leost, M., Ryan, X. P., Vonica, C. A., Brivanlou, A., Dajani, R., et al. (2003). GSK-3-selective inhibitors derived from Tyrian purple indirubins. Chem Biol 10, 1255–1266.

Miyajima, M., Nornes, H. O., and Neuman, T. (1995). Cyclin E is expressed in neurons and forms complexes with cdk5. Neuroreport 6, 1130–1132.

Nebreda, A. R. (2006). CDK activation by non-cyclin proteins. Curr Opin Cell Biol 18, 192–198.

Nishizawa, M., Kanaya, Y., and Toh, E. A. (1999). Mouse cyclin-dependent kinase (Cdk) 5 is a functional homologue of a yeast Cdk, pho85 kinase. J Biol Chem 274, 33859–33862.

Noble, M. E., Endicott, J. A., Brown, N. R., and Johnson, L. N. (1997). The cyclin box fold: protein recognition in cell-cycle and transcription control. Trends Biochem Sci 22, 482–487.

Nolen, B., Taylor, S., and Ghosh, G. (2004). Regulation of protein kinases: controlling activity through activation segment conformation. Mol Cell 15, 661–675.

Otyepka, M., Bartova, I., Kriz, Z., and Koca, J. (2006). Different mechanisms of CDK5 and CDK2 activation as revealed by CDK5/p25 and CDK2/cyclin A dynamics. J Biol Chem 281, 7271–7281.

Pavletich, N. P. (1999). Mechanisms of cyclin-dependent kinase regulation: structures of Cdks, their cyclin activators, and Cip and INK4 inhibitors. J Mol Biol 287, 821–828.

Pawson, T., and Nash, P. (2003). Assembly of cell regulatory systems through protein interaction domains. Science 300, 445–452.

Pellicena, P., and Kuriyan, J. (2006). Protein-protein interactions in the allosteric regulation of protein kinases. Curr Opin Struct Biol 16, 702–709.

Polychronopoulos, P., Magiatis, P., Skaltsounis, A. L., Myrianthopoulos, V., Mikros, E., Tarricone, A., Musacchio, A., Roe, S. M., Pearl, L., Leost, M., et al. (2004). Structural basis for the synthesis of indirubins as potent and selective inhibitors of glycogen synthase kinase-3 and cyclin-dependent kinases. J Med Chem 47, 935–946.

Poon, R. Y., Lew, J., and Hunter, T. (1997). Identification of functional domains in the neuronal Cdk5 activator protein. J Biol Chem 272, 5703–5708.

Qi, Z., Huang, Q. Q., Lee, K. Y., Lew, J., and Wang, J. H. (1995). Reconstitution of neuronal Cdc2-like kinase from bacteria-expressed Cdk5 and an active fragment of the brain-specific activator. Kinase activation in the absence of Cdk5 phosphorylation. J Biol Chem 270, 10847–10854.

Reindl, C., and Spiekermann, K. (2006). From kinases to cancer: leakiness, loss of autoinhibition and leukemia. Cell Cycle 5, 599–602.

Remenyi, A., Good, M. C., and Lim, W. A. (2006). Docking interactions in protein kinase and phosphatase networks. Curr Opin Struct Biol 16, 676–685.

Russo, A. A., Jeffrey, P. D., Patten, A. K., Massague, J., and Pavletich, N. P. (1996a). Crystal structure of the p27Kip1 cyclin-dependent-kinase inhibitor bound to the cyclin A-Cdk2 complex. Nature 382, 325–331.

Russo, A. A., Jeffrey, P. D., and Pavletich, N. P. (1996b). Structural basis of cyclin-dependent kinase activation by phosphorylation. Nat Struct Biol 3, 696–700.

Scheeff, E. D., and Bourne, P. E. (2005). Structural evolution of the protein kinase-like superfamily. PLoS Comput Biol 1, e49.

Schindler, T., Bornmann, W., Pellicena, P., Miller, W. T., Clarkson, B., and Kuriyan, J. (2000). Structural mechanism for STI-571 inhibition of abelson tyrosine kinase. Science *289*, 1938–1942.

Schulman, B. A., Lindstrom, D. L., and Harlow, E. (1998). Substrate recruitment to cyclin-dependent kinase 2 by a multipurpose docking site on cyclin A. Proc Natl Acad Sci USA *95*, 10453–10458.

Schulze-Gahmen, U., and Kim, S. H. (2002). Structural basis for CDK6 activation by a virus-encoded cyclin. Nat Struct Biol *9*, 177–181.

Seet, B. T., Dikic, I., Zhou, M. M., and Pawson, T. (2006). Reading protein modifications with interaction domains. Nat Rev Mol Cell Biol *7*, 473–483.

Serber, Z., and Ferrell, Jr., J. E. (2007). Tuning bulk electrostatics to regulate protein function. Cell *128*, 441–444.

Sessa, F., Mapelli, M., Ciferri, C., Tarricone, C., Areces, L. B., Schneider, T. R., Stukenberg, P. T., and Musacchio, A. (2005). Mechanism of Aurora B activation by INCENP and inhibition by hesperadin. Mol Cell *18*, 379–391.

Sharma, P., Sharma, M., Amin, N. D., Albers, R. W., and Pant, H. C. (1999). Regulation of cyclin-dependent kinase 5 catalytic activity by phosphorylation. Proc Natl Acad Sci USA *96*, 11156–11160.

Sim, K. L., and Creamer, T. P. (2002). Abundance and distributions of eukaryote protein simple sequences. Mol Cell Proteomics *1*, 983–995.

Solomon, M. J., and Kaldis, P. (1998). Regulation of CDKs by phosphorylation. Results Probl Cell Differ *22*, 79–109.

Sridhar, J., Akula, N., and Pattabiraman, N. (2006). Selectivity and potency of cyclin-dependent kinase inhibitors. Aaps J *8*, E204–221.

Takizawa, C. G., and Morgan, D. O. (2000). Control of mitosis by changes in the subcellular location of cyclin-B1-Cdk1 and Cdc25C. Curr Opin Cell Biol *12*, 658–665.

Tan, E. K., and Skipper, L. M. (2007). Pathogenic mutations in Parkinson disease. Hum Mutat *28*, 641–653.

Tang, D., Yeung, J., Lee, K. Y., Matsushita, M., Matsui, H., Tomizawa, K., Hatase, O., and Wang, J. H. (1995). An isoform of the neuronal cyclin-dependent kinase 5 (Cdk5) activator. J Biol Chem *270*, 26897–26903.

Tarricone, C., Dhavan, R., Peng, J., Areces, L. B., Tsai, L., and Musacchio, A. (2001). Structure and regulation of the cdk5-p25(nck5a) complex. Mol Cell *8*, 657–669.

ter Haar, E., Coll, J. T., Austen, D. A., Hsiao, H. M., Swenson, L., and Jain, J. (2001). Structure of GSK3beta reveals a primed phosphorylation mechanism. Nat Struct Biol *8*, 593–596.

Ubersax, J. A., and Ferrell, Jr., J. E. (2007). Mechanisms of specificity in protein phosphorylation. Nat Rev Mol Cell Biol *8*, 530–541.

Welburn, J. P., Tucker, J. A., Johnson, T., Lindert, L., Morgan, M., Willis, A., Noble, M. E., and Endicott, J. A. (2007). How tyrosine 15 phosphorylation inhibits the activity of cyclin-dependent kinase 2-cyclin A. J Biol Chem *282*, 3173–3181.

Wood-Kaczmar, A., Gandhi, S., and Wood, N. W. (2006). Understanding the molecular causes of Parkinson's disease. Trends Mol Med *12*, 521–528.

Xiong, Y., Zhang, H., and Beach, D. (1992). D type cyclins associate with multiple protein kinases and the DNA replication and repair factor PCNA. Cell *71*, 505–514.

Yaffe, M. B., and Smerdon, S. J. (2004). The use of in vitro peptide-library screens in the analysis of phosphoserine/threonine-binding domain structure and function. Annu Rev Biophys Biomol Struct *33*, 225–244.

Zhang, B. F., Peng, F. F., Zhang, J. Z., and Wu, D. C. (2003). Staurosporine induces apoptosis in NG108-15 cells. Acta Pharmacol Sin *24*, 663–669.

Zhang, F., Strand, A., Robbins, D., Cobb, M. H., and Goldsmith, E. J. (1994). Atomic structure of the MAP kinase ERK2 at 2.3 A resolution. Nature *367*, 704–711.

Zheng, Y. L., Kesavapany, S., Gravell, M., Hamilton, R. S., Schubert, M., Amin, N., Albers, W., Grant, P., and Pant, H. C. (2005). A Cdk5 inhibitory peptide reduces tau hyperphosphorylation and apoptosis in neurons. Embo J *24*, 209–220.

Zukerberg, L. R., Patrick, G. N., Nikolic, M., Humbert, S., Wu, C. L., Lanier, L. M., Gertler, F. B., Vidal, M., Van Etten, R. A., and Tsai, L. H. (2000). Cables links Cdk5 and c-Abl and facilitates Cdk5 tyrosine phosphorylation, kinase upregulation, and neurite outgrowth. Neuron *26*, 633–646.

Cdk5, a Journey from Brain to Pain: Lessons from Gene Targeting

Tej K. Pareek and Ashok B. Kulkarni

Abstract Cyclin-dependent kinase 5 (Cdk5) is a ubiquitously expressed proline-directed serine/threonine kinase. The monomeric form of Cdk5 is inactive and requires binding with its activator p35 and/or p39 to execute its kinase activity. Cdk5 was initially identified by purification from bovine brain extract and termed "Cdk5" because of its nucleotide sequence homology with human CDC2. Since the discovery of this kinase in 1992, it has been extensively studied by different laboratories to gain insights into its multiple roles in many important physiological systems. It is interesting to note that this kinase, which was initially considered as a postmitotic, neuron-specific kinase, has also been recognized as a key molecule in many cellular functions in non-neuronal tissues. We have now determined that this kinase was not only misnamed, as it neither requires binding to cyclin for activation, nor is it critically essential in the cell cycle, but was also not specific to neurons, as previously thought. Gene targeting is a very powerful tool for understanding the function of genes in human development and disease; in 2007, the Nobel Prize in Physiology or Medicine was awarded for introducing principles of specific gene modifications in mice by the use of embryonic stem cells. After the generation of the first gene-targeted mouse, our knowledge of specific gene functions has been immensely augmented by making use of gene-targeting techniques. In the last decade, we and others have employed functional genomics tools for a better understanding of Cdk5 biology. In this chapter, we will discuss the lessons learned from different strategies undertaken to understand Cdk5 biology.

Cyclin-dependent kinase 5 (Cdk5) is an ubiquitously expressed proline-directed serine/threonine kinase. The monomeric form of Cdk5 is inactive and requires binding with its activator p35 and/or p39 to execute its kinase activity. Cdk5 was initially identified by purification from bovine brain extract and termed "Cdk5" because of its nucleotide sequence homology with human

A.B. Kulkarni
Functional Genomics Section, Laboratory of Cell and Developmental Biology,
National Institute of Dental and Craniofacial Research, National Institutes of Health,
Bethesda, MD, 20892, USA
e-mail: ak40m@nih.gov

N.Y. Ip, L.-H. Tsai (eds.), *Cyclin Dependent Kinase 5 (Cdk5)*,
DOI: 10.1007/978-0-387-78887-6_15, © Springer Science+Business Media, LLC 2008

CDC2 [1,2]. Since the discovery of this kinase in 1992, it has been extensively studied by different laboratories to gain insights into its multiple roles and its involvement in molecular mechanisms in many important physiological systems. It is interesting to note this kinase, which was initially considered as a postmitotic, neuron-specific kinase [3], has also been recognized as a key molecule in many cellular functions in non-neuronal tissues [4]. We have now determined that this kinase was not only misnamed, as it neither requires binding to cyclin for activation, nor is it critically essential in the cell cycle, but also cast aside the myth of its neuronal specificity.

Due to the colossal power of gene targeting in understanding the gene function and its importance in studying human health and disease, this year's (2007) Nobel prize in Physiology or Medicine has been jointly awarded to Drs. Mario R. Capecchi, Martin J. Evans, and Oliver Smithies for introducing principles of specific gene modifications in mice by the use of embryonic stem cells. After the generation of first gene-targeted mouse [5,6] in their labs, our knowledge of understanding a specific gene function has immensely augmented by making use of gene-targeting techniques. In the last decade, we and others have employed functional genomics tools for a better understanding of Cdk5 biology. In this chapter we will discuss the lessons learned from different strategies undertaken to understand Cdk5 biology.

Cdk5 Knockout (Cdk5–/–) Mice

To delineate the molecular function of Cdk5, we developed a conventional gene knockout mouse of Cdk5, which exhibits unique lesions in the central nervous system associated with prenatal mortality [7]. More than 60% of Cdk5–/– mice die in utero, and the newborns become weak and die within 12 h after birth. The Cdk5–/– mouse brain lacks cortical laminar structure and cerebellar foliation. Cdk5–/– mice display migration defects in facial branchiomotor and inferior olive neurons of brain stem [8]. Currently, the neuronal pathology in Cdk5–/– mice is considered to be mainly due to perturbations of neuronal migration. In addition, the large neurons in the brain stem and spinal cord show chromatolytic changes with accumulation of neurofilament immunoreactivity, suggestive of neuronal response to axonal injury. These findings indicate that Cdk5 is an important molecule for brain development and neuronal migration and also suggest that Cdk5 may play critical roles in neuronal cytoskeleton structure and organization. Initially it was proposed that degenerative changes in large neurons of the brain stem, including motor neurons in the lower cranial nerve nuclei and spinal cord, may be the cause of the early lethality observed in Cdk5–/– mice. This mouse model became an essential tool to understand Cdk5 function. Histological analysis of Cdk5–/– mouse by Carl Herrup's group revealed that the cerebral cortex of these mice is abnormal in its structure, with a neurogenic gradient that is inverted from the normal inside-out development. Despite this, the early pre-plate layer separates correctly, and neurons with a normal pyramidal morphology can be found between the true marginal zone and sub-plate [9]. Among the defects in these animals, a disruption of the normal pattern of cell migrations in the cerebellum was particularly apparent, including a pronounced abnormality in the cerebellar Purkinje cells. These results prompted studies to delineate the molecular mechanism of Cdk5-mediated cerebral cortical development. For a long time, this ambiguity was unclear until the yeast two-hybrid screen isolated β-catenin as a p35-binding protein [10,11]. Studies from these authors suggest that β-catenin is a substrate of Cdk5 and that active Cdk5 dissociates β-catenin from N-cadherin, which is accompanied by a loss of cell adhesion. These results indicate that Cdk5 might control neuronal migration by regulation of the cadherin–catenin complex. Further studies from Gleeson's group indicate that Cdk5/p35-mediated doublecortin phosphorylation is important for cell migration [12].

Cdk5–/– mice also become an important tool in identifying its role in development of motor axons and neuromuscular synapses [13]. Studies from Nancy IP's group suggest that Cdk5–/– mice not only display morphological abnormalities at the pre- and post-synaptic neuromuscular junction but also exhibit profuse intramuscular nerve projections and anomalous branching patterns with wider central band of acetylcholine receptor (AChR) clusters. Furthermore, using Cdk5–/– mice it has been described that ephrin-A1 promotes EphA4-dependent spine retraction through the activation of Cdk5 and ephexin1, which in turn modulates actin-cytoskeletal dynamics. Recently this group reported that Cdk5 mediated phosphorylation of TrkB at 478 at the intracellular juxtamembrane region of TrkB, which is required for brain-derived neurotrophic factor, triggered dendritic growth in primary hippocampal neurons. These results reveal an unanticipated role of Cdk5 in TrkB-mediated regulation of dendritic growth through modulation of BDNF-induced Cdc42 activation [14].

Cdk5 Chimeric Mice (Cdk5–/–↔Cdk5+/+)

Complete analysis of the brain is hindered in Cdk5–/– mice because most cerebellar morphogenesis occurs after birth. To overcome this disadvantage, we generated chimeric mice by injecting Cdk5–/– embryonic stem cells into Cdk5+/+ host blastocysts [15]. Analysis of the cerebellum from the resulting Cdk5 chimeric mice shows that the abnormal location of the mutant Purkinje cells is a cell-autonomous defect. In addition, significant numbers of granule cells remain in the molecular layer, suggesting a failure to complete migration from the external to the internal granule cell layer. In contrast to the Purkinje and granule cell populations, all three deep cerebellar nuclear cell groupings formed correctly and are composed of cells of both mutant and wild-type genotypes. Together, the data further supported the hypothesis that Cdk5 activity is required for specific aspects of neuronal migration that are differentially required by various neuronal cell types and even by a single neuronal cell type at different developmental stages. Since the Cdk5–/– mouse phenotype mimics the reeler and scrambler/yotari mutant phenotype [16–24] and Cdk5 gene is close to the reeler gene on the chromosome [25–27], it was obvious to ask if reeler and scrambler gene alteration could account for the cerebellar phenotype of Cdk5–/– mice. To address this, we analyzed the expression levels and patterns of reelin and disabled-1 mRNA in wild-type and Cdk5–/– mouse brains and found no difference, suggesting that the migration defect of Cdk5–/– neurons is not mediated by changes in the expression of either of these two phenotypically related genes.

Cdk5 Overexpressing Transgenic (TgCdk5) Mice

After learning that Cdk5 has an indispensable role in proper brain development, we next asked what the consequences of Cdk5 overexpression would be. We generated transgenic Cdk5 overexpressing (TgCdk5) mice in which the p35 promoter drives the expression of Cdk5 [28]. These mice showed increased levels of Cdk5 in neurons but not in astrocytes. These mice are grossly normal without any noticeable neuronal phenotype. Unexpectedly these mice showed a marked reduction in Cdk5 kinase activity in the brain compared to that of their wild-type littermate controls. These results were surprising, as one would expect a higher level of Cdk5 activity in these mice, not a decreased level. The first question we asked was whether the Cdk5 transgene is functional *in vivo*. The functional activity of Cdk5 transgene was confirmed by performing *in vitro* kinase assays by adding active p35 or p25, which resulted in enhanced Cdk5 kinase activity. The other possibility is that the higher level of transgenic Cdk5 protein may interfere with the transcription of p35 and reduce its production, which may contribute to decreased Cdk5 activity. However, p35 protein levels were unaltered in TgCdk5 mice. One another possible explanation for this reduced Cdk5 activity reflected in kinase assay is that further overexpression of Cdk5 competes for the already rate-limiting p35, resulting in smaller fraction of Cdk5 molecules bound with a p35 cofactor. If the immunoprecipitation in the *in vitro* kinase assay is not quantitative (not a complete pull-down), it may result in an apparent and paradoxical reduction in immunoprecipitated kinase activity. In fact, the dramatic increase in kinase activity after re-introduction of p35 shown in the same study supports this hypothesis. However, the reduction in Cdk5 kinase activity in these mice did not contribute to any noticeable phenotype.

Cdk5 Overexpression Under the Control of p35 Promoter in Cdk5-null (TgKO) Mice

Since Cdk5 is expressed in both neurons and astrocytes, it was unclear whether the phenotype of Cdk5–/– mice was primarily attributable to defects in neurons or astrocytes. To determine whether the expression of Cdk5 in p35-expressing areas was sufficient for reversing the lethality of Cdk5-null mice, we crossed TgCdk5 mice with Cdk5 heterozygous mice to generate mice lacking endogenous Cdk5 but with the Cdk5 transgene under the control of the p35 promoter (TgKO mice) [28]. These mice were not only viable and fertile but were also free of the defects in the nervous system. These results clearly demonstrate that Cdk5 activity is necessary for normal development and survival of neurons. These gene manipulation studies have answered the cause of lethality to certain extent. We may soon learn the underlying pathology responsible for this lethality with the growing knowledge of Cdk5 biology.

p35 Knockout (p35–/–) Mice

Cdk5/p35 binding is required for Cdk5 kinase activity [29,30]. The *in vitro* functional studies of p35 and Cdk5 suggest that the Cdk5/p35 kinase plays a role in dynamic, developmental neuronal processes. In order to analyze the role of p35 *in vivo*, p35–/– (p35–/–) mice were generated [31]. These mice lack the p35 protein and have undetectable Cdk5 kinase activity. In order to elucidate the relationship between Cdk5 and p35 expression and their functional consequences alone or in partnership with other genes, in brain development we also generated p35–/– mice [32]. These mice exhibit substantial residual Cdk5 kinase activity, corresponding to 10 and 20% in the cerebral cortex and the cerebellum, respectively, compared with their wild-type controls. One another independent development of p35 knockout mice depict 38% depletion in Cdk5 activity in adult brain [33]. Unlike Cdk5–/– mice, these mice are not embryonically lethal but prone to develop fatal seizures. These mice display defects mostly confined to the forebrain, with cortical lamination defects similar to those observed in Cdk5–/– mice. Unlike Cdk5–/– mice, p35–/– mice show typical foliation and tripartite layering in the cerebellum and a completely normal brain stem and spinal cord. The results obtained from p35–/– mice elucidate a critical role of p35 in the proper organization of cortical structures. Histological examination of these mice revealed major defects in the developing neocortex, where the normal lamination pattern of the cortical neurons is disrupted. In addition, certain axonal trajectories and dendritic structures are altered in p35 mutant mice. It was suggested that abnormal patterning of the cortex is due to a defect in neuronal migration. The phenotype of p35 mutant mice provides the first evidence for an intracellular kinase activity regulating the migration and hence the laminar fate of neurons during cortical development. Studies from Vincent's group on p35–/– mice illustrate that p35 regulates the sub-cellular distribution of Cdk5 and cytoskeleton proteins in neurons and play a hierarchical role in regulating the phosphorylation and function of cytoskeleton proteins [33].

p35 Knockout and Cdk5 Heterozygous Mice (p35–/–; Cdk5+/–)

In order to understand if residual Cdk5 kinase activity contributes to mild phenotype of p35–/– mice, we generated a p35–/–; Cdk5+/– mouse line by crossing of p35–/– mice with Cdk5+/– mice, and compared its histological phenotype with that of p35–/– mice [32]. The p35–/–;Cdk5+/– mice showed additional abnormalities in the cerebellum. In the cerebellum of p35–/– mice, the alignment of Purkinje cells in the Purkinje cell layer was disturbed in some areas at which Purkinje cells form 2 to 3 cell-width layers. Additionally, significant numbers of granule cells are found in the molecular cell layer. In the cerebellum of p35–/–;Cdk5+/– mice, extensive migration defects of Purkinje and granule cells are present. Ectopic Purkinje cells are found frequently in the granule cell layer, and greater numbers of granule cells are observed in the molecular layer in the cerebellum of p35–/–; Cdk5+/– mice.

p35 Overexpressing Transgenic (Tgp35) Mice

Since overexpression of Cdk5 does not show any noticeable phenotype, we wondered if p35 overexpression would increase Cdk5 activity and if it would have any consequences in brain development. In order to address this issue, we generated p35 overexpressing mice under p35 promoter [34]. Unlike TgCdk5 mice, Tgp35 mice reveal an approximately 2-fold increase in Cdk5 activity. This phenomenon suggests that p35 is a limiting factor for *in vivo* Cdk5 activity. These mice are completely normal and fertile and do not exhibit any noticeable brain abnormalities. Interestingly, comprehensive behavior analyses showed hyperactivity in these mice. Further examination of TgCdk5 or Tgp35 revealed that p35 but not Cdk5 overexpression leads to attenuation of cocaine-mediated dopamine signaling. These results support the idea that Cdk5 activity is involved in altered gene expression after chronic exposure to cocaine and hence impacts the long-lasting changes in neuronal function underlying cocaine addiction [34].

p35 Overexpression Under a p35 Promoter in p35-null (Tgp35;p35–/–) Mice

The next step was to determine whether re-introduction of the p35 gene in p35-null mice can rescue its phenotype. In order to address this issue, we generated double transgenic mice (Tgp35; p35–/–) in which p35 expression was driven only from the transgene [34]. p35 expression in Tgp35;p35–/– mice was observed predominantly in the brain, where the spatial expression pattern was similar to that of wild-type mice. A lack of p35 has been shown to result in abnormal layering structure in the cerebral cortex and hippocampus of mice; however, the Tgp35;p35–/– mice show a complete rescue of the p35–/– brain phenotype.

p25 Expressing Transgenic (p25-tg) Mice

An N-terminally truncated form of p35, termed p25, is generated through cleavage by the Ca^{2+}-dependent protease calpain. Cdk5/p25 has been shown to have a higher half-life and more stability than Cdk5/p35 and is produced and accumulates in the brain of patients with neurodegenerative disorders such as Alzheimer's disease (AD). Additionally, Cdk5/p25 displays a different sub-cellular localization than Cdk5/p35. It is still debatable whether or not p25 is essential for normal biological function or is a poor activator of Cdk5. To understand the potential role of Cdk5/p25-mediated Cdk5 activity, human p25 was overexpressed in the mouse brain under the regulation of the neuronal-specific enolase (NSE) promoter [35]. Except in cerebellum, these p25-tg mice show at least 2-fold increase in Cdk5 activity, resulting in hyperphosphorylated tau and neurofilament with disturbances in neuronal cytoskeletal organization. These changes are localized predominantly in the amygdala, thalamus/hypothalamus, and cortex. Additionally, these mice have increased sponta-neous locomotor activity like Tgp35 mice and differ from controls in their performance on the elevated plus-maze test.

Inducible Transgenic Mice Overexpressing p25 in the Postnatal Forebrain

It has been shown that *in vitro* p25 causes apoptosis and tau hyperphosphoryla-tion, and it is also implicated in neurodegenerative disorders such as AD. Since the p25-tg mouse does not depict any of these phenotypes, a new strategy was adopted to generate p25-overexpressing mice. Tsai's group generated inducible transgenic mouse lines overexpressing p25 (CK-p25) in the postnatal forebrain under the control of the CamKII-α promoter [36]. For this purpose, tetracy-cline-controlled transactivator (tTA) system was used to generate bitransgenic mice that have inducible overexpression of human p25 that drives high trans-gene expression in the forebrain. Another unique feature of these mice is the capability to turn transgene expression on and off. In the presence of the tetracycline derivative doxycycline, expression of the p25 transgene can be inhibited in these mice. Induction of p25 preferentially directed Cdk5 to patho-logical substrates.

These animals exhibit neuronal loss in the cortex and hippocampus, accom-panied by forebrain atrophy, astrogliosis, and caspase-3 activation. Endogen-ous tau is hyperphosphorylated at many epitopes, aggregated tau accumulated, and neurofibrillary pathology developed progressively in these animals. This mouse model provides the first *in vivo* evidence suggesting that deregulation of Cdk5 by p25 plays a causative role in neurodegeneration and the development of neurofibrillary pathology. Moreover, prolonged p25 expression dramatically impaired hippocampal LTP and memory with accompanying synaptic and neuronal loss, whereas transient p25 expression profoundly enhanced hippocampal LTP and facilitated learning and memory without causing neurodegeneration [37]. Thus, a prolonged expression of p25 may turn a physiological action of Cdk5 into a pathological one. In contrast, recent studies from Vincent's group reported that lack of p35/p25 does not slow down the onset or progression or improve the neuropathology of Niemann–Pick type C disease (NPC) [38]. Although these studies concentrate on NPC and hypothesize to be the same mechanism of neurodegeneration in AD, but still these results need to be analyzed more carefully. In another recent report, it has been shown that CK-p25 mice severely impaired extinction in contextual fear-conditioning procedure, as indicated by the persistence of aversive freezing behavior of these mice [39].

These results provide the requirement for generation of such inhibitors which can inhibit Cdk5/p25 activity without altering Cdk5/p35 or any other kinase activity. The first Cdk5/p25-specific small-inhibitory peptide (CIP) were developed in Dr. Pant's lab at NIH [40,41]. The *in vivo* viral infection of this peptide in rat brain cortical neurons has shown to selectively inhibit Cdk5/p25 activity without altering the activity of Cdk5/p35 or other Cdks such as Cdc2, Cdk2, Cdk4, and Cdk6 [42]. Generation of this peptide provides the hope of development of CIP transgenic mice. It would be interesting to see if double transgenic of p25-tg and CIP-tg would be able to rescue the neurodegenerative phenotype of p25-tg mice.

Low Expression of the p25 Transgene Under the NSE Promoter in p35-null (p35–/–;p25-tg) Mice

Results obtained from Tgp35 and p25-tg mice suggest that overexpression of either p35 or p25 leads to hyperactivation of Cdk5. Tgp35 mice do not portray any noticeable neuropathology; however, p25-tg mice show disturbances in neuronal cytoskeleton architecture and develop neurodegeneration. These observations indicate that when Cdk5 is activated with p25, it gains some toxic function compared to Cdk5/p35. However, it is still unclear if Cdk5/p25 activity is required for normal neuronal functions parallel to Cdk5/p35. Although it has been considered that p25 is a byproduct of p35, it is not clear that p25 is produced because of stress or toxicity or because of loss of function of p35. To clarify these issues, p25 transgenic mice in a p35-null background (p35–/–;p25-tg) were generated [43]. It was interesting to notice that, like Tgp35;p35–/–, low levels of p25 in p35–/–;tg-p25 mice, during development induce a partial rescue of the p35–/– phenotype in several brain regions, including rescue of cell positioning of a subset of neurons in the neocortex. In accordance with the partial rescue of brain anatomy, phosphorylation of the Cdk5 substrate mouse disabled-1 (mDAB1) is partially restored but other Cdk5/p35 substrates such as NUDEL and PSD-95 is not phosphorylated.

p39 Knockout (p39−/−) Mice

Although p35 was identified as the first regulatory activator of Cdk5, p39 was later reported as a second activator of Cdk5, and shares 57% amino acid identity with p35 [44]. Like p35, p39 interacts with and activates Cdk5, although p39 expression is mainly postnatal [45]. In addition, the temporal and spatial expression of p39 indicates a possible role of the Cdk5/p39 kinase at the synapse. To investigate the *in vivo* function of p39, a targeted deletion in the p39 locus was created in mice [46]. Surprisingly, the p39−/− mice are viable and fertile and do not display any obvious phenotypes. These mice do not exhibit detectable abnormalities in neuronal positioning in the nervous system nor produce any significant difference in the levels of p35 and Cdk5. Similarly, p35- and Cdk5-associated histone H1 kinase activity is unaltered in these mice. These results indicate that although p39 binding can activate Cdk5, it is not an absolute requirement for Cdk5 function. Therefore, the lack of phenotypic change from the loss of p39 may be due to an unequal distribution of p35 and p39 or p35 may be able to take over the function of p39.

p35 and p39 Double Knockout (p35−/−;p39−/−) Mice

Prominent differences in the phenotypic severity of Cdk5, p35, and p39−/− mice raised several basic questions about Cdk5 regulation. Is p39 a dispensable activator of Cdk5? Does p35 alone take over all the functions of p39, and if so why do p35−/− mice still survive even after similar abnormalities in cortical development as Cdk5−/− mice, and why do they die at a later age? In order to answer these questions, compound-mutant mice were generated via a two- or three-stage cross of p35−/− and p39−/− mice [47]. Surprisingly, p35−/− and p39−/− mice showed a phenotype exactly similar to Cdk5−/− mice. These findings suggest that both p35 and p39 are essential for proper Cdk5 functioning. The developmental profiling of Cdk5, p35, and p39 in various brain regions suggests that p39 expression is higher during embryonic development in the cerebellum, brain stem, and spinal cord, where the pathological defects are mostly absent in p35−/− mice. However, the level of p39 expression is very low in the embryonic cerebral cortex, where p35−/− mice have lamination defects. These results suggest that how Cdk5/p39 may compensate for a lack of Cdk5/p35 activity in p35−/− mice, indicating the overlapping roles of p35 and p39. Furthermore, the fact that p39−/− mice do not show any noticeable abnormalities is also compatible with the observation that the p35 expression level is high in all regions of the brain during embryonic development, suggesting a compensatory role of p35 in the absence of p39 [48].

Cdk5 Floxed (Cdk5f/f) Mice

Though the classical gene-targeting strategy can successfully introduce specific mutations into the mouse genome, there are several disadvantages in employing this approach. Embryonic lethality in the absence of one gene and the limitation of addressing tissue-specific roles of a targeted gene are a few of them. Site-specific recombination systems, such as Cre/LoxP from bacteriophage P1 and the yeast Flp/Frt systems have enabled scientists to achieve tissue-specific gene targeting. These systems have been successful in expressing a transgene in transgenic mice or ES cells; its translation is blocked by the presence of the artificial stop codon, which is designed to be deleted by Cre/Flp recombinase [49,50]. In more recent years, scientists have exploited this technology extensively in generating conditional deletions (tissue- and age specific) of genes of interest. Embryonic lethality of Cdk5−/− mice hampered the process of delineating its role in adult brain. To circumvent the embryonic lethality associated with Cdk5−/− mice, we took advantage of a Cre/LoxP strategy to induce deletion of the Cdk5 gene in a temporally and spatially regulated manner. We generated mice that are homozygous for the Cdk5 allele flanked by LoxP motifs (Cdk5f/f) [51]. These mice are healthy and fertile without any obvious phenotype. These mice provide an extraordinary tool to analyze the role of Cdk5 in the adult brain and other non-neuronal tissues. We have generated conditional deletions of Cdk5 in different regions of the brain and peripheral sensory neurons by crossing Cdk5-loxP flanked mice (Cdk5f/f) mice with Cre transgenic mice.

The rapid development and characterization of the mouse genome sequence has opened several new ways to understand the complexity of biological system, but at the same time it has generated new challenges for functional genomics studies. The first draft of the sequence of the mouse genome (Mouse genome Sequencing Consortium, 2002) has facilitated the hopes of manipulating any gene of interest. Increasing knowledge of the mouse genome and the fast-paced development of gene mapping have taught us the lessons in gene clustering and overlapping of these genes on chromosomes. Newly developed tools at *NCBI Map Viewer* and several other resources make it possible to analyze the exact location of a gene and identify its neighboring gene maps. We now know that Cdk5 is located on chromosome 5 and is closely associated with two neighboring genes Accn3 (408 bp from the 3′ end of Cdk5) and Slc4a2 (1.78 bp from the 5′ end of Cdk5). The method we used to generate the Cdk5-LoxP mice enabled us to delete the entire 11.28 Kb fragment including the 5.28 Kb fragment of Cdk5 by Cre recombination. This strategy generates a risk of deletion of non-coding sequences in neighboring genes. To overcome this problem and develop a specific deletion of Cdk5, we developed a new strategy for Cdk5 conditional deletion using the Cdk5f/f mouse. We first cross Cdk5f/f mice with Cdk5+/− mice to generate Cdk5f/− mice. Since these mice lack one Cdk5 allele and one allele is flanked with the Lox-P sequence, Cre recombination in this mouse assures the specific deletion of Cdk5 without significantly altering neighboring gene function.

We first addressed the question of whether or not perinatal deletion of Cdk5 will allow viable offspring. In order to test this hypothesis, we crossed Cdk5f/f mice with the murine neurofilament-heavy chain promoter-driven Cre mice (NFH-Cre) [51]. This crossing resulted in significant downregulation of Cdk5 in different regions of the brain beginning around embryonic day 16.5. Unlike Cdk5−/− mice, these Cdk5 conditional knockout mice are viable and fertile. Abrogated Cdk5 expression in these mice was associated with neuronal migration defects in the cerebral cortex, where the defects were restricted to the later-generated cortical neurons and in the olfactory bulb and cerebellar cortex, where neuronal migration continues through the perinatal period. This mouse model offers further opportunities to investigate the molecular roles of Cdk5 during postnatal developmental stages. Recently, studies from Ohshima's group suggest that cortex-restricted deletion of Cdk5 (by crossing Cdk5f/f with Emx1-Cre mice) results in abnormal dendritic development and defective axonal trajectories of pyramidal neurons in the postnatal cerebral cortex [52]. Results obtained from these studies suggest that transition of migrating neurons into a bipolar shape is essential to form the pyramidal morphology (in which neurons extend their single apical dendrite toward the pia and extend single axons toward the bottom of the cortex) and that this radial axis is brought about by the active radial migration of neurons in a Cdk5-dependent manner.

Recently James Bibb's group developed another Cdk5f/f mice where exons encoding vital Cdk5 catalytic-domain components were flanked with loxP elements [53]. In order to generate inducible conditional Cdk5 knockdown, these mice were crossed with the prion protein promoter–driven Cre-ERT mice. These mice showed improved performance in spatial learning tasks and enhanced hippocampal long-term potentiation and NMDA receptor (NMDAR)-mediated excitatory postsynaptic currents. This mouse model revealed a previously unknown role for Cdk5 in regulating glutamate receptor degradation.

Cdk5 and Pain

Pain is a combination of sensory (discriminative) and affective (emotional) components. The sensory component of pain is defined as nociception and is required for survival and maintenance of the integrity of the organism. Nociception results from the activation of molecular and cellular mechanisms in damaged tissue, sensory neurons, and the spinal cord. Many cellular pathways have been implicated in nociceptive signaling, but their precise molecular mechanisms have not been clearly defined. The results obtained from behavioral phenotype analysis of the p35 transgenic and knockout mice prompted us to analyze the role of Cdk5 in nociceptive signaling. An in-depth analysis of the p35–/– mouse showed abnormalities in pain signaling. When p35–/– mice were exposed to hot water bath, their reflex action was delayed compared to that of their counter-littermate controls. Inversely, Tgp35 mice showed hyperalgesia [54]. These mice also showed altered morphine tolerance during

Table 1 Cdk5 transgenic and knockout mouse models

Mouse Model	Genotype	Year	Reference	Foremost phenotype
Cdk5 knockout mice	Cdk5–/–	1996	[7]	Embryonic lethality, improper neuronal migration, abnormal neuromuscular junction, unresponsive to noxious cutaneous pinching
Cdk5 chimeric mice	Cdk5–/–↔Cdk5 +/+	1999	[15]	Neuronal migration defects
Cdk5 overexpresser mice	Tg-Cdk5	2001	[28]	No obvious phenotype
Cdk5 overexpression in Cdk5 null mice	TgKO	2001	[28]	Rescue of Cdk5–/– phenotype
p35 knockout mice	p35–/–	1997, 2001, 2003	[31, 32, 33]	Reduced Cdk5 activity, abnormal neuronal patterning, improper brain development, prone to develop fatal seizures, hypoalgesia
p35 knockout and Cdk5 heterozygous mice	p35–/–;Cdk5+/–	2002	[32]	Extensive neuronal migration defects
p35 overexpresser mice	Tg-p35	2005	[34]	Increased Cdk5 activity, hyperactive, attenuation of cocaine-mediated dopamine signaling, hyperalgesia

Table 1 (continued)

Mouse Model	Genotype	Year	Reference	Foremost phenotype
p35 overexpression in p35 null background	Tg-p35;p35–/–	2005	[34]	Rescue of p35–/– phenotype
p25 expresser mice	p25-tg	2000	[35]	Increased Cdk5 activity, increased locomotor activity, distressed neuronal cytoarchitecture
p25 inducible overexpression	CK-p25	2003	[36]	Neurodegeneration, impaired hippocampal LTP and memory
p25 overexpression under p35 null background	p35–/–;p25-tg	2003	[43]	Partial rescue of p35–/– phenotype
p39 knockout mice	p39–/–	2000	[44]	No obvious phenotype
p35 and p39 double knockout mice	p35–/–;p39–/–	2001	[47]	Similar to Cdk5–/– mice
Lox-P flanked Cdk5 mice	Cdk5 f/f	2004, 2007	[51, 53]	No obvious phenotype
Cdk5 conditional deletion in adult brain using NFH-Cre	Cdk5 f/-;NF-H Cre	2004	[51]	Downregulated Cdk5 activity in brain, defective neuronal migration
Cdk5 conditional deletion in adult brain using Emx-Cre	Cdk5 f/-;Emx-Cre	2007	[52]	Abnormal dendritic development and defective axonal trajectories of pyramidal neurons
Inducible conditional Cdk5 knockdown	Cdk5f/f;Cre-ERT	2007	[53]	Improved performance in spatial learning tasks, enhanced hippocampal long-term potentiation
Cdk5 conditional deletion in peripheral nervous system using SNS-Cre	Cdk5 f/-;SNS-Cre	2007	[56]	Hypoalgesia

inflammatory pain [55]. Additionally, Cdk5–/– pups are unresponsive to noxious cutaneous pinching [13], but this observation may be influenced by the likelihood that Cdk5 and p35–/– mice suffer from secondary neuronal dysfunctions. To address these concerns, we generated primary nociceptor-specific Cdk5 conditional knockout (Cdk5-CoKO) mice by crossing Cdk5f/f mice with SNS-Cre mice [56]. These Cdk5-CoKO mice lack Cdk5 expression predominantly in C-fiber neurons. We observed abrogation of TRPV1 phosphorylation at Thr407 in the DRG of these mice. Moreover, these mice showed significant hypoalgesia in the basal thermal response. This phenotype of Cdk5-CoKO mice is similar to that observed in p35 and TRPV1–/– mice, which suggests that Cdk5-mediated TRPV1 phosphorylation is important in nociceptive signaling. Thus, Cdk5-mediated TRPV1 phosphorylation is important for heat- and capsaicin-induced activation and provides insight into Cdk5-mediated pain signaling.

Since the discovery of Cdk5 in 1992, we have come a long way in understanding its biology, cellular functions, and molecular roles in multiple physiological systems, and the functional genomics-based analysis proved to be a powerful tool in this process. A significant progress has been made in the development of numerous transgenic and knockout mouse models to understand Cdk5 function (Table 1). However, there is still a long way to go to delineate all the vital functions of this kinase. The accumulating evidence indicates that Cdk5 is a key regulator of important functions in neuronal as well as non-neuronal cells, and it is now considered as a therapeutic target to treat various diseases. However, several biological questions need to be carefully addressed before one can begin to develop therapeutic strategies based on Cdk5 functions. We believe numerous mouse models with altered expression of Cdk5 and/or its activators will accelerate this process, resulting in speedy development of such therapeutic strategies.

Acknowledgments We would like to thank Drs. Roscoe Brady, Harish Pant, Elias Utreras, Akira Futatsugi, Veeranna, and Vinod Yaragudri for a critical reading of this chapter and significant contribution to our Cdk5 studies described in this chapter, and Harry Grant for editorial assistance. This work was supported by funds from the Divisions of Intramural Research of the National Institute of Dental and Craniofacial Research and the National Institute of Neurological Disorders and Stroke.

References

1. Lew, J., Beaudette, K., Litwin, C. M., & Wang, J. H. (1992) *J Biol Chem* **267,** 13383–13390.
2. Meyerson, M., Enders, G. H., Wu, C. L., Su, L. K., Gorka, C., Nelson, C., Harlow, E., & Tsai, L. H. (1992) *Embo J* **11,** 2909–2917.
3. Dhavan, R. & Tsai, L. H. (2001) *Nat Rev Mol Cell Biol* **2,** 749–759.
4. Rosales, J. L. & Lee, K. Y. (2006) *Bioessays* **28,** 1023–1034.
5. Doetschman, T., Gregg, R. G., Maeda, N., Hooper, M. L., Melton, D. W., Thompson, S., & Smithies, O. (1987) *Nature* **330,** 576–578.
6. Thomas, K. R. & Capecchi, M. R. (1987) *Cell* **51,** 503–512.
7. Ohshima, T., Ward, J. M., Huh, C. G., Longenecker, G., Veeranna, Pant, H. C., Brady, R. O., Martin, L. J., & Kulkarni, A. B. (1996) *Proceedings of the National Academy of Sciences of the United States of America* **93,** 11173–11178.

8. Ohshima, T., Ogawa, M., Takeuchi, K., Takahashi, S., Kulkarni, A. B., & Mikoshiba, K. (2002) *J Neurosci* **22,** 4036–4044.

9. Gilmore, E. C., Ohshima, T., Goffinet, A. M., Kulkarni, A. B., & Herrup, K. (1998) *J Neurosci* **18,** 6370–6377.

10. Kesavapany, S., Lau, K. F., McLoughlin, D. M., Brownlees, J., Ackerley, S., Leigh, P. N., Shaw, C. E., & Miller, C. C. (2001) *Eur J Neurosci* **13,** 241–247.

11. Kwon, Y. T., Gupta, A., Zhou, Y., Nikolic, M., & Tsai, L. H. (2000) *Curr Biol* **10,** 363–372.

12. Tanaka, T., Serneo, F. F., Tseng, H. C., Kulkarni, A. B., Tsai, L. H., & Gleeson, J. G. (2004) *Neuron* **41,** 215–227.

13. Fu, A. K., Ip, F. C., Fu, W. Y., Cheung, J., Wang, J. H., Yung, W. H., & Ip, N. Y. (2005) *Proceedings of the National Academy of Sciences of the United States of America* **102,** 15224–15229.

14. Cheung, Z. H., Chin, W. H., Chen, Y., Ng, Y. P., & Ip, N. Y. (2007) *PLoS Biol* **5,** e63.

15. Ohshima, T., Gilmore, E. C., Longenecker, G., Jacobowitz, D. M., Brady, R. O., Herrup, K., & Kulkarni, A. B. (1999) *J Neurosci* **19,** 6017–6026.

16. Gilmore, E. C. & Herrup, K. (1997) *Curr Biol* **7,** R231–R234.

17. Goldowitz, D., Cushing, R. C., Laywell, E., D'Arcangelo, G., Sheldon, M., Sweet, H. O., Davisson, M., Steindler, D., & Curran, T. (1997) *J Neurosci* **17,** 8767–8777.

18. Gonzalez, J. L., Russo, C. J., Goldowitz, D., Sweet, H. O., Davisson, M. T., & Walsh, C. A. (1997) *J Neurosci* **17,** 9204–9211.

19. Howell, B. W., Hawkes, R., Soriano, P., & Cooper, J. A. (1997) *Nature* **389,** 733–737.

20. Ogawa, M., Miyata, T., Nakajima, K., Yagyu, K., Seike, M., Ikenaka, K., Yamamoto, H., & Mikoshiba, K. (1995) *Neuron* **14,** 899–912.

21. Sheldon, M., Rice, D. S., D'Arcangelo, G., Yoneshima, H., Nakajima, K., Mikoshiba, K., Howell, B. W., Cooper, J. A., Goldowitz, D., & Curran, T. (1997) *Nature* **389,** 730–733.

22. Sweet, H. O., Bronson, R. T., Johnson, K. R., Cook, S. A., & Davisson, M. T. (1996) *Mamm Genome* **7,** 798–802.

23. Ware, M. L., Fox, J. W., Gonzalez, J. L., Davis, N. M., Lambert de Rouvroit, C., Russo, C. J., Chua, S. C., Jr., Goffinet, A. M., & Walsh, C. A. (1997) *Neuron* **19,** 239–249.

24. Yoneshima, H., Nagata, E., Matsumoto, M., Yamada, M., Nakajima, K., Miyata, T., Ogawa, M., & Mikoshiba, K. (1997) *Neurosci Res* **29,** 217–223.

25. Dernoncourt, C., Ruelle, D., & Goffinet, A. M. (1991) *Genomics* **11,** 1167–1169.

26. Goffinet, A. M. & Dernoncourt, C. (1991) *Mamm Genome* **1,** 100–103.

27. Ohshima, T., Nagle, J. W., Pant, H. C., Joshi, J. B., Kozak, C. A., Brady, R. O., & Kulkarni, A. B. (1995) *Genomics* **28,** 585–588.

28. Tanaka, T., Veeranna, Ohshima, T., Rajan, P., Amin, N. D., Cho, A., Sreenath, T., Pant, H. C., Brady, R. O., & Kulkarni, A. B. (2001) *J Neurosci* **21,** 550–558.

29. Lew, J., Huang, Q. Q., Qi, Z., Winkfein, R. J., Aebersold, R., Hunt, T., & Wang, J. H. (1994) *Nature* **371,** 423–426.

30. Tsai, L. H., Delalle, I., Caviness, V. S., Jr., Chae, T., & Harlow, E. (1994) *Nature* **371,** 419–423.

31. Chae, T., Kwon, Y. T., Bronson, R., Dikkes, P., Li, E., & Tsai, L. H. (1997) *Neuron* **18,** 29–42.

32. Ohshima, T., Ogawa, M., Veeranna, Hirasawa, M., Longenecker, G., Ishiguro, K., Pant, H. C., Brady, R. O., Kulkarni, A. B., & Mikoshiba, K. (2001) *Proceedings of the National Academy of Sciences of the United States of America* **98,** 2764–2769.

33. Hallows, J. L., Chen, K., DePinho, R. A., & Vincent, I. (2003) *J Neurosci* **23,** 10633–10644.

34. Takahashi, S., Ohshima, T., Cho, A., Sreenath, T., Iadarola, M. J., Pant, H. C., Kim, Y., Nairn, A. C., Brady, R. O., Greengard, P.*, et al.* (2005) *Proceedings of the National Academy of Sciences of the United States of America* **102,** 1737–1742.

35. Ahlijanian, M. K., Barrezueta, N. X., Williams, R. D., Jakowski, A., Kowsz, K. P., McCarthy, S., Coskran, T., Carlo, A., Seymour, P. A., Burkhardt, J. E.*, et al.* (2000)

Proceedings of the National Academy of Sciences of the United States of America **97,** 2910–2915.

36. Cruz, J. C., Tseng, H. C., Goldman, J. A., Shih, H., & Tsai, L. H. (2003) *Neuron* **40,** 471–483.
37. Fischer, A., Sananbenesi, F., Pang, P. T., Lu, B., & Tsai, L. H. (2005) *Neuron* **48,** 825–838.
38. Hallows, J. L., Iosif, R. E., Biasell, R. D., & Vincent, I. (2006) *J Neurosci* **26,** 2738–2744.
39. Sananbenesi, F., Fischer, A., Wang, X., Schrick, C., Neve, R., Radulovic, J., & Tsai, L. H. (2007) *Nature neuroscience* **10,** 1012–1019.
40. Amin, N. D., Albers, W., & Pant, H. C. (2002) *J Neurosci Res* **67,** 354–362.
41. Zheng, Y. L., Li, B. S., Amin, N. D., Albers, W., & Pant, H. C. (2002) *Eur J Biochem* **269,** 4427–4434.
42. Zheng, Y. L., Kesavapany, S., Gravell, M., Hamilton, R. S., Schubert, M., Amin, N., Albers, W., Grant, P., & Pant, H. C. (2005) *Embo J* **24,** 209–220.
43. Patzke, H., Maddineni, U., Ayala, R., Morabito, M., Volker, J., Dikkes, P., Ahlijanian, M. K., & Tsai, L. H. (2003) *J Neurosci* **23,** 2769–2778.
44. Tang, D., Yeung, J., Lee, K. Y., Matsushita, M., Matsui, H., Tomizawa, K., Hatase, O., & Wang, J. H. (1995) *J Biol Chem* **270,** 26897–26903.
45. Humbert, S., Dhavan, R., & Tsai, L. (2000) *J Cell Sci* **113 (Pt 6),** 975–983.
46. Humbert, S., Lanier, L. M., & Tsai, L. H. (2000) *Neuroreport* **11,** 2213–2216.
47. Ko, J., Humbert, S., Bronson, R. T., Takahashi, S., Kulkarni, A. B., Li, E., & Tsai, L. H. (2001) *J Neurosci* **21,** 6758–6771.
48. Takahashi, S., Saito, T., Hisanaga, S., Pant, H. C., & Kulkarni, A. B. (2003) *J Biol Chem* **278,** 10506–10515.
49. Dymecki, S. M. (1996) *Proceedings of the National Academy of Sciences of the United States of America* **93,** 6191–6196.
50. Orban, P. C., Chui, D., & Marth, J. D. (1992) *Proceedings of the National Academy of Sciences of the United States of America* **89,** 6861–6865.
51. Hirasawa, M., Ohshima, T., Takahashi, S., Longenecker, G., Honjo, Y., Veeranna, Pant, H. C., Mikoshiba, K., Brady, R. O., & Kulkarni, A. B. (2004) *Proceedings of the National Academy of Sciences of the United States of America* **101,** 6249–6254.
52. Ohshima, T., Hirasawa, M., Tabata, H., Mutoh, T., Adachi, T., Suzuki, H., Saruta, K., Iwasato, T., Itohara, S., Hashimoto, M., et al. (2007) *Development* **134,** 2273–2282.
53. Hawasli, A. H., Benavides, D. R., Nguyen, C., Kansy, J. W., Hayashi, K., Chambon, P., Greengard, P., Powell, C. M., Cooper, D. C., & Bibb, J. A. (2007) *Nature neuroscience* **10,** 880–886.
54. Pareek, T. K., Keller, J., Kesavapany, S., Pant, H. C., Iadarola, M. J., Brady, R. O., & Kulkarni, A. B. (2006) *Proceedings of the National Academy of Sciences of the United States of America* **103,** 791–796.
55. Pareek, T. K. & Kulkarni, A. B. (2006) *Cell Cycle* **5,** 585–588.
56. Pareek, T. K., Keller, J., Kesavapany, S., Agarwal, N., Kuner, R., Pant, H. C., Iadarola, M. J., Brady, R. O., & Kulkarni, A. B. (2007) *Proceedings of the National Academy of Sciences of the United States of America* **104,** 660–665.

Involvement of Cdk5 in Synaptic Plasticity, and Learning and Memory

Florian Plattner, K. Peter Giese, and Marco Angelo

Abstract Cdk5 has been demonstrated to be one of the most diversely functional kinases within neurons. It is unsurprising then that recent advances implicate Cdk5 in synaptic plasticity, and learning and memory. In this chapter, we summarize the data that reveal the involvement of Cdk5 in mnemonic processes on molecular as well as cellular levels and relate these findings to its emerging function in learning and memory. From amongst the impressive range of candidate mechanisms by which Cdk5 might influence mnemonic processes, we pay particular attention to mechanisms with well-established function in both, synaptic plasticity, and learning and memory, including NMDA receptor modulation, transcriptional regulation and organization of synaptic structures. We aim to show that Cdk5 is uniquely placed amongst kinases to orchestrate the multi-level processes inherent in learning and memory owing to its integral role in many neuronal functions.

Introduction

Over the last two decades, the understanding of the molecular and cellular basis of synaptic plasticity, and learning and memory has progressed significantly, and hundreds of molecules have been implicated in these phenomena. Essential neuronal functions such as synaptic transmitter release and ion channel modulation rely on highly specialized pre- and postsynaptic molecular complexes. The precise organization of these complexes and their proper functioning requires an extensive degree of regulation. Protein phosphorylation has been found to be one of the fundamental mechanisms for such regulation, and consequently numerous protein kinases and phosphatases are now known to have functions in synaptic plasticity and memory processes.

As highlighted in other chapters, Cdk5 is probably one of the most diversely functional kinases within neurons. Cdk5 is intimately implicated in the

F. Plattner
Institute of Neurology, University College London, London, UK
e-mail: f.plattner@ucl.ac.uk

N.Y. Ip, L.-H. Tsai (eds.), *Cyclin Dependent Kinase 5 (Cdk5)*,
DOI: 10.1007/978-0-387-78887-6_16, © Springer Science+Business Media, LLC 2008

regulation of numerous neuronal processes including axonal outgrowth, CNS development and neuronal cell death. Hence, it is not surprising that recent advances suggest a role of Cdk5 as a regulator of mnemonic functions. However, it is precisely this diversity of function that has complicated analysis of Cdk5 in synaptic plasticity and memory. In particular, the dramatic effects of Cdk5 in developmental structuring of the CNS have confounded analysis of many genetically modified mouse models with altered Cdk5 activity (Giese et al. 2005). A further confounding factor precluding detailed analysis of Cdk5-mediated effects is the lack of highly specific pharmacological Cdk5 inhibitors. Despite these obvious drawbacks, a growing number of molecular, electrophysiological, and behavioural studies implicate Cdk5 in synaptic plasticity, and learning and memory (for review see Angelo et al. 2006).

In this chapter, we summarize the data that reveal the involvement of Cdk5 in mnemonic processes from molecular and cellular levels through to learning and memory. By exploring how Cdk5 is linked to synaptic plasticity and memory, we aim to show that Cdk5 is uniquely placed amongst kinases to orchestrate such activities owing to its integral role in many neuronal functions.

Molecular Mechanisms of Synaptic Plasticity

The physiological description and molecular characterization of synaptic processes have greatly furthered understanding of neuronal functions and their relationships to cognition. Experiments on the physiology of neuronal networks have unearthed various activity-dependent changes in synaptic transmission termed synaptic plasticity (Malenka & Bear 2004). These changes in synaptic efficacy can be classified according to their anatomical, molecular, physiological and temporal characteristics. Molecular studies have shown the involvement of pre-, post- and extra-synaptic processes in synaptic plasticity. Pre-synaptic components of synaptic plasticity are influenced by parameters including the shape of the action potential, Ca^{2+} channel conductance and Ca^{2+} levels (Zucker & Regehr 2002). These parameters control pre-synaptic vesicle release and thereby regulate synaptic transmission. Transmitter release in turn triggers postsynaptic mechanisms via the activation of receptors and ion channels. Postsynaptic processes of synaptic plasticity involve alteration of ion channel conductance, modulation of intracellular signalling and reorganization of synaptic structures. Events external to the synapse may also impinge on plasticity through axonal transport, cell adhesion, gene transcription and protein synthesis amongst others.

Cdk5 function has been associated with an ever-increasing number of molecules with pre-, post- and extra-synaptic localization, many of which are recognized to be important for synaptic plasticity. Numerous studies implicate Cdk5 in the direct phosphorylation of such substrates relevant for synaptic plasticity (Table 1–3). Other data demonstrate that Cdk5 and its activators are linked to

Table 1 Pre-synaptic substrates of Cdk5 with putative functions in synaptic plasticity

Cellular Process	Cdk5 Substrate	Phospho-site	Substrate Function	Plasticity Affected by Substrate	Effect of Cdk5 Phosphorylation on Substrate	References
Endocytosis and exocytosis	Munc-18	Thr574	SNARE complex formation and vesicle fusion	BST, STP (?)	Involvement in exocytosis (negative/positive?)	Fletcher et al. 1999; Barclay et al. 2004
	Trio (?)	Ser1720 Ser2358 Ser2454	Activator of Rac, involvement in exocytosis	BST, STP (?)	Positive regulation of exocytosis (?)	Xin et al. 2004
	Sept5	Ser17	SNARE complex formation	BST, STP (?)	Positive regulation of exocytosis via syntaxin1A (?)	Taniguchi et al. 2007
	Dynamin I	Ser774/778 Ser857	Pinching off vesicles during endocytosis	BST, STP (?)	Implicated in endocytosis (negative/positive?)	Tan et al. 2003; Tomizawa et al. 2003; Graham et al. 2007
	Amphiphysin I	Ser262/272 Ser276/285 Thr310	Binds dynamin I and involvement in endocytosis	BST, STP (?)	Regulation of endocytosis (negative/positive?)	Floyd et al. 2001; Tomizawa et al. 2003; Liang et al. 2007; but see Tan et al. 2003
	Synaptojanin	Ser1144	Synthesis of $PI(4,5)P_2$,	BST, STP (?)	Inhibits synaptojanin	Lee et al. 2004
	PIPKIγ	Ser650	$PI(4,5)P_2$ regulates dynamin I function	BST, STP (?)	Activates PIPKIγ (role in endocytosis unclear)	Lee et al. 2005
Vesicle pool dynamics	Synapsin I	Ser551/553 (?)	Regulation of vesicle availability	BST, STP (?)	Vesicle recruitment (?)	Matsubara et al. 1996
Ion influx	L-type VDCC; P/Q-type VDCC	Ser783(?)	Ca^{2+} influx; interaction with SNARE protein complex	BST, STP, LTP	Reduces VDCC conductance	Tomizawa et al. 2002; Wei et al. 2005a
Transmitter synthesis	Tyrosine hydroxylase	Thr31	Synthesis of dopamine	BST	Stabilizes tyrosine hydroxylase and increases its activity	Kansy et al. 2004; Moy & Tsai 2001
Synaptic maturation	CASK	Ser51; Ser395	Regulates αCaMKII autophosphorylation and synaptogenesis via liprin-α	All (?)	Regulates membrane recruitment of CASK and interaction of CASK and liprin-α	Samuels et al. 2007

Abbreviations: αCaMKII, alpha Ca^{2+}/calmodulin-dependent kinase II; BST, basal synaptic transmis-sion; $PI(4,5)P_2$, phosphatidylinositol-4,5- biphosphate; PIPKIγ, phosphatidylinositol phosphate kinase type Iγ; SNARE, soluble NSF (N-ethylmaleimide-sensitive fusion protein) accessory protein SNAP receptor; STP, short-term plasticity; VDCC, voltage-dependent calcium channel. (*Table adapted from Angelo et al. 2006*).

Table 2 Post-synaptic substrates of Cdk5 with putative functions in synaptic plasticity

Cellular Process	Cdk5 Substrate	Phospho-site	Substrate Function	Plasticity Affected by Substrate	Effect of Cdk5 Phosphorylation on Substrate	References
Ion influx	NR2A	Ser1232	Ca^{2+} influx, NMDA-mediated currents	LTP, LTD	Increases NMDAR conductance	Li et al. 2001; Wang et al. 2003
Protein clustering	PSD-95	Thr19, Ser25/35	Clustering of synaptic proteins including NMDAR	LTP, LTD	Negative regulation of PSD-95 clustering	Morabito et al. 2004; Roselli et al. 2005
Intracellular signaling	MEK1	Thr286	Regulation of MAPK pathway	LTP, LTD	Inhibition of MEK1 and in turn ERK activity	Sharma et al. 2002
	DARPP-32	Thr75	Regulation of PKA and PP1 activity	LTP, LTD	Inhibition of PKA and subsequent activation of PP1	Bibb et al. 1999; Nishi et al. 2000
	PP1 inhibitor 1 (I1); PP1 inhibitor 2 (I2)	Ser67; Thr72	Regulation of PP1 activity	LTP, LTD	Involvement in the regulation of PP1 (negative/positive?)	Huang & Paudel 2000; Agarwal-Mawal & Paudel 2001; Bibb et al. 2001; Nguyen et al. 2007
	Dab1	Ser400/491	Effector in Reelin signaling pathway	LTP (?), LTD (?)	Alters Tyr phosphorylation of Dab1 and thus Dab1 function (isoform-specific)	Keshvara et al. 2002; Ohshima et al. 2007; Sato et al. 2007
Post-synaptic architecture	TrkB	Ser478	Promotes dendritic growth in response to BNDF	LTP	Stimulates TrkB receptor; promotes BNDF-triggered dendritic growth	Cheung et al. 2007
	Spinophilin	Thr17	Both bind PP1 and regulate spine morphology (may act antagonistically via G-protein-coupled receptors)	LTP, LTD	Effect unknown	Futter et al. 2005
	Neurabin-1	Ser95		LTP, LTD	Reduces binding to actin and so alters neuronal morphology	Causeret et al. 2007
	RasGRF1; RasGRF2	Ser731; Ser737	Regulates GTPase activity of Rac	LTP, LTD	Reduction of Rac activity and in turn ERK activity through Cdk5 phosphorylation of RasGRF2	Kesavapany et al. 2004; Kesavapany et al. 2006
	Pak1	Thr212	Regulator of dendrite and spine morphology, probably through actin remodeling, Rac effector	LTP, LTD, MP	Inhibits Pak1 and relocalizes it from membrane to cytosol	Nikolic et al. 1998; Rashid et al. 2001

Table 2 (continued)

Cellular Process	Cdk5 Substrate	Phospho-site	Substrate Function	Plasticity Affected by Substrate	Effect of Cdk5 Phosphorylation on Substrate	References
	Ephexin1	Thr41/47(?), Ser139(?)	Growth cone collapse and spine morphogenesis (via RhoA)	LTP (?), LTD (?)	Leads to enhanced RhoA activity and thus decreased spine density	Fu et al. 2007
	CRMP-1; CRMP-2	Thr509/ Ser522; Ser522	Growth cone collapse; spine development	LTP, LTD (?)	Modulation of CRMP signaling (?); Reduces binding of CRMP-2 to tubulin	Cole et al. 2006; Brown et al. 2004; Yamashita et al. 2007
	WAVE1	Ser310/397; Ser441	Positive regulation of actin-dependent dendritic spine maturation	LTP, LTD (?)	Inhibits WAVE1 and thus decreases numbers of mature spines	Kim et al. 2006
Other	APP	Thr668	Regulator of vesicular transport and membrane dynamics (?) Precursor to several peptides	LTP, LTD (?), MP (?)	Alters APP-localization and processing (?)	Iijima et al. 2000; Liu et al. 2003; Cruz et al. 2006

Abbreviations: APP, amyloid precursor protein; BDNF, brain-derived neurotrophic factor; CRMP, collapsin response mediating proteins; DARPP-32, dopamine and cyclic AMP-regulated phosphoprotein molecular mass 32 kDa; ERK, extracellular signal-regulated kinase; LTD, long-term depression; LTP, long-term potentiation; MAPK, mitogen-activated protein kinase; MEK1, MAPK/ERK kinase 1; MP, metaplasticity; NMDAR, N-methyl-D-aspartate receptor; NR2A, NMDAR subunit 2A; PAK1, p21-activated kinase-1; PKA, cyclic AMP-dependent protein kinase; PP1, protein kinase 1; PSD-95, post-synaptic density 95 kDa; RasGRF1/2, Ras guanine nucleotide-releasing factor 1/2; WAVE1, Wiskott-Aldrich syndrome protein verprolin homologous-1. (*Table adapted from Angelo et al. 2006*).

Table 3 Extra-synaptic substrates of Cdk5 with putative functions in synaptic plasticity

Cellular Process	Cdk5 Substrate	Phospho-site	Substrate Function	Plasticity Affected by Substrate	Effect of Cdk5 Phosphorylation on Substrate	References
Cytoskeletal architecture (axonal transport)	Stathmin	Ser25(?)/38	Regulation of MT	LTP, LTD (?)	Alters binding to MT	Hayashi et al. 2006
	Tau	Ser205/235	Organization of axonal architecture (via MT binding), axonal transport (?)	LTP (?)	Altered MT binding and protein-protein interaction (?)	Sengupta et al. 2006
	SCLIP	Ser68/73	Inhibitor of RasGRF1	LTP (?), LTD (?)	Effect unknown	Horiuchi et al. 2006
Transcription	GR	Ser203/211	Hormone-dependent transcription factor	L-LTP	Suppression of GR, transcriptional activation	Kino et al. 2007
	STAT3	Ser727	Transcription factor, MT regulation	L-LTP (?)	Enhances STAT3-mediated transcription	Fu et al. 2004
	p53	Ser15/33; Ser46	Transcription factor	L-LTP (?)	Stabilizes p53 and promotes p53-induced transcription	Zhang et al. 2002; Lee et al. 2007
	MEF2A; MEF2D	Ser408; Ser444	Transcription factors linked to cell death	L-LTP (?)	Reduces MEF2-mediated transcription, enhances cell death	Gong et al. 2003
	mSds3	Ser228	Modulation of histone acetylation	L-LTP (?)	Enhances mSds3-mediated transcriptional repression	Li et al. 2004b
	JNK3	Thr131	Transcriptional regulation, cell survival	L-LTP (?)	Inhibition of JNK3 activity, promotes cell survival	Li et al. 2002
Translation	S6K1	Ser411	Critical regulator of translation	L-LTP (?)	Activates S6K1, promotes S6K1-mediated translation	Hou et al. 2007

Abbreviations: GR, glucocorticoid receptor; JNK, c-Jun N-terminal kinase; LTD, long-term depression; LTP, long-term potentiation; L-LTP, late-phase LTP; MEF2A/D, myocyte enhancer factor 2A/D; mSds3, mammalian suppressor of defective silencing 3; MT, microtubules; RasGRF1, Ras guanine nucleotide-releasing factor 1; S6K1, ribosomal protein S6 kinase 1; SCLIP, SCG10-like protein; STAT3, signal transducer and activator of transcription 3. (*Table adapted from Angelo et al. 2006*).

plasticity-related molecules via protein–protein interactions or indirect actions (Table 4). A new line of evidence implies that numerous signalling pathways thought to participate in plasticity converge on Cdk5 (Table 5). For ease of comprehension, the molecular functions of Cdk5 with relevance for synaptic plasticity are listed according to their likely pre-, post- or extra-synaptic sites of action.

Pre-synaptic Mechanisms

Involvement of Cdk5 in a number of pre-synaptic molecular mechanisms suggests its probable contribution to synaptic plasticity by affecting ion channel conductance, vesicle endocytosis and exocytosis, vesicle pool dynamics and pre-synaptic components in synapse organization and maturation (Table 1 and 4).

The conductance of P/Q-type and L-type voltage-dependent calcium channels (VDCC) can be reduced via their phosphorylation by Cdk5 (Tomizawa et al. 2002; Wei et al. 2005a). Reduction in Cdk5 activity was found to decrease Cdk5-directed VDCC phosphorylation and synchronously influence exocytosis. These observations support the idea that Cdk5 affects exocytosis via the regulation of VDCC, possibly by modulation of pre-synaptic Ca^{2+} signals. A further association of Cdk5 and exocytosis comes from studies in chromaffin cells and pancreatic β-cells (Fletcher et al. 1999; Lilja et al. 2004; Barclay et al. 2004). This work implicates Cdk5 in exocytosis via the direct phosphorylation of Munc-18. Finally, phosphorylation of Trio might link Cdk5 function to the exocytotic process (Xin et al. 2004).

Cdk5 has been linked to clathrin-mediated endocytosis via the phosphorylation of dynamin I and amphiphysin I (Tan et al. 2003; Tomizawa et al. 2003). Furthermore, Cdk5 might affect endocytosis via modulation of phospholipid signalling pathways, as it can regulate synaptojanin, a phosphatidylinositol(4,5)-bisphosphate [PI(4,5)P$_2$] phosphatase and PIP kinase type Iγ (PIPKIγ) (Lee et al. 2004, 2005). Compelling evidence for a role of Cdk5 in slow synaptic endocytosis comes from studies employing pharmacological and molecular ablation of Cdk5 activity (Tan et al. 2003; Evans & Cousin 2007). These data indicate that Cdk5 is required for multiple rounds of endocytosis induced through strong stimulation by re-phosphorylation of proteins dephosphorylated by calcineurin. Whether Cdk5 participates in fast vesicle endocytosis is however still contested (Barclay et al. 2004; Evans & Cousin 2007). In line with a role in slow endocytosis, Cdk5 has been linked to vesicle pool dynamics via the phosphorylation of synapsin I, a protein regulating availability of vesicles for release and the interaction of the Cdk5 activator, p39, with myosin light chain, a component of the molecular motor myosin (Matsubara et al. 1996; Ledee et al. 2007). Furthermore, Cdk5 has been reported to regulate neurotransmitter synthesis, as it stabilizes and increases the activity of tyrosine hydroxylase, an enzyme involved in the synthesis of dopamine (Kansy et al. 2004; Moy & Tsai 2004).

Table 4 Synaptic-plasticity-related molecules with links to Cdk5 holoenzyme

Cellular Process	Plasticity-Related Molecule	Substrate Function	Plasticity Affected	Link to Cdk5	Effects of Cdk5 Holoenzyme	References
Pre-synaptic						
Vesicle pool dynamics	MLC	Component of molecular motor myosin	STP; LTP (?)	Interacts with p39	(?)	Ledee et al. 2007
Post-synaptic						
Ion influx	NR2B	Ca^{2+} influx, NMDA-mediated current	LTP; LTD	NR2B in complex with Cdk5 and calpain	Regulation of NR2B degradation via calpain	Hawasli et al. 2007
	NR1	Ca^{2+} influx, NMDA-mediated current	LTP; LTD	NR1 phosphorylation at Ser 897 is indirectly affected by Cdk5 activity	Indirect negative regulation of NMDAR conductance possibly via dopaminergic input	Chergui et al. 2004
Intracellular signaling	αCaMKII	Protein kinase essential for CA1 LTP induction	LTP; LTD	CaMKII is in complex with p35 and p39	Suppression of CaMKII activation	Dhavan et al. 2002; Hosokawa et al. 2006
	GSK3	Multifunctional protein kinase, transcriptional control	LTP, LTD	GSK3 is in complex with Cdk5, indirect negative control over GSK3 activity	Regulation of axonal transport (?)	Morfini et al. 2004; Plattner et al. 2006
	PI3 K and AKT	Protein kinases	LTP; LTD	Indirect (?)	Indirect positive regulation of PI3 K and AKT activity via neuregulin	Li et al. 2003; Sarker & Lee 2004
	Src	Protein kinase	LTP; (LTD)	Direct phosphorylation of Src	(?)	Kato & Maeda 1999
	PP2A	Multifunctional protein phosphatase	LTP; LTD	PP2A is in complex with Cdk5 and GSK3	Inhibition of PP2A (?)	Plattner et al. 2006
Protein cleavage	Presenilin 1 (PS 1)	Secretase	LTP; (LTD)	Direct phosphorylation at carboxy-terminus	Regulation of PS 1 carboxy-terminal fragment stability	Kesavapany et al. 2001; Lau et al. 2002
	Calpain	Ca^{2+}-dependent protease	LTP; (LTD)	Calpain in complex with Cdk5 and p35	Localization of calpain to substrates	Hawasli et al. 2007
Extra-synaptic						
Cell adhesion	N-cadherin	Neuronal cell adhesion molecule; Interacts with catenins	LTP	N-cadherin in complex with Cdk5, p35, and p39	Negative regulation of cadherin-mediated cell adhesion	Kwon et al. 2000; Negash et al. 2002

Table 4 (continued)

Cellular Process	Plasticity-Related Molecule	Substrate Function	Plasticity Affected	Link to Cdk5	Effects of Cdk5 Holoenzyme	References
Protein clustering, transcription	β-catenin	Binds to N-cadherin; transcription factor	LTP	β-catenin in complex with Cdk5, p35, and p39	Regulation of localization and protein interactions of β-catenin (?)	Kwon et al. 2000; Schuman & Murase 2003
Transcription	CREB	Transcription factor	L-LTP	(?): indirect via MAPK pathway	CREB activity is regulated by Cdk5 via MEK1	Takahashi et al. 2005; Sharma et al. 2007
Nuclear import	Importins	Transport of proteins from/ to nucleus	L-LTP (?)	Importin–β/–5/–7 bind to p35	Cdk5 is not associated with importin/p35 complex	Fu et al. 2006

Abbreviations: αCaMKII, alpha Ca^{2+}/calmodulin-dependent kinase II; BST, basal synaptic transmission; CA1 LTP, NMDA receptor-dependent LTP in CA1 neurons; CREB, cAMP responsive element-binding protein; GSK3, glycogen synthase kinase 3; L-LTP, late-phase LTP; LTD, long-term depression; LTP, long-term potentiation; MAPK, mitogen-activated protein kinase; MEK1, MAPK/ERK kinase 1; MLC, myosin light chain; NR1, NMDAR subunit 1; NR2B, NMDAR subunit 2B; P13 K, phosphoinositide-3 kinase; PP2A, protein phosphatase 2A; STP, short-term plasticity.

Table 5 Synaptic plasticity-related extracellular ligands known to modify Cdk5 activity

Extracellular Ligand	Receptor	Effect of Receptor Activation on Cdk5 Activity	Is Receptor Cdk5 Substrate?; Cdk5 Effect on Receptor	Plasticity Affected by Ligand	Effect of Cdk5 on Plasticity Through Modulation of Ligand Signaling	Remarks	Reference
Neurotransmitter							
Glutamate	NR2A	Upregulation (?), p25 generation	Yes; enhanced conductance	LTP, LTD	Complex bidirectional effects of Cdk5 on NMDAR, relative contributions of NR2A and NR2B to these effects is unclear	Phosphorylation of NR2A greater by p25/Cdk5 than by p35/Cdk5	Li et al. 2001; Wang et al. 2003; Kerokoski et al. 2004
	NR2B	Upregulation (?), p25 generation	No (?); interacts with p35/Cdk5	LTP, LTD		Decreases in Cdk5 lead to reduced NR2B degradation, thereby increasing NMDAR currents	Kerokoski et al. 2004; Hawasli et al. 2007
	mGluR	Upregulation	Unknown	LTP, LTD	Differential effects according to brain region (?)		Liu et al. 2001, 2002
Dopamine	D1R/D2R	Different effects by receptor type and brain region	Unknown	LTP, LTD (according to synapse type)	Differential effects according to receptor type expressed	D1R agonists inhibit Cdk5 in striatum, whereas striatal D2Rs stimulate Cdk5	Nishi et al. 2000; Chergui et al. 2004; Zhen et al. 2004
Acetylcholine	Nicotinic AChR	Upregulation	Unknown	STP, LTP (?), MP	(?) Positive enhancement of AChR-mediated effects	Nicotinic AChRs are expressed both pre- and post-synaptically	Lin et al. 2005; Fu et al. 2005
Trophic factors							
BDNF	TrkB	Upregulation	Yes; activation	Augmentation of LTP	Enhancement of BDNF-mediated augmentation of LTP	TrkB phosphorylation by Cdk5 affects dendritic outgrowth	Cheung et al. 2007
NGF	?TrkA/B ?p75NTR	Upregulation (?)	Unknown	Augmentation of LTP	(?)Enhancement of NGF-mediated augmentation of LTP	NGF can induce p35 expression	Harada et al. 2001; Zheng et al. 2001
GDNF	GFRα1 and NCAM (?)	Upregulation	Unknown	Augmentation of BST (?)	(?)Enhancement of GDNF-mediated augmentation of BST		Paratcha et al. 2006; Ledda et al. 2006

Table 5 (continued)

Extracellular Ligand	Receptor	Effect of Receptor Activation on Cdk5 Activity	Is Receptor Cdk5 Substrate?; Cdk5 Effect on Receptor	Plasticity Affected by Ligand	Effect of Cdk5 on Plasticity Through Modulation of Ligand Signaling	Remarks	Reference
Neuregulin	ErbB2/3/4	Upregulation	Yes (for ErbB2/3); activation	Reduction in CA1 LTP	(?)Enhancement of neuregulin-mediated depotentiation of potentiated synapses		Fu et al. 2001, 2005; Li et al. 2003
Others							
TNFα	TNF-R1/R2	Downregulation	Unknown	Synaptic scaling	(?) Downregulation of TNFα-mediated synaptic scaling	Downstream effects through activation of NF-κB	Orellana et al. 2006
Ephrin	EphA4/?EphA2	Upregulation	Unknown	Required for early CA1 LTP	(?) Enhancement of ephrin-mediated LTP component		Fu et al. 2001; Cheng et al. 2003; Fu et al. 2007
Corticosterone	GR	Upregulation (?)	Yes; suppresses GR-mediated transcription	MP (?)	(?) Cdk5 decreases GR-mediated transcription, thereby affecting L-LTP	Cdk5 expression and activity in CNS may be upregulated by stress	Kino et al. 2007
Aβ	Unknown	Upregulation (?)	Unknown	LTP, MP (?)	Inhibition of Cdk5 ablates block of LTP by Aβ	Complex reciprocal effects of APP/Aβ on Cdk5 and its activators	Wang et al. 2004
Reelin	ApoER2/VLDLR	(?)	Unknown	Augmentation of CA1 LTP	Genetic ablation of p35 or p39 prevents augmentation of LTP by reelin	Phosphorylation of ApoER2/VLDLR adaptor protein Dab1 can modulate reelin signaling	Beffert et al. 2004; Ohshima et al. 2007

Abbreviations: Aβ, amyloid beta peptide; ACh, acetylcholine; APP, amyloid precursor protein; BDNF, brain-derived neurotrophic factor; BST, basal synaptic transmission; CA1 LTP, NMDA receptor-dependent LTP in CA1 neurons; D1R/D2R, dopamine receptor 1/2; GDNF, glial-derived neurotrophic factor; GR, glucocorticoid receptor; LTD, long-term depression; LTP, long-term potentiation; mGluR, metabotropic glutamate receptors; MP, metaplasticity; NCAM, neuronal cell-adhesion molecule; NGF, nerve growth factor; NR2A/B, NMDAR subunit 2A/B; p75NTR, p75 neurotrophin receptor; R, receptor; STP, short-term plasticity; TNFα, tumor necrosis factor alpha; VLDLR, very-low-density lipoprotein receptor.

Finally, Cdk5 may be linked to the structural organization of the pre-synapse. A recent paper suggests that Cdk5-dependent phosphorylation of CASK may influence synapse maturation through regulation of its association with liprin-α (Samuels et al. 2007).

Post-synaptic Mechanisms

A growing body of evidence links Cdk5 to postsynaptic processes relevant for synaptic plasticity such as modulation of ion channel conductance, protein clustering, regulation of kinase and phosphatase activities and postsynaptic structural changes (Table 2 and 4). Recent advances indicate that there is an intricate relationship between Cdk5 and the NMDA receptor (NMDAR), a molecular coincidence detector essential for the induction of some forms of synaptic plasticity such as NMDAR-dependent long-term potentiation (LTP) and long-term depression (LTD) (Malenka & Bear 2004). In contrast, Cdk5 seems not to regulate AMPA receptors (AMPAR), as Cdk5 activity does not influence AMPAR currents (Li et al. 2001; Wang et al. 2003; Chergui et al. 2004; Hawasli et al. 2007). Cdk5 phosphorylates the NMDAR subunit, NR2A, which leads to an increased NMDAR conductance (Li et al. 2001; Wang et al. 2003). A study in conditional Cdk5 hypomorphic 'null' mutant mice finds that NR2B, another NMDAR subunit, occurs in a complex with Cdk5, p35 and calpain and concludes that this complex is involved in calpain-mediated NR2B degradation (Hawasli et al. 2007). Hence, Cdk5 can influence NMDAR conductance by regulating NR2B levels in a kinase activity–independent manner. Experiments in medium spiny neurons in the striatum demonstrate that Cdk5 indirectly modulates NMDAR conductance via the regulation of NR1, a subunit of both striatal and hippocampal NMDARs (Chergui et al. 2004). Another important observation is that NMDAR-mediated Ca^{2+} influx triggers the proteolytic cleavage of the Cdk5 activator p35 into p25 via the Ca^{2+}-dependent protease calpain (Kerokoski et al. 2004; Wei et al. 2005b). Interestingly, the quantity of p25 formed depends on the magnitude of Ca^{2+} influx. Lower levels of Ca^{2+} influx do not result in p25 generation, but in fact are found to promote brief periods of p35-mediated, Ca^{2+}/calmodulin-dependent Cdk5 activity (Wei et al. 2005b). Furthermore, p35 can be protected from calpain cleavage via phosphorylation by Cdk5, which ultimately can direct p35 for degradation via the proteosomal pathway (Saito et al. 2003; Kamei et al. 2007). Together, these data start to give insight into the complex regulation of Cdk5, which is intricately linked to Ca^{2+}-signalling, and suggest that Cdk5 can act as a Ca^{2+}-responsive enzyme implicated in synaptic plasticity and memory (Angelo et al. 2006). These findings further support the idea that the evolutionarily conserved ability to generate p25 may itself be a physiological, non-pathological process, with a role in mnemonic functions (Angelo et al. 2006). An additional association between NMDARs and Cdk5 is the Ca^{2+}-dependent formation of postsynaptic protein complexes. Cdk5-mediated phosphorylation regulates the clustering properties of postsynaptic density-95 (PSD-95), a scaffolding protein involved in NMDAR binding (Morabito et al. 2004). Phosphorylation of PSD-95 by Cdk5 can also promote proteosomal degradation of PSD-95 in the presence of amyloid beta peptide (Roselli et al. 2005). Importantly, glutamatergic signalling stimulates the association of the Cdk5 holoenzymes, p35/Cdk5 and p39/Cdk5, with alpha Ca^{2+}/calmodulin-dependent kinase II (αCaMKII) together with α-actinin-1, partially via NMDAR-mediated Ca^{2+} influx (Dhavan et al. 2002). In line with these data, inhibition of Cdk5 enhances CaMKII autophosphorylation and subsequently increased CaMKII activation (Hosokawa et al. 2006). This observation is of likely relevance for mnemonic processes, as CaMKII activation is an essential requirement for some forms of plasticity and memory (Irvine et al. 2006). Besides CaMKII, numerous other protein kinases and phosphatases have been implicated in mnemonic functions. Hence, regulatory cross-talk of Cdk5 with such intracellular signalling pathways may constitute a link to plasticity and memory processes. Several studies implicate Cdk5 in the regulation of the MAPK pathway. Firstly, Cdk5 phosphorylates and inhibits MAPK/ERK kinase 1 (MEK1) directly, which subsequently leads to a reduction in ERK activity (Sharma et al. 2002). Secondly, Cdk5 phosphorylates the guanine exchange factor (GEF) RasGRF2 (Ras guanine nucleotide releasing factor 2), which downregulates the GTPase activity of Rac. This in turn reduces MEK1 activity and thus also ERK activity (Kesavapany et al. 2004).

The phosphorylation of dopamine and cyclic AMP-regulated phospho-protein, molecular mass 32 kDa (DARPP-32), links Cdk5 to PKA signalling and protein phosphatase 1 (PP1) in striatum (Bibb et al. 1999). A number of studies provide further evidence associating Cdk5 and phosphatase function. Work in cortical neurons observed that PP1 is inhibited by Cdk5 via an indirect mechanism (Morfini et al. 2004). In line with this result, Cdk5 was found to regulate PP1 via the phosphorylation of the homologous regulatory subunits inhibitor-1 (I-1), inhibitor-2 (I-2) and DARPP-32 (Huang & Paudel 2000; Agarwal-Mawal & Paudel 2001; Bibb et al. 1999; Ohshima et al. 2005; Bibb et al. 2001; Nguyen et al. 2007). In addition, Cdk5 can reportedly phosphorylate both spinophilin and neurabin, scaffolding proteins thought to target PP1 to membrane receptors and the cytoskeleton (Futter et al. 2005; Causeret et al. 2007). Moreover, experiments in p25 transgenic mice find Cdk5 in a complex with PP2A and glycogen synthase kinase 3 (GSK3), suggesting a functional interaction (Plattner et al. 2006). This study reveals that Cdk5 negatively controls GSK3 activity, possibly indirectly via phosphatases and suggests that this regulatory control may have important implications for the hyperphosphorylation of tau that is the hallmark of neurodegenerative tauopathies. Additionally, Cdk5 may be involved in the regulation of PI3K and AKT signalling (Li et al. 2003; Sarker & Lee 2004).

Activity-dependent morphological changes of dendritic spines, the postsynaptic compartment of most excitatory synapses, have recently excited much interest as a putative mechanism underlying synaptic plasticity and memory. Dendritic spine growth and changes in morphology are mediated by the modulation of the postsynaptic cytoskeleton, which is mainly formed by actin filaments. Dynamic changes of the actin cytoskeleton are effected by multiple signalling pathways including Rho and Ras family small GTPases. Many proteins of the Rho and Ras GTPase pathways have themselves been implicated in synaptic plasticity and memory processes.

Mounting evidence indicates that Cdk5 may play an important role in the formation and morphogenesis of dendritic spines. In line with this notion, Cdk5 activity has been shown to affect spine density (Norrholm et al. 2003; Fischer et al. 2005). Numerous studies implicate Cdk5 in molecular mechanisms underlying activity-dependent changes of the actin cytoskeleton at the postsynapse. This regulation of actin dynamics by Cdk5 appears in part to occur through the Rho and Ras GTPase signalling pathways. Cdk5 was found to influence signalling through Rho GTPases, including Rac1, RhoA and Cdc42 (cell-division cycle 42). Cdk5 can decrease the activity of Rac through phosphorylation of RasGRF2 (Kesavapany et al. 2004). Moreover, Cdk5 phosphorylates RasGRF1, a GEF that controls Rac1 activity (Kesavapany et al. 2006). Interestingly, a recent study suggests that RasGRF1 and RasGRF2 may act antagonistically to favour induction of LTD and LTP, respectively (Li et al. 2006). A further GEF, ephexin1, is phosphorylated by Cdk5 in response to ephrin A1 signalling, which results in increased RhoA activity. Enhanced RhoA activity in this case occurs concomitantly with decreased spine density (Fu et al. 2007). Brain-derived neurotrophic factor (BDNF)-induced phosphorylation of TrkB receptor by Cdk5 can trigger activation of Cdc42, which has been correlated with dendritic growth (Cheung et al. 2007).

Cdk5 also phosphorylates a number of Rac signalling effectors, including WAVE-1 (Wiskott-Aldrich syndrome protein verprolin homologous-1), Pak1 (p21-activated kinase-1), STAT3 and S6 kinase (Kim et al. 2006; Nikolic et al. 1998; Fu et al. 2004; Hou et al. 2007). Phosphorylation of WAVE-1 by Cdk5 results in a decreased number of mature spines (Kim et al. 2006). Pak1 is inhibited through Rac-dependent phosphorylation at Thr212 by Cdk5 (Nikolic et al. 1998). Reduction in Pak1 activity has been correlated with reduced spine density, but an increased proportion of larger, mature spines (Hayashi et al. 2004). Reduced Pak1 activity enhances cortical LTP, impairs LTD, diminishes long-term contextual fear memory consolidation and reduces extinction of contextual fear memories (Hayashi et al. 2004; Sananbenesi et al. 2007). Cdk5 can phosphorylate a number of collapsin response mediating proteins (CRMPs) including CRMP-1 and CRMP-2, which have been implicated in regulation of dendritic spine formation possibly by acting downstream of RhoA (Cole et al. 2006; Brown et al. 2004; Yamashita et al. 2007). Interestingly, phosphorylation of CRMP-2 by Cdk5 and/or RhoA prevents CRMP-2 from binding to tubulin, but not actin (Arimura et al. 2005). Additional substrates of Cdk5 with a functional link to spine formation and morphogenesis include c-Jun N-terminal kinase (JNK), MEK1, neurabin, PSD-95 and spinophilin

(Li et al. 2002; Kesavapany et al. 2004; Causeret et al. 2007; Morabito et al. 2004; Futter et al. 2005).

Extra-Synaptic Mechanisms

Cdk5 is engaged in processes located outside the pre- and postsynapse relevant for synaptic plasticity including cell adhesion, cytoskeletal dynamics, axonal transport, gene transcription and protein synthesis (Table 3 and 4). Cdk5 has been associated with cell adhesion via regulation of N-cadherin and β-catenin (e.g. Kwon et al. 2000; Negash et al. 2002; Murase et al. 2002). In this context, Cdk5 is thought to promote β-catenin phosphorylation possibly via the tyrosine kinase Src, thereby affecting cadherin-mediated cell adhesion (Lilien & Balsamo 2005; Kato & Maeda 1999). It is also becoming clear that Cdk5 can influence cytoskeletal architecture and thereby neuronal morphology via regulation of the actin cytoskeleton, as well as organization of microtubules (MT). Cdk5 is recognized to phosphorylate various proteins regulating polymerization of tubulin into MT, including tau, stathmin, SCLIP (SCG10-like protein), TPPP (tubulin polymerization-promoting protein) and STAT3 (signal transducer and activator of transcription 3) (Sengupta et al. 2006; Hayashi et al. 2006; Hlavanda et al. 2007; Horiuchi et al. 2006; Fu et al. 2004). In addition to structural functions, MT also play a role in axonal transport. Hence, it is conceivable that the involvement of Cdk5 in axonal transport (Morfini et al. 2004) may be mediated via its regulation of MT.

Later stages of synaptic plasticity and memory critically rely on gene transcription and protein synthesis. Cdk5 has been linked to transcriptional regulation via the phosphorylation of transcription factors including STAT3, p53 and MEF2A/2D (Fu et al. 2004; Zhang et al. 2002; Lee et al. 2007; Gong et al. 2003). Cdk5 also phosphorylates the glucocorticoid receptor which leads to the suppression of its transcriptional activation and thereby possibly modulation of stress-related glucocorticoid responses (Kino et al. 2007). Moreover, Cdk5 has been implicated in CREB-mediated- and possibly β-catenin-mediated transcription (Takahashi et al. 2005; Sharma et al. 2007). The regulation of JNK-3 and GSK3, protein kinases shown to be important for transcriptional control, might constitute another link between Cdk5 and transcription (Li et al. 2002; Plattner et al. 2006).

Additional evidence for a critical role of Cdk5 in nuclear functions comes from a number of studies. Cdk5 and its activators show nuclear localization. In fact, p35 can be transported into the nucleus via an importin-mediated mechanism (Fu et al. 2006). Another study shows that p25 can become sequestered in the nucleus (O'Hare et al. 2005). Cdk5 is involved in neuronal differentiation, since its absence leads to the arrest of the cell cycle in precursor cells (Cicero & Herrup 2005). Moreover, Cdk5 and its activators have been linked to the regulation of histone acetylation, thereby affecting the chromatin structure and subsequently transcription. Cdk5 directly phosphorylates mSds3, a modulator of the histone deacetylase (HDAC) co-repressor complex (Li et al. 2004b). Another study suggesting that p25 may influence transcription, possibly via the regulation of histone acetylation levels and chromatin structure, observes that a decrease in level of synaptic proteins induced by prolonged p25 expression can be attenuated by application of an HDAC inhibitor (Fischer et al. 2007). In addition, other data also associate Cdk5 with protein synthesis. Phosphorylation of S6 kinase1 by Cdk5 can activate this key regulator of translation (Hou et al. 2007). Furthermore, Cdk5 can enhance local protein synthesis in response to semaphorin-3 signalling (Li et al. 2004a).

Besides the numerous downstream functions of Cdk5 described above, it has become apparent that Cdk5 itself can be regulated by a diverse array of signalling pathways. A number of extracellular ligands with importance to synaptic plasticity and memory processes have been reported to alter Cdk5 activity via their respective receptors (Table 5). Trophic factors including BDNF, glial-derived neurotrophic factor (GDNF), nerve growth factor (NGF) and neuregulin are shown to elevate Cdk5 activity (e.g. Cheung et al. 2007; Harada et al. 2001; Ledda et al. 2002; Fu et al. 2001). Furthermore, neurotransmitters including acetylcholine, dopamine and glutamate may upregulate Cdk5 activity (e.g. Lin et al. 2005; Chergui et al. 2004; Liu et al. 2001; Kerokoski et al. 2004). In addition, other extracellular ligands and signalling molecules upstream of Cdk5 may affect its activity including $A\beta$ peptide, corticosterone, ephrin, reelin and tumor necrosis factor alpha ($TNF\alpha$) (e.g. Wang et al. 2004; Kino et al. 2007; Fu et al. 2007; Beffert et al. 2004; Orellana et al. 2007). Interestingly, several receptors of these ligands including NR2A, NR2B and TrkB are themselves known either to be phosphorylated by Cdk5 or else to interact with the kinase (Li et al. 2001; Hawasli et al. 2007; Cheung et al. 2007).

Cdk5 in Cellular Mechanisms of Synaptic Plasticity

In light of the involvement of Cdk5 in the molecular mechanisms of vesicle cycling, ion channel conductance, intracellular signalling and transcription, it is conceivable that Cdk5 might also exert an important influence in cellular mechanisms of synaptic plasticity (Angelo et al. 2006). Changes in synaptic efficacy may persist from milliseconds to hours or even years and according to their temporal characteristics are grouped as short-term or long-term synaptic plasticity.

A prerequisite for normal short-term and long-term plasticity is the basic functioning of processes underlying synaptic transmission. To assess whether synaptic transmission is affected by Cdk5, several studies have investigated basal synaptic transmission as measured by input–output curves in genetically modified mouse models with altered Cdk5 activity (Table 6). Most of the studies find that alterations in Cdk5 activity have no effect on basal synaptic

Table 6 Involvement of Cdk5 in synaptic transmission and short-term synaptic plasticity

Plasticity	Synapse Type	Preparation	Effect of Cdk5 Modification On Plasticity	Remarks	References
BST	SC–CA1 Synapses	Hpc slices from p25 TG mice	No effect	Results depend on level and duration of p25 expression and gender	Fischer et al. 2005; Ris et al. 2005
		Hpc slices from Cdk5 conditional KO mice	No effect		Hawasli et al. 2007
		Hpc slices from p35(−/−) mice	No effect	Developmental abnormalities in p35(−/−) mice	Ohshima et al. 2005
		Hpc slices from p35(−/−) and p39(−/−) mice	Enhanced BST	Developmental abnormalities in p35(−/−) mice	Beffert et al. 2004
		Hpc slices from rat with viral p25 expression	Enhanced BST		Wang et al. 2003
PPF	SC–CA1 Synapses	Hpc slices from rat with Cdk5 inhibitor	Reduced PPF	Roscovitine used	Tomizawa et al. 2002
		Hpc slices from p35(−/−) and p39(−/−) mice	Reduced PPF	Developmental abnormalities in p35(−/−) mice	Beffert et al. 2004
		Hpc slices from p25 TG mice	No changes or elevated PPF	Results depend on level and duration of p25 expression and gender	Fischer et al. 2005; Ris et al. 2005
		Hpc slices from Cdk5 conditional KO mice	No changes		Hawasli et al. 2007
	Striatal medium spiny neurons	Coronal brain slices with Cdk5 inhibitors	No changes	Roscovitine and butyrolactone I used	Chergui et al. 2004
PTP	SC–CA1 Synapses	Hpc slices from rat with Cdk5 inhibitor	No effect	Roscovitine used	Li et al. 2001
		Hpc slices from p35 (−/−) and p39(−/−) mice	Reduction in PTP	Two trains of HFS	Beffert et al. 2004
		Hpc slices from p35(−/−) mice	No effect	One train of HFS	Ohshima et al. 2005
		Hpc slices from p35(−/−) mice	Elevated PTP	Different TBS protocol used; stronger TBS has no effect	Wei et al. 2005b
		Hpc slices from Cdk5 conditional KO mice	Elevated PTP	Elevated with stronger TBS, weak TBS has no effect	Hawasli et al. 2007
		Hpc slices from p25 TG mice	Elevated PTP or no effect	Results depend on level and duration of p25 expression and gender	Fischer et al. 2005; Ris et al. 2005

Abbreviations: BST, basal synaptic transmission; DG, dentate gyrus; HFS, high-frequency stimulation; Hpc, hippocampus; KO, knockout; PPF, paired-pulse facilitation; PTP, post-tetanic potentiation; SC–CA1, Schaffer collateral–hippocampal area CA1; TBS, theta burst stimulation; TG, transgenic.

transmission at Schaffer collateral to hippocampal area CA1 synapses (Ris et al. 2005; Fischer et al. 2005; Ohshima et al. 2005; Hawasli et al. 2007). In contrast, work in p35- and p39-null mutant mouse lines, in which Cdk5 activity is significantly reduced, observe enhancements in basal synaptic transmission (Beffert et al. 2004). However, studies in p35 and p39 null mutants might be confounded by developmental disturbances of their hippocampal circuitry (Wenzel et al. 2001; Patel et al. 2004).

Short-Term Plasticity

Several studies have investigated the involvement of Cdk5 in short-term plasticity, such as synaptic facilitation and post-tetanic potentiation (PTP) (Table 6). Fundamental neuronal processes underlying short-term synaptic plasticity rely on pre-synaptic Ca^{2+} influx, which influences critical parameters including Ca^{2+} level and neurotransmitter release. Both facilitation as well as PTP are at least partially pre-synaptic in origin but may also reflect changes in postsynaptic responsiveness (Zucker & Regehr 2002).

Facilitation refers to a synaptic enhancement within a timescale of hundreds of milliseconds and is commonly assessed via the measurement of paired-pulse facilitation (PPF). PPF is a transient enhancement of synaptic responses, so that an initial stimulus facilitates transmitter release triggered by a subsequent stimulus, due to residual pre-synaptic calcium. Electrophysiological studies at the Schaffer collateral–CA1 synapses observe no changes in PPF in some mutant mouse models with altered Cdk5 activity (Ris et al. 2005; Fischer et al. 2005; Hawasli et al. 2007) (Table 6). In line with these data, application of Cdk5 inhibitors do not alter PPF in striatal medium spiny neurons (Chergui et al. 2004). Other results suggest that Cdk5 activity might regulate PPF positively. Pharmacological inhibition of Cdk5 with roscovitine reduces PPF at Schaffer collateral–CA1 synapses in rat hippocampal slices (Tomizawa et al. 2002), whilst hippocampal slices of p35- and p39-null mutant mice also display PPF deficits (Beffert et al. 2004).

PTP describes the initial synaptic potentiation after tetanic induction, which persists for 30 seconds to several minutes. Pharmacological inhibition of Cdk5 using roscovitine or butyrolactone I has no impact on PTP (Li et al. 2001; Wang et al. 2004). In agreement with this finding, no changes in PTP have been observed in a p35-null mutant mouse line (Ohshima et al. 2005). In contrast, work on another p35-null mutant line and on p39-null mutant mice reveals a reduction in PTP (Beffert et al. 2004). One factor complicating the comparison of the various results on PTP is the application of different stimulation protocols. In fact, some of the studies indicate that there are differential effects of Cdk5 on PTP according to the stimulation protocol employed. Weak theta burst stimulation enhances PTP, whereas stronger theta burst or tetanic stimulation has no significant effect on PTP in a p35-null mutant line (Wei et al.

2005b; Ohshima et al. 2005). Analysis of conditional Cdk5-null mutant mice show that PTP is elevated using a stronger, but not weaker, TBS protocol (Hawasli et al. 2007).

Despite good evidence linking Cdk5 to pre- and postsynaptic molecular mechanisms relevant for synaptic plasticity, the data from electrophysiological experiments remain inconclusive as to an involvement of Cdk5 in short-term plasticity. As both facilitation and PTP are probably mainly pre-synaptic in origin, it can be speculated that the pre-synaptic functions of Cdk5 play a subordinate role in synaptic plasticity or can be compensated for. However, as most of the studies on short-term plasticity have looked at Schaffer collateral–CA1 synapses, it is possible that Cdk5 exerts a more pronounced effect on short-term plasticity at other synapses, at which the pre-synaptic contribution is of higher importance to the induction of plasticity.

Long-Term Plasticity

A number of studies have implicated Cdk5 in long-term synaptic plasticity (Table 7). Long-lasting forms of activity-dependent changes in synaptic transmission may persist from minutes to months and include LTP and LTD (Malenka & Bear 2004). LTP is an enhancement of synaptic efficacy at excitatory synapses and is broadly classed as NMDAR dependent or NMDAR independent. NMDAR-dependent LTP is thought to have induction, expression and maintenance phases, of which the former two probably occur mainly at the postsynapse. The induction phase relies on NMDAR activation with resultant Ca^{2+} influx that triggers signalling pathways involving CaMKII and possibly Cdk5. The expression phase, which involves AMPAR insertion into the postsynaptic membrane, is less well understood. The maintenance phase (late-phase LTP, L-LTP) requires gene transcription and protein synthesis.

The initial association of Cdk5 with LTP came from a study claiming that LTP at Schaffer collateral–CA1 synapses (CA1 LTP) is impaired by application of roscovitine (Li et al. 2001). However, the interpretation of these data is now known to be problematic through potential confounding actions of roscovitine including alteration of conductance of several potassium and calcium channels relevant for LTP (Yan et al. 2002; Buraei et al. 2005, 2007). Nevertheless, compelling evidence implicating Cdk5 in LTP originates from the analysis of several genetically modified mouse lines with altered Cdk5 activity. Work on a conditional Cdk5 null mutant mouse model, which exhibits significantly reduced Cdk5 activity, observed an increase in CA1 LTP (Hawasli et al. 2007). The enhancement in LTP was linked to increased NMDAR-mediated synaptic transmission and elevated NR2B levels in the Cdk5 hypomorphic mutant mice. An enhancement in CA1 LTP has also been reported in two transgenic mouse lines expressing the Cdk5 activator p25 (Ris et al. 2005; Fischer et al. 2005). In both p25 mouse lines,

Table 7 Involvement of Cdk5 in long-term synaptic plasticity

Plasticity	Synapse Type	Preparation	Effect of Cdk5 Modification on Plasticity	Remarks	References
LTP	SC–CA1 synapses	Hpc slices from rat with Cdk5 inhibitor	Impaired LTP	Roscovitine used	Li et al. 2001
		Hpc slices from p35 (−/−) and p39 (−/−) mice	No effect	HFS protocol used; developmental abnormalities	Beffert et al. 2004; Ohshima et al. 2005
		Hpc slices from p35 (−/−) mice	LTP threshold is reduced with weak TBS	No differences in LTP with stronger TBS	Wei et al. 2005b
		Hpc slices from p25 TG mice	Enhanced CA1 LTP; reduced LTP threshold	Gender-dependent effect; enhancement persists in L-LTP phase	Ris et al. 2005
		Hpc slices from inducible p25 TG mice	Enhanced CA1 LTP with 50 Hz tetanus	Results depend on level and duration of p25 expression	Fischer et al. 2005
		Hpc slices from Cdk5 conditional KO mice	Enhanced CA1 LTP; reduced LTP induction threshold	2xTBS and 50 Hz tetanus readily induce LTP; no differences in LTP with stronger stimulation	Hawasli et al. 2007
	PP–DG synapses	Hpc slices from rat with Cdk5 inhibitor	No effect	Using roscovitine and butyrolactone I	Wang et al. 2004
		In vivo LTP in anaesthetized p25 TG mice	No effect		Ris et al. 2005
LTD	SC–CA1 synapses	Hpc slices from p35 (−/−) mice	Defective induction of LTD	Depotentiation is impaired	Ohshima et al. 2005

Abbreviations: CA1 LTP, LTP induced at SC–CA1 synapses; HFS, high-frequenzy stimulation; Hpc, hippocampus; KO, knockout; L-LTP, late-phase LTP; LTD, long-term depression; LTP, long-term potentiation; PP-DG, perforant path-dentate gyrus; SC–CA1, Schaffer collateral-hippocampal area CA1; TBS, theta burst stimulation; TG, transgenic.

transgene expression is limited in a spatio-temporal fashion and induces a significant increase in Cdk5 activity (Angelo et al. 2003; Fischer et al. 2005; Plattner et al. 2006). Interestingly, in these p25 transgenic mice, CA1 LTP was differently affected depending on p25 level, duration of p25 expression and sex (Ris et al. 2005; Fischer et al. 2005).

One important observation concerns the effect of the Cdk5 holoenzymes on the induction threshold of CA1 LTP. Several different mutant mouse lines with alterations of Cdk5 activity display reduction of the CA1 LTP induction threshold. In one p35-null mutant line, weak theta burst stimulation readily induces CA1 LTP in mutants, but not wild-type (WT) controls (Wei et al. 2005b). In contrast, stronger theta burst or tetanic stimulation likewise potentiates synapses in both mutants and controls (Wei et al. 2005b; Ohshima et al. 2005). Accordingly, in conditional Cdk5-null mutant mice, CA1 LTP can be readily induced using weak theta burst or tetanic stimulation of 50 Hz, whereas stronger stimulation protocols induce comparable levels of LTP in mutant and WT mice (Hawasli et al. 2007). Analysis of p25-expressing mouse lines points to a similar effect on LTP induction. In low-level-expressing p25 mice, CA1 LTP is improved using one train of high-frequency stimulation, whilst four trains trigger comparable LTP in TG and WT mice (Ris et al. 2005). In addition, tetanic stimulation of 50 Hz readily induces CA1 LTP in inducible p25 TG mice, but fails to potentiate WT controls (Fischer et al. 2005). Together, these results indicate that Cdk5 has a modulatory, perhaps homeostatic, role in CA1 LTP by regulating the induction threshold of LTP.

The modulatory effect of Cdk5 on LTP is apparent only for a certain range of stimulation intensities; stronger stimulation protocols abrogate this effect indicating that Cdk5 is not required under these conditions. More difficult to rationalize is the apparent discrepancy that both Cdk5- and p35-null mutants, as well p25 TG mice, can reduce the induction threshold. One explanation for the parallels is that the different genetic modifications in these mouse lines affect similar or even identical molecular mechanisms. This seems plausible as the modifications in all these lines concern Cdk5 level and/or activity. However, Cdk5 activity is dramatically reduced in conditional Cdk5-null mutant mice, but significantly increased in the p25 TG mouse lines. One plausible hypothesis, which unifies these apparently contradictory results, is that p25 redistributes Cdk5 within neurons in p25 TG mice, thereby inducing similar effects on plasticity as observed in Cdk5 null mutants. The p25-induced redistribution of Cdk5 would result in a reduction in or loss of p35-mediated Cdk5 functions, with membrane-located processes being specifically susceptible. This is in line with data suggesting that p25 redistributes Cdk5 within neurons and that p25 can substitute some, but not all, p35-medaited functions (Patrick et al. 1999; Patzke et al. 2003). In Cdk5 hypomorphic mutants, LTP enhancement has been linked to reduced degradation of NR2B (Hawasli et al. 2007). As NR2B degradation is modulated by a protein complex involving p35 and Cdk5, it can be speculated that p25 might disturb this p35/Cdk5-mediated function resulting in increased numbers of synaptic NR2B subunits and hence increased plasticity.

Another interesting observation in p25 transgenic mouse lines is that the CA1 LTP enhancement extends into the transcription/translation-dependent late phase of LTP (Ris et al. 2005). This may be of relevance in light of the emerging roles of Cdk5 in transcriptional and translational regulation. Furthermore, analysis of LTP at other synapses reveals that Cdk5 might have differential effects at distinct synapses. Accordingly, Cdk5 inhibitors do not perturb LTP at perforant path-dentate gyrus synapses (DG LTP) in rat slices (Wang et al. 2004). DG LTP is also unaffected in p25 TG mice *in vivo* (Ris et al. 2005). Moreover, synaptic transmission at perforant path-dentate gyrus synapses is unaltered in p35-null mutant slices (Patel et al. 2004). The absence of any significant effect of altered Cdk5 activity in DG LTP may indicate that these synapses do not ordinarily employ Cdk5 in LTP. With respect to other long-term plasticities, one study in p35-null mutants reports impaired LTD induction as well as an inability to depotentiate synapses that have undergone LTP (Ohshima et al. 2005). Examination of the role of p39 in synaptic plasticity has so far failed to reveal any function, but more comprehensive studies are required (Beffert et al. 2004).

Cdk5 in Learning and Memory

The role of Cdk5 and its activators in learning and memory has only just started to be explored, but has already provided exciting evidence for the participation of this kinase and its activators in this higher CNS function. Data come from the study of pharmacological inhibition of Cdk5 activity and genetically modified mouse models with altered levels of Cdk5 or its activators. Effects of these manipulations have been studied in hippocampal- and amygdala-dependent forms of learning and memory, which will be explored in this chapter. Cdk5 is also implicated in behavioural sensitization, a striatum-dependent process that is considered a paradigm for addiction. Arguably some of the most significant advances in understanding learning and memory have been made in the study of hippocampus- and amygdala-dependent memory, principally due to the versatility of rodents as model systems. Hippocampal function in mice and rats is commonly assessed in the Morris water maze (MWM) and contextual fear conditioning, whilst amygdala function is tested by use of contextual and cued (tone) fear conditioning (D'Hooge & De Deyn 2001; Anagnostaras et al. 2001).

An initial insight that Cdk5 might have a role in associative learning and memory came with experiments analyzing the effects on fear conditioning of infusing a Cdk5 inhibitor into various CNS structures (Fischer et al. 2002). Injection of butyrolactone I into hippocampus, ventricles or septum, immediately before or after training, resulted in impairment of contextual, but not cued fear conditioning responses. In contrast, infusions at later time points, including immediately before a 24 h memory test, had no effect (Fischer et al. 2002; Fischer et al. 2003). It would appear therefore that Cdk5 activity is required for hippocampus-dependent learning at around the time of conditioning, but not for longer-term memory storage or retrieval, or indeed for learning occurring solely in the amygdala. However, it is important to note that some of the behavioural effects of butyrolactone I might also result from inhibition of Cdk5 in other neuroanatomical substrates, Cdk5 inhibition in non-neuronal cells or inhibition of other enzymes such as other Cdks and MAPK. In line with these pharmacological studies is work on p35-null mutant mice in both contextual fear conditioning and MWM. Loss of this Cdk5 activator severely impairs contextual memory 24 h after conditioning in homozygous p35-null mice (Fischer et al. 2005). Interestingly, heterozygous p35-null mutant mice also display reduced freezing behaviour in comparison to controls. However, the reduction is less pronounced than in homozygous p35 mutants, suggesting a dose-dependent effect (Fischer et al. 2005). MWM studies in another p35-null mutant line reveal increased escape latencies and non-selective searching for the platform during the probe trial, indicative of deficient spatial learning (Ohshima et al. 2005). However, severely disordered brain development, including disturbances in hippocampal cytoarchitecture, may confound the analysis of memory in adult p35-null mutants (Wenzel et al. 2001; Patel et al. 2004). Interestingly, recent research has hinted at a functional significance of p35 expression in human cognitive capacity. Mutations in the coding region or the 3' untranslated region of the p35 gene correlate with non-syndromic mental retardation, the most usual cause of cognitive impairment not linked to dementia (Venturin et al. 2006).

One simple interpretation of these data would be that reduction in or loss of Cdk5 activity is detrimental for memory formation. In apparent contrast to these observations, an important study shows that reduction in Cdk5 levels in brain enhances hippocampal memory formation. This study employed an inducible conditional Cdk5 hypomorphic 'null' mutant mouse model with significantly reduced Cdk5 expression and activity on the order of 70% in several brain structures including cerebral cortex and hippocampus (Hawasli et al. 2007). Surprisingly, the reduction in Cdk5 coincides with enhanced reversal learning in the MWM and improved contextual conditioning. These conditional Cdk5 hypomorphic mice additionally exhibit increased extinction of fear memories concomitantly with reduced Cdk5 activity, a finding recently supported by murine *in vivo* studies with Cdk5 inhibitors (Sananbenesi et al. 2007). Importantly, these data show striking parallels with work from two p25 TG mouse lines which reveal similar enhancements in learning and memory (Angelo et al. 2003; Ris et al. 2005; Fischer et al. 2005). In these TG mouse lines, p25 expression is restricted to postnatal excitatory forebrain neurons and leads to a significant increase in Cdk5 activity. Expression of p25 results in enhanced spatial learning as shown by improved target quadrant selectivity during the MWM probe trial. Moreover, reversal learning in the MWM is improved in the p25 TG mice (Angelo et al. 2003). In line with the MWM findings, data show that p25 may enhance fear conditioning (Angelo et al. 2003; Ris et al. 2005; Fischer et al. 2005). Interestingly, even a brief period of p25 expression seems sufficient to enhance hippocampus-dependent memory, as well as LTP, at later times (Fischer et al. 2005). It has also been recently shown that expression of high levels of p25 appears to inhibit extinction of contextual fear memories (Sananbenesi et al. 2007). The effects of p25 on learning and memory are preserved in different genetic backgrounds, but may vary according to gender, p25 level, training history of the animals and spatio-temporal characteristics of p25 expression including the advent of neurode-generation with prolonged high-level p25 expression (Angelo et al. 2003; Ris et al. 2005; Fischer et al. 2005; Mizuno et al. 2006). Together, these results reveal an unexpected role of p25 and point to an involvement of the p25/Cdk5 holoenzyme in mnemonic functions.

As observed for synaptic plasticity, the p25 TG mouse lines and the conditional Cdk5 hypomorphic mutant mice exhibit similar phenotypic alterations in learning and memory. This finding further supports the idea that the phenotypic parallels are due to alterations in similar or identical molecular mechanisms in the Cdk5 hypomorphic mice and p25 TG mouse lines. In the Cdk5 hypomorphic mutants, the enhancement in learning and memory was associated with an increased number of synaptic NR2B subunits (Hawasli et al. 2007). This result is consistent with a study demonstrating that increased levels of NR2B enhance spatial learning in the MWM, fear conditioning and extinction of fear memories (Tang et al. 1999). Therefore, it is conceivable that the improvements in learning and memory in the p25 TG mice are also linked to increased levels of NMDAR subunits. This notion is supported by the finding that NMDA currents are increased in p25 TG mice (Fischer et al. 2005).

The findings presented here indicate that the two Cdk5 holoenzymes, p35/Cdk5 and p25/Cdk5, differently affect plasticity and memory processes. In fact, the dramatic effect of p25 expression on synaptic plasticity, and learning and memory suggest to us that p25 formation may be a physiological process involved in mnemonic functions. The p25-induced improvements in learning and memory are also of special interest in light of the apparent generation of p25 in neurodegenerative diseases. We have previously developed a theory that p25 might be formed as a compensatory response by pathologically 'distressed' neurons, as for example in Alzheimer's disease, to stave off early memory deficits, but that eventually p25 itself contributes to cognitive decline once its levels are sufficiently high (Angelo et al. 2003, 2006; Giese et al. 2005). However, neither pharmacological studies using Cdk5 inhibitors nor analysis of p35- and Cdk5-null mutants are currently capable of differentiating effects mediated by p25/Cdk5 or p35/Cdk5. Hence, novel approaches will be required to study the roles of the different Cdk5 holoenzymes in synaptic plasticity and memory.

Putative Mechanisms Underlying Cdk5-Mediated Effects in Synaptic Plasticity, and Learning and Memory

This review outlines a multitude of potential molecular mechanisms through which Cdk5 might conceivably influence synaptic plasticity, and learning and memory. However, currently only a few of these mechanisms have good evidence linking them to mnemonic processes via Cdk5.

One candidate mechanism, which ties Cdk5 directly to the initial phase of synaptic plasticity and memory, is the modulation of NMDAR conductance. A number of studies consistently observe that reduction in Cdk5 activity results in enhanced NMDAR-dependent currents (Chergui et al. 2004; Hawasli et al. 2007). The enhancement in NMDAR currents was attributed to a reduced degradation of NR2B, which is regulated by a Cdk5-dependent mechanism (Hawasli et al. 2007). The Cdk5-related enhancement in NMDAR transmission was found to increase synaptic plasticity and coincided with improved memory. Together these findings reveal an important functional relationship between Cdk5 and NMDAR, which clearly is of importance during the initial (induction) phase of synaptic plasticity, and possibly learning and memory.

The functional relevance of Cdk5 in later (protein synthesis–dependent) phases of synaptic plasticity and memory is less certain. This is both a reflection of the uncertainty of the mechanisms underlying these processes and a lack of definitive data from Cdk5 studies. Nevertheless, some electrophysiological and behavioural studies point to a function of Cdk5 and its activators in late-stage processes. Firstly, the increase in CA1 LTP persists well into the transcription/translation-dependent late-phase of LTP in a low-level p25-expressing mouse line (Ris et al. 2005). Secondly, the enhancements in LTP and memory induced by p25 are still observed even after 4 weeks of repression of transgene expression in inducible p25 TG mice (Fischer et al. 2005). Candidate mechanisms by which Cdk5 might influence the late-stage of plasticity and memory include gene transcription, protein synthesis and synaptic structural changes.

With respect to transcription, although Cdk5 has been reported to affect several transcription factors, none of these mechanisms have yet been studied in plasticity or memory. However, evidence for an involvement of Cdk5 and its activators in transcriptional regulation has now emerged. Work in an inducible p25 TG mouse line correlated changes in protein levels with altered learning and memory (Fischer et al. 2007). Application of an HDAC inhibitor was found to attenuate the behavioural impairments and the decrease in synaptic proteins induced by prolonged high-level p25 expression. These findings suggest that p25 and/or the p25/Cdk5 holoenzyme can negatively modulate transcription possibly by affecting histone acetylation levels and chromatin structure. Further work will be required to assess if HDAC inhibitors can also rescue plasticity deficits associated with prolonged high-level p25 expression. As low-level p25-expressing mice exhibit neither reductions in synaptic protein levels nor significant neuronal loss (Angelo et al. 2003), it suggests that low levels of p25 do not affect nuclear functions to the same degree as high p25 levels.

Lastly, Cdk5 has been found to influence structural changes at the postsynapse including regulation of spine formation. Analysis of an inducible p25 TG mouse line has correlated enhancements in synaptic plasticity, and learning and memory with increased dendritic spine density and synapse number (Fischer et al. 2005). However, there is currently no good evidence that changes in spine density are causally related to synaptic plasticity and memory formation. Many non-mnemonic processes including stages of estrus cycle, anoxia and hypoglycaemia can also increase spine density (Segal 2005; Nikonenko et al. 2002). Recent advances suggest that changes in spine morphology rather than spine density are critical for synaptic plasticity and mnemonic functions (Segal 2005). A potential mechanism, by which Cdk5 may induce activity-dependent structural changes of the synapse is the regulation of the actin cytoskeleton. Cdk5 has been implicated in actin organization through a multitude of substrates in Rho GTPase signalling pathways. Interestingly, a recent study has observed that a decline in Rac1-mediated membrane localization of p35/Cdk5 coincides with extinction of contextual fear memories (Sananbenesi et al. 2007). The latter occurs simultaneously with a decrease in membrane-localized phospho-Thr212 Pak1, a form of Pak1 inactivated through phosphorylation by Cdk5. As pharmacological inhibition of either Rac1 or Cdk5, reduction of Cdk5 expression levels and expression of dominant negative Pak1 all impair extinction, so the Rac1-Cdk5-Pak1 pathway may be important for extinction of fear memories (Sananbenesi et al. 2007; Hawasli et al. 2007). The question as to whether this occurs through some membrane-localized mechanism remains elusive, as expression of p25, which relocates Cdk5 away from the membrane, appears also to impair extinction contrary to this Rac1-Cdk5-Pak1 hypothesis (Patrick et al. 1999; Sananbenesi et al. 2007). Else it may indicate that p25 confers its effect on extinction through some other, perhaps Cdk5-independent, mechanisms.

Conclusion

An increasing number of studies support a role of Cdk5 in synaptic plasticity, and learning and memory. On a molecular level, Cdk5 has been implicated in numerous neuronal processes relevant for synaptic plasticity and memory such as vesicle cycling, ion channel modulation and spine morphogenesis. Accordingly, cellular studies indicate that Cdk5 affects synaptic transmission and modulates various forms of plasticity. Behavioural analysis of genetically modified mouse models has associated alterations in levels of Cdk5 or its activators with distinct learning and memory changes. Moreover, these behavioural changes are correlated with isomorphic effects on synaptic plasticity.

Even though the precise contributions of Cdk5 and its activators to synaptic plasticity and memory are still elusive, recent advances have singled out strong candidate mechanisms including NMDAR modulation, transcriptional and translational regulation and organization of synaptic structures. However, it remains to be determined which of the many potential plasticity and memory processes affected by Cdk5 are ultimately relevant, and in turn whether Cdk5 is involved globally through multiple processes or via discrete mechanisms.

References

Agarwal-Mawal A, Paudel HK. (2001) Neuronal Cdc2-like protein kinase (Cdk5/p25) is associated with protein phosphatase 1 and phosphorylates inhibitor-2. J. Biol. Chem. 276: 23712–23718.

Anagnostaras SG, Gale GD, Fanselow MS. (2001) Hippocampus and contextual fear conditioning: recent controversies and advances. Hippocampus 11: 8–17.

Angelo M, Plattner F, Irvine EE, Giese KP. (2003) Improved reversal learning and altered fear conditioning in transgenic mice with regionally restricted p25 expression. Eur. J. Neurosci. 18: 423–431.

Angelo M, Plattner F, Giese KP. (2006) Cyclin-dependent kinase 5 in synaptic plasticity, learning and memory. J. Neurochem. 99: 353–370.

Arimura N, Ménager C, Kawano Y, Yoshimura T, Kawabata S, Hattori A, Fukata Y, Amano M, Goshima Y, Inagaki M, Morone N, Usukura J, Kaibuchi K. (2005) Phosphorylation by Rho kinase regulates CRMP-2 activity in growth cones. Mol. Cell. Biol. 25:9973–9984.

Barclay JW, Aldea M, Craig TJ, Morgan A, Burgoyne RD. (2004) Regulation of the fusion pore conductance during exocytosis by cyclin-dependent kinase 5. J. Biol. Chem. 279: 41495–41503.

Beffert U, Weeber EJ, Morfini G, Ko J, Brady ST, Tsai LH, Sweatt JD, Herz J. (2004) Reelin and cyclin-dependent kinase 5-dependent signals cooperate in regulating neuronal migration and synaptic transmission. J. Neurosci. 24: 1897–1906.

Bibb JA, Snyder GL, Nishi A, et al. (1999) Phosphorylation of DARPP-32 by Cdk5 modulates dopamine signalling in neurons. Nature 402: 669–671.

Bibb JA, Nishi A, O'Callaghan JP, et al. (2001) Phosphorylation of protein phosphatase inhibitor-1 by Cdk5. J. Biol. Chem. 276: 14490–14497.

Brown M, Jacobs T, Eickholt B, Ferrari G, Teo M, Monfries C, Qi RZ, Leung T, Lim L, Hall C. (2004) Alpha2-chimaerin, cyclin-dependent Kinase 5/p35, and its target collapsin response mediator protein-2 are essential components in semaphorin 3A-induced growth-cone collapse. J. Neurosci. 24: 8994–9004.

Buraei Z, Anghelescu M, Elmslie KS. (2005) Slowed N-type calcium channel (CaV2.2) deactivation by the cyclin-dependent kinase inhibitor roscovitine. Biophys. J. 89: 1681–1691.

Buraei Z, Schofield G, Elmslie KS (2007) Roscovitine differentially affects CaV2 and Kv channels by binding to the open state. Neuropharmacology. 52: 883–894.

Causeret F, Jacobs T, Terao M, Heath O, Hoshino M, Nikolic M. (2007) Neurabin-I is phosphorylated by Cdk5: implications for neuronal morphogenesis and cortical migration. Mol. Biol. Cell. 18: 4327–4342.

Cheng Q, Sasaki Y, Shoji M, Sugiyama Y, Tanaka H, Nakayama T, Mizuki N, Nakamura F, Takei K, Goshima Y. (2003) Cdk5/p35 and Rho-kinase mediate ephrin-A5-induced signaling in retinal ganglion cells. Mol. Cell. Neurosci. 24: 632–645.

Chergui K, Svenningsson P, Greengard P. (2004) Cyclin-dependent kinase 5 regulates dopa-minergic and glutamatergic transmission in the striatum. Proc. Natl. Acad. Sci. U.S.A. 101: 2191–2196.

Cheung ZH, Chin WH, Chen Y, Ng YP, Ip NY (2007) Cdk5 is involved in BDNF-stimulated dendritic growth in hippocampal neurons. PLoS Biol. 5: e63

Cicero S, Herrup K. (2005) Cyclin-dependent kinase 5 is essential for neuronal cell cycle arrest and differentiation. J. Neurosci. 25: 9658–9668.

Cole AR, Causeret F, Yadirgi G, Hastie CJ, McLauchlan H, McManus EJ, Hernández F, Eickholt BJ, Naikolic M, Sutherland C. (2006) Distinct priming kinases contribute to differential regulation of collapsin response mediator proteins by glycogen synthase kinase-3 in vivo. J. Biol. Chem. 281: 16591–16598.

Cruz JC, Kim D, Moy LY, Dobbin MM, Sun X, Bronson RT, Tsai LH. (2006) p25/cyclin-dependent kinase 5 induces production and intraneuronal accumulation of amyloid beta in vivo. J. Neurosci. 26: 10536–10541.

Dhavan R, Greer PL, Morabito MA, Orlando LR, Tsai LH. (2002) The cyclin-dependent kinase 5 activators p35 and p39 interact with the alpha-subunit of Ca2 + /calmodulin-dependent protein kinase II and alpha-actinin-1 in a calcium-dependent manner. J. Neurosci. 22: 7879–7891.

D'Hooge R, De Deyn P.P. (2001) Applications of the Morris water maze in the study of learning and memory. Brain. Res. Brain. Res. Rev. 36: 60–90.

Evans GJO, Cousin MA. (2007) Activity-dependent control of slow synaptic vesicle endocy-tosis by cyclin-dependent kinase 5. J. Neurosci. 27: 401–411.

Fischer A, Sananbenesi F, Schrick C, Spiess J, Radulovic J. (2002) Cyclin-dependent kinase 5 is required for associative learning. J. Neurosci. 22: 3700–3707.

Fischer A, Sananbenesi F, Schrick C, Spiess J, Radulovic J. (2003) Regulation of contextual fear conditioning by baseline and inducible septo-hippocampal cyclin-dependent kinase 5. Neuropharmacology 44: 1089–1099.

Fischer A, Sananbenesi F, Pang PT, Lu B, Tsai LH. (2005) Opposing Roles of Transient and prolonged expression of p25 in synaptic plasticity and hippocampus-dependent memory. Neuron 48: 825–838.

Fischer A, Sananbenesi F, Wang X, Dobbin M, Tsai L.H. (2007) Recovery of learning and memory is associated with chromatin remodelling. Nature 447: 178–182.

Fletcher AI, Shuang R, Giovannucci DR, Zhang L, Bittner MA, Stuenkel EL. (1999) Regulation of exocytosis by cyclin-dependent kinase 5 via phosphorylation of Munc18. J. Biol. Chem. 274: 4027–4035.

Floyd SR, Porro EB, Slepnev VI, Ochoa GC, Tsai LH, De Camilli P. (2001) Amphiphysin 1 binds the cyclin-dependent kinase (Cdk) 5 regulatory subunit p35 and is phosphorylated by cdk5 and cdc2. J. Biol. Chem. 276: 8104–8110.

Fu AKY, Fu WY, Cheung J, Tsim KW, IP FC, Wang J.H, Ip NY (2001) Cdk5 is involved in neuregulin-induced AchR expression at the neuromuscular junction. Nat Neurosci 4: 374–381.

Fu AK., Fu WY, Ng AK, Chien WW, Ng YP, Wang JH, Ip NY. (2004) Cyclin-dependent kinase 5 phosphorylates signal transducer and activator of transcription 3 and regulates its transcriptional activity. Proc. Natl. Acad. Sci. U.S.A. 101: 6728–6733.

Fu AK, Ip FC, Fu WY, Cheung J, Wang JH, Yung WH, Ip NY. (2005) Aberrant motor axon projection, acetylcholine receptor clustering, and neurotransmission in cyclin-dependent kinase 5 null mice. Proc. Natl. Acad. Sci. U.S.A. 102: 15224–15229.

Fu X, Choi YK, Qu D, Yu Y, Cheung NS, Qi RZ. (2006) Identification of nuclear import mechanisms for the neuronal Cdk5 activator. J. Biol. Chem. 281: 39014–39021.

Fu WY, Chen Y, Sahin M, Zhao XS, Shi L, Bikoff JB, Lai KO, Yung WH, Fu AK, Greenberg ME, Ip NY (2007) Cdk5 regulates EphA4-mediated dendritic spine retraction through an ephexin1-dependent mechanism. Nat. Neurosci. 10: 67–76.

Futter M, Uematsu K, Bullock SA, Kim Y, Hemmings HC Jr, Nishi A, Greengard P, Nairn AC. (2005) Phosphorylation of spinophilin by ERK and cyclin-dependent PK 5 (Cdk5). Proc. Natl. Acad. Sci. U.S.A. 102: 3489–3494.

Giese KP, Ris L, Plattner F. (2005) Is there a role of the cyclin-dependent kinase 5 activator p25 in Alzheimer's disease? Neuroreport 16: 1725–1730.

Gong X, Tang X, Wiedmann M, Wang X, Peng J, Zheng D, Blair LA, Marshall J, Mao Z. (2003) Cdk5-mediated inhibition of the protective effects of transcription factor MEF2 in neurotoxicity-induced apoptosis. Neuron 38: 33–46.

Graham ME, Anggono V, Bache N, Larsen MR, Craft GE, Robinson PJ. (2007) The in vivo phosphorylation sites of rat brain dynamin I. J. Biol. Chem. 282: 14695–14707.

Harada T, Morooka T, Ogawa S, Nishida E. (2001) ERK induces p35, a neuron-specific activator of Cdk5, through induction of Egr1. Nat Cell Biol. 3: 453–459.

Hawasli AH, Benavides DR, Nguyen C, Kansy JW, Hayashi K, Chambon P, Greengard P, Powell CM, Cooper DC, Bibb JA. (2007) Cyclin-dependent kinase 5 governs learning and synaptic plasticity via control of NMDAR degradation. Nat. Neurosci. 10: 880–886.

Hayashi ML, Choi SY, Rao BS, Jung HY, Lee HK, Zhang D, Chattarji S, Kirkwood A, Tonegawa S. (2004) Altered cortical synaptic morphology and impaired memory consolidation in forebrain- specific dominant-negative PAK transgenic mice. Neuron 42: 773–787.

Hayashi K, Pan Y, Shu H, Ohshima T, Kansy JW, White CL 3rd, Tamminga CA, Sobel A, Curmi PA, Mikoshiba K, Bibb JA. (2006) Phosphorylation of the tubulin-binding protein, stathmin, by Cdk5 and MAP kinases in the brain. J. Neurochem. 99: 237–250.

Hlavanda E, Klement E, Kókai E, Kovács J, Vincze O, Tökési N, Orosz F, Medzihradszky KF, Dombrádi V, Ovádi J. (2007) Phosphorylation blocks the activity of tubulin polymerization-promoting protein (TPPP): identification of sites targeted by different kinases. J. Biol. Chem. 282: 29531–29539.

Horiuchi Y, Asada A, Hisanaga S, Toh-e A, Nishizawa M. (2006) Identifying novel substrates for mouse Cdk5 kinase using the yeast Saccharomyces cerevisiae. Genes Cells. 11: 1393–1404.

Hosokawa T, Saito T, Asada A, Ohshima T, Itakura M, Takahashi M, Fukunaga K, Hisanaga S. (2006) Enhanced activation of $Ca2+$/calmodulin-dependent protein kinase II upon downregulation of cyclin-dependent kinase 5-p35. J. Neurosci. Res. 84: 747–754.

Hou Z, He L, Qi RZ. (2007) Regulation of s6 kinase 1 activation by phosphorylation at ser-411. J. Biol. Chem. 282: 6922–6928.

Huang KX, Paudel HK. (2000) Ser67-phosphorylated inhibitor 1 is a potent protein phosphatase 1 inhibitor. Proc. Natl. Acad. Sci. U.S.A. 97: 5824–5829.

Iijima K, Ando K, Takeda S, Satoh Y, Seki T, Itohara S, Greengard P, Kirino Y, Nairn AC, Suzuki T. (2000) Neuron-specific phosphorylation of Alzheimer's beta-amyloid precursor protein by cyclin-dependent kinase 5. J. Neurochem. 75: 1085–1091.

Irvine EE, von Hertzen LSJ, Plattner F, Giese KP. (2006) αCaMKII autophosphorylation: a fast track to memory. Trends Neurosci. 29: 459–465.

Kamei H, Saito T, Ozawa M, Fujita Y, Asada A, Bibb JA, Saido TC, Sorimachi H, Hisanaga S. (2007) Suppression of calpain-dependent cleavage of the CDK5 activator p35 to p25 by site-specific phosphorylation. J. Biol. Chem. 282: 1687–1694.

Kansy JW, Daubner SC, Nishi A et al. (2004) Identification of tyrosine hydroxylase as a physiological substrate for Cdk5. J. Neurochem. 91: 374–384.

Kato G, Maeda S. (1999) Neuron-specific Cdk5 kinase is responsible for mitosis-independent phosphorylation of c-Src at Ser75 in human Y79 retinoblastoma cells. J. Biochem. (Tokyo) 126: 957–961.

Kerokoski P, Suuronen T, Salminen A, Soininen H, Pirttila T. (2004) Both N-methyl-D-aspartate (NMDA) and non-NMDA receptors mediate glutamate-induced cleavage of the cyclin-dependent kinase 5 (Cdk5) activator p35 in cultured rat hippocampal neurons. Neurosci. Lett. 368: 181–185.

Kesavapany S, Lau KF, McLoughlin DM, Brownlees J, Ackerley S, Leigh PN, Shaw CE, Miller CC (2001) p35/cdk5 binds and phosphorylates beta-catenin and regulates beta-catenin/presenilin-1 interaction. Eur. J. Neurosci. 13: 241–247.

Kesavapany S, Amin N, Zheng YL et al. (2004) p35/cyclin-dependent kinase 5 phosphorylation of Ras guanine nucleotide releasing factor 2 (RasGRF2) mediates Rac-dependent Extracellular signal-regulated kinase 1/2 activity, altering RasGRF2 and microtubule-associated protein 1b distribution in neurons. J. Neurosci. 24: 4421–4431.

Kesavapany S, Pareek TK, Zheng YL, Amin N, Gutkind JS, Ma W, Kulkarni AB, Grant P, Pant HC. (2006/7) Neuronal nuclear organization is controlled by cyclin-dependent kinase 5 phosphorylation of Ras Guanine nucleotide releasing factor-1. Neurosignals. 15: 157–173.

Keshvara L, Magdaleno S, Benhayon D, Curran T. (2002) Cyclin-dependent kinase 5 phosphorylates disabled 1 independently of Reelin signaling. J. Neurosci. 22: 4869–4877.

Kim Y, Sung JY, Ceglia I, Lee KW, Ahn JH, Halford JM, Kim AM, Kwak SP, Park JB, Ho Ryu S, Schenck A, Bardoni B, Scott JD, Nairn AC, Greengard P. (2006) Phosphorylation of WAVE1 regulates actin polymerization and dendritic spine morphology. Nature 442: 814–817.

Kino T, Ichijo T, Amin ND, Kesavapany S, Wang Y, Kim N, RaoS, Player A, Zheng Y, Garabedian MJ, Kawasaki E, Pant HC, Chrousos GP (2007) Cyclin-dependent kinase 5 differentially regulates the transcriptional activity of the glucocorticoid receptor through phosphorylation: Clinical implications for the Nervous system response to glucocorticoids and stress. Mol. Endocrinology 21: 1552–1568.

Kwon YT, Gupta A, Zhou Y, Nikolic M, Tsai LH. (2000) Regulation of N-cadherin-mediated adhesion by the p35/Cdk5 kinase. Curr. Biol. 10: 363–372.

Lau KF, Howlett DR, Kesavapany S, Standen CL, Dingwall C, McLoughlin DM, Miller CC (2002) Cyclin-dependent kinase-5/p35 phosphorylates Presenilin 1 to regulate carboxy-terminal fragment stability. Mol. Cell. Neurosci. 20: 13–20.

Ledda F, Paratcha G, Ibanez CF. (2002) Target-derived GFRalpha1 as an attractive guidance signal for developing sensory and sympathetic axons via activation of Cdk5. Neuron 36: 387–401.

Ledee DR, Tripathi BK, Zelenka PS. (2007) The CDK5 activator, p39, binds specifically to myosin essential light chain. Biochem. Biophys. Res. Commun. 354: 1034–1039.

Lee SY, Wenk MR, Kim Y, Nairn AC, De Camilli P. (2004) Regulation of synaptojanin 1 by cyclin-dependent kinase 5 at synapses. Proc. Natl. Acad. Sci. U.S.A. 101: 546–551.

Lee SY, Voronov S, Letinic K, Nairn AC, Di Paolo G, De Camilli P. (2005) Regulation of the interaction between PIPKI gamma and talin by proline-directed protein kinases. J. Cell. Biol. 168: 789–799.

Lee JH, Kim HS, Lee SJ, Kim KT. (2007) Stabilization and activation of p53 induced by Cdk5 contributes to neuronal cell death. J. Cell. Sci. 120: 2259–2271.

Li BS, Sun MK, Zhang L, Takahashi S, Ma W, Vinade L, Kulkarni AB, Brady RO, Pant HC. (2001) Regulation of NMDA receptors by cyclin-dependent kinase-5. Proc. Natl. Acad. Sci. U.S.A. 98: 12742–12747.

Li BS, Zhang L, Takahashi S, Ma W, Jaffe H, Kulkarni AB, Pant HC. (2002). Cyclin-dependent kinase 5 prevents neuronal apoptosis by negative regulation of c-Jun N-terminal kinase 3. EMBO J. 21: 324–333.

Li BS, Ma W, Jaffe H, Zheng Y, Takahashi S, Zhang L, Kulkarni AB, Pant HC. (2003) Cyclin-dependent kinase-5 is involved in neuregulin-dependent activation of phosphatidylinositol 3-kinase and Akt activity mediating neuronal survival. J. Biol. Chem. 278: 35702–35709.

Li C, Sasaki Y, Takei K, Yamamoto H, Shouji M, Sugiyama Y, Kawakami T, Nakamura F, Yagi T, Ohshima T, Goshima Y. (2004a) Correlation between semaphorin3A-induced facilitation of axonal transport and local activation of a translation initiation factor eukaryotic translation initiation factor 4E. J. Neurosci. 24: 6161–6170.

Li Z, David G, Hung KW, DePinho RA, Fu AK, Ip N.Y. (2004b) Cdk5/p35 phosphorylates mSds3 and regulates mSds3-mediated repression of transcription. J. Biol. Chem. 279: 54438–54444.

Li S, Tian X, Hartley DM, Feig LA. (2006) Distinct roles for Ras-guanine nucleotide-releasing factor 1 (Ras-GRF1) and Ras-GRF2 in the induction of long-term potentiation and long-term depression. J. Neurosci. 26: 1721–1729.

Liang S, Wei FY, Wu YM, Tanabe K, Abe T, Oda Y, Yoshida Y, Yamada H, Matsui H, Tomizawa K, Takei K. (2007) Major Cdk5-dependent phosphorylation sites of amphiphysin 1 are implicated in the regulation of the membrane binding and endocytosis. J. Neurochem. (doi:10.1111/j.1471-4159.2007.04507.x)

Lilien J, Balsamo J. (2005) The regulation of cadherin-mediated adhesion by tyrosine phosphorylation/dephosphorylation of beta-catenin. Curr Opin Cell Biol. 17: 459–465.

Lilja L, Johansson JU, Gromada J, Mandic SA, Fried G, Berggren PO, Bark C. (2004) Cyclin-dependent kinase 5 associated with p39 promotes Munc18-1 phosphorylation and Ca(2+)-dependent exocytosis. J. Biol. Chem. 279: 29534–29541.

Lin W, Dominguez B, Yang J, Aryal P, Brandon EP, Gage FH, Lee KF. (2005) Neurotransmitter acetylcholine negatively regulates neuromuscular synapse formation by a Cdk5-dependent mechanism. Neuron 46: 569–579.

Liu F, Ma XH, Ule J, Bibb JA, Nishi A, DeMaggio AJ, Yan Z, Nairn AC, Greengard P. (2001) Regulation of cyclin-dependent kinase 5 and casein kinase 1 by metabotropic glutamate receptors. Proc. Natl. Acad. Sci. U.S.A. 98: 11062–11068.

Liu F, Virshup DM, Nairn AC, Greengard P. (2002) Mechanism of regulation of casein kinase I activity by group I metabotropic glutamate receptors. J. Biol. Chem. 277: 45393–45399.

Liu F, Su Y, Li B, Zhou Y, Ryder J, Gonzalez-DeWhitt P, May PC, Ni B. (2003) Regulation of amyloid precursor protein (APP) phosphorylation and processing by p35/Cdk5 and p25/Cdk5. FEBS Lett. 547: 193–196.

Malenka RC, Bear MF. (2004) LTP and LTD: an embarrassment of riches. Neuron 44: 5–21.

Matsubara M, Kusubata M, Ishiguro K, Uchida T, Titani K, Taniguchi H. (1996) Site-specific phosphorylation of synapsin I by mitogen-activated protein kinase and Cdk5 and its effects on physiological functions. J. Biol. Chem. 271: 21108–21113.

Mizuno K, Plattner F, Giese KP. (2006) Expression of p25 impairs contextual learning but not latent inhibition in mice. Neuroreport. 17: 1903–1905.

Morabito MA, Sheng M, Tsai LH. (2004) Cyclin-dependent kinase 5 phosphorylates the N-terminal domain of the postsynaptic density protein PSD-95 in neurons. J. Neurosci. 24: 865–876.

Morfini G, Szebenyi G, Brown H, Pant HC, Pigino G, DeBoer S, Beffert U, Brady ST. (2004) A novel CDK5-dependent pathway for regulating GSK3 activity and kinesin-driven motility in neurons. EMBO J. 23: 2235–2245.

Moy LY, Tsai LH. (2004) Cyclin-dependent kinase 5 phosphorylates serine 31 of tyrosine hydroxylase and regulates its stability. J. Biol. Chem. 279: 54487–54493.

Murase S, Mosser E, Schuman EM. (2002) Depolarization drives beta-Catenin into neuronal spines promoting changes in synaptic structure and function. Neuron 35: 91–105.

Negash S, Wang HS, Gao C, Ledee D, Zelenka P. (2002) Cdk5 regulates cell-matrix and cell-cell adhesion in lens epithelial cells. J. Cell. Sci. 115: 2109–2117.

Nguyen C, Nishi A, Kansy JW, Fernandez J, Hayashi K, Gillardon F, Hemmings HC Jr, Nairn AC, Bibb JA. (2007) Regulation of protein phosphatase inhibitor-1 by cyclin-dependent kinase 5. J. Biol. Chem. 282: 16511–16520.

Nikolic M, Chou MM, Lu W, Mayer BJ, Tsai LH. (1998) The p35/Cdk5 kinase is a neuron-specific Rac effector that inhibits Pak1 activity. Nature. 395: 194–198.

Nikonenko I, Jourdain P, Alberi S, Toni N, Muller D. (2002) Activity-induced changes of spine morphology. Hippocampus 12, 585–591.

Nishi A, Bibb JA, Snyder GL, Higashi H, Nairn AC, Greengard P. (2000) Amplification of dopaminergic signaling by a positive feedback loop. Proc. Natl. Acad. Sci. U.S.A. 97: 12840–12845.

Norrholm SD, Bibb JA, Nestler EJ, Ouimet CC, Taylor JR, Greengard P. (2003) Cocaine-induced proliferation of dendritic spines in nucleus accumbens is dependent on the activity of cyclin-dependent kinase-5. Neuroscience 116: 19–22.

O'Hare MJ, Kushwaha N, Zhang Y, Aleyasin H, Callaghan SM, Slack RS, Albert PR, Vincent I, Park D.S. (2005) Differential roles of nuclear and cytoplasmic cyclin-dependent kinase 5 in apoptotic and excitotoxic neuronal death. J. Neurosci. 25: 8954–8966.

Ohshima T, Ogura H, Tomizawa K et al. (2005) Impairment of hippocampal long-term depression and defective spatial learning and memory in p35 mice. J. Neurochem. 94: 917–925.

Ohshima T, Suzuki H, Morimura T, Ogawa M, Mikoshiba K (2007) Modulation of reelin signaling by cyclin-dependent kinase 5. Brain Res. 1140: 84–95.

Orellana DI, Quintanilla RA, Maccioni RB. (2007) Neuroprotective effect of TNFalpha against the beta-amyloid neurotoxicity mediated by CDK5 kinase. Biochim Biophys Acta. 1773: 254–263.

Paratcha G, Ibanez CF, Ledda F. (2006) GDNF is a chemoattractant factor for neuronal precursor cells in the rostral migratory stream. Mol Cell Neurosci. 31: 505–514.

Patel LS, Wenzel HJ, Schwartzkroin PA. (2004) Physiological and morphological character-ization of dentate granule cells in the p35 knock-out mouse hippocampus: evidence for an epileptic circuit. J. Neurosci. 24: 9005–9014.

Patrick GN, Zukerberg L, Nikolic M, de la Monte S, Dikkes P, Tsai LH. (1999) Conversion of p35 to p25 deregulates Cdk5 activity and promotes neurodegeneration. Nature 402: 615–622.

Patzke H, Maddineni U, Ayala R, Morabito M, Volker J, Dikkes P, Ahlijanian MK, Tsai LH. (2003) Partial rescue of the p35-/- brain phenotype by low expression of a neuronal-specific enolase p25 transgene. J Neurosci 23: 2769–2778.

Plattner F, Angelo M, Giese KP. (2006). The roles of Cdk5 and GSK3 in tau hyperpho-sphorylation. J. Biol. Chem. 281: 25457–25466.

Rashid T, Banerjee M, Nikolic M. (2001) Phosphorylation of Pak1 by the p35/Cdk5 kinase affects neuronal morphology. J. Biol. Chem. 276: 49043–49052.

Ris L, Angelo M, Plattner F, Capron B, Errington ML, Bliss TV, Godaux E, Giese KP. (2005) Sexual dimorphisms in the effect of low-level p25 expression on synaptic plasticity and memory. Eur. J. Neurosci. 21: 3023–3033.

Roselli F, Tirard M, Lu J, Hutzler P, Lamberti P, Livrea P, Morabito M, Almeida OF. (2005) Soluble beta-amyloid1-40 induces NMDA-dependent degradation of postsynaptic den-sity-95 at glutamatergic synapses. J. Neurosci. 25: 11061–11070.

Saito T, Onuki R, Fujita Y, Kusakawa G, Ishiguro K, Bibb JA, Kishimoto T, Hisanaga S. (2003) Developmental regulation of the proteolysis of the p35 cyclin-dependent kinase 5 activator by phosphorylation. J. Neurosci. 23: 1189–1197.

Samuels BA, Hsueh YP, Shu T, Liang H, Tseng HC, Hong CJ, Su SC, Volker J, Neve RL, Yue DT, Tsai LH. (2007) Cdk5 Promotes Synaptogenesis by Regulating the Subcellular Distribution of the MAGUK Family Member CASK. Neuron 56: 823–837.

Sananbenesi F, Fischer A, Wang X, Schrick C, Neve R, Radulovic J, Tsai LH. (2007) A hippocampal Cdk5 pathway regulates extinction of contextual fear. Nat. Neurosci. 10: 1012–1019.

Sarker KP, Lee KY. (2004) L6 myoblast differentiation is modulated by Cdk5 via the PI3 K-AKT-p70S6 K signaling pathway. Oncogene 23: 6064–6070.

Sato Y, Taoka M, Sugiyama N, Kubo K, Fuchigami T, Asada A, Saito T, Nakajima K, Isobe T, Hisanaga S. (2007) Regulation of the interaction of Dis-abled-1 with CIN85 by phosphorylation with Cyclin-dependent kinase 5. Genes Cells. 12: 1315–1327.

Schuman EM, Murase S. (2003) Cadherins and synaptic plasticity: activity-dependent cyclin-dependent kinase 5 regulation of synaptic beta-catenin-cadherin interactions. Philos. Trans. R. Soc. Lond. B Biol. Sci. 358: 749–756.

Segal M. (2005) Dendritic spines and long-term plasticity. Nat. Rev. Neurosci. 6: 277–284.

Sengupta A, Novak M, Grundke-Iqbal I, Iqbal K. (2006) Regulation of phosphorylation of tau by cyclin-dependent kinase 5 and glycogen synthase kinase-3 at substrate level. FEBS Lett. 580: 5925–5933.

Sharma P, Veeranna, Sharma M, Amin ND, Sihag RK, Grant P, Ahn N, Kulkarni AB, Pant HC. (2002) Phosphorylation of MEK1 by cdk5/p35 down-regulates the mitogen-activated protein kinase pathway. J. Biol. Chem. 277: 528–534.

Sharma M, Hanchate NK, Tyagi RK, Sharma P. (2007) Cyclin dependent kinase 5 (Cdk5) mediated inhibition of the MAP kinase pathway results in CREB down regulation and apoptosis in PC12 cells. Biochem. Biophys. Res. Commun. 358: 379–384.

Takahashi S, Ohshima T, Cho A, et al. (2005) Increased activity of cyclin-dependent kinase 5 leads to attenuation of cocaine-mediated dopamine signaling. Proc. Natl. Acad. Sci. U.S.A. 102: 1737–1742.

Tan TC, Valova VA, Malladi CS, et al. (2003) Cdk5 is essential for synaptic vesicle endocytosis. Nat. Cell. Biol. 5: 701–710.

Tang YP, Shimizu E, Dube GR, Rampon C, Kerchner GA, Zhuo M, Liu G, Tsien JZ. (1999). Genetic enhancement of learning and memory in mice. Nature 401: 63–69.

Taniguchi M, Taoka M, Itakura M, Asada A, Saito T, Kinoshita M, Takahashi M, Isobe T, Hisanaga S. (2007) Phosphorylation of adult type Sept5 (CDCrel-1) by cyclin-dependent kinase 5 inhibits interaction with syntaxin-1. J. Biol. Chem. 282: 7869–7876.

Tomizawa K, Ohta J, Matsushita M, Moriwaki A, Li ST, Takei K, Matsui H. (2002) Cdk5/p35 regulates neurotransmitter release through phosphorylation and downregulation of P/Q-type voltage-dependent calcium channel activity. J. Neurosci. 22: 2590–2597.

Tomizawa K, Sunada S, Lu YF, et al. (2003) Cophosphorylation of amphiphysin I and dynamin I by Cdk5 regulates clathrin-mediated endocytosis of synaptic vesicles. J. Cell. Biol. 163: 813–824.

Venturin M, Moncini S, Villa V, Russo S, Bonati MT, Larizza L, Riva P. (2006) Mutations and novel polymorphisms in coding regions and UTRs of CDK5R1 and OMG genes in patients with non-syndromic mental retardation. Neurogenetics 7: 59–66.

Wang J, Liu S, Fu Y, Wang JH, Lu Y. (2003) Cdk5 activation induces hippocampal CA1 cell death by directly phosphorylating NMDA receptors. Nat. Neurosci. 6: 1039–1047.

Wang Q, Walsh DM, Rowan MJ, Selkoe DJ, Anwyl R. (2004) Block of long-term potentiation by naturally secreted and synthetic amyloid beta-peptide in hippocampal slices is mediated via activation of the kinases c-Jun N-terminal kinase, cyclin-dependent kinase 5, and p38 mitogen-activated protein kinase as well as metabotropic glutamate receptor type 5. J. Neurosci. 24: 3370–3378.

Wei FY, Nagashima K, Ohshima T et al. (2005a) Cdk5-dependent regulation of glucose-stimulated insulin secretion. Nat. Med. 11: 1104–1108.

Wei FY, Tomizawa K, Ohshima T, et al. (2005b). Control of cyclin-dependent kinase 5 (Cdk5) activity by glutamatergic regulation of p35 stability. J. Neurochem. 93: 502–512.

Wenzel HJ. Robbins CA, Tsai LH, Schwartzkroin PA. (2001) Abnormal morphological and functional organization of the hippocampus in a p35 mutant model of cortical dysplasia associated with spontaneous seizures. J Neurosci 21: 983–998.

Xin X, Ferraro F, Back N, Eipper BA, Mains RE. (2004) Cdk5 and Trio modulate endocrine cell exocytosis. J. Cell. Sci. 117: 4739–4748.

Yan Z, Chi P, Bibb JA, Ryan TA, Greengard P. (2002) Roscovitine: a novel regulator of P/Q-type calcium channels and transmitter release in central neurons. J. Physiol. 540: 761–770.

Yamashita N, Morita A, Uchida Y, Nakamura F, Usui H, Ohshima T, Taniguchi M, Honnorat J, Thomasset N, Takei K, Takahashi T, Kolattukudy P, Goshima Y. (2007)

Regulation of spine development by semaphorin3A through cyclin-dependent kinase 5 phosphorylation of collapsin response mediator protein 1. J. Neurosci. 27: 12546–12554.

Zhang J, Krishnamurthy PK, Johnson GV. (2002) Cdk5 phosphorylates p53 and regulates its activity. J. Neurochem. 81: 307–313.

Zhen X, Goswami S, Abdali SA, Gil M, Bakshi K, Friedman E. (2004) Regulation of cyclin-dependent kinase 5 and calcium/calmodulin-dependent protein kinase II by phosphatidylinositol-linked dopamine receptor in rat brain Mol. Pharmacol. 66: 1500–1507.

Zheng YL, Li BS, Kanungo J, Kesavapany S, Amin N, Grant P, Pant HC. (2007) Cdk5 Modulation of mitogen-activated protein kinase signaling regulates neuronal survival. Mol. Biol. Cell. 18: 404–413.

Zucker RS, Regehr WG. (2002) Short-term synaptic plasticity. Annu. Rev. Physiol. 64: 355–405.

Cyclin-Dependent Kinase 5 (Cdk5): Linking Synaptic Plasticity and Neurodegeneration

Andre Fischer and Li-Huei Tsai

Abstract It is well established that cyclin-dependent kinase 5 (Cdk5) is critically involved in neurodevelopmental processes. In addition, recent data point toward an important role of Cdk5 in regulating synaptic plasticity, learning, and memory in the adult brain. However, aberrant Cdk5 activity has been implicated in various neurodegenerative diseases such as Alzheimer's disease. Deregulation of Cdk5 has been attributed to calpain-mediated cleavage of the Cdk5 activator p35 to the N-terminally truncated p25 protein. p25 levels are elevated in many neurodegenerative diseases and implicated in neuronal cell death *in vitro* and *in vivo*. More importantly, p25/Cdk5 causes hyperphosphorylation of tau and affects processing of APP, leading to increased levels of toxic Aβ-peptides. Surprisingly, recent data indicate that *in vivo* p25 is not toxic per se but that a transient increase in p25 levels may even facilitate neuroplasticity. Here we will review these recent developments and propose a scenario in which p25 generation during aging and Alzheimer's disease might initially be a compensatory phenomenon to enhance neuroplasticity but eventually contributes to the pathogenesis of Alzheimer's disease when chronically elevated.

Synaptic Plasticity, Learning, and Memory

The molecular mechanisms underlying the acquisition, consolidation, and retrieval of memories are a challenging but fascinating aspect of neurobiology. This research is of course driven by the long-standing curiosity that scientists try to understand how the human brain achieves the miraculous task of coordinating communication among the approximately 100 billion neurons to form memories, emotions, and conscience, for example. However, deregulation of neuronal processes is known to cause devastating consequences such as Alzheimer's

A. Fischer
European Neuroscience Institute (ENI), Department for Experimental Neuropathology, Medical School University Goettingen, Max Planck Society, Germany
e-mail: andre.fischer@mpi-mail.mpg.de

N.Y. Ip, L.-H. Tsai (eds.), *Cyclin Dependent Kinase 5 (Cdk5)*,
DOI: 10.1007/978-0-387-78887-6_17, © Springer Science+Business Media, LLC 2008

disease (AD), schizophrenia, anxiety disorders, and phobias to name a few. Such diseases pose a huge emotional and economical burden to our societies, and suitable therapeutic strategies are not yet available. Understanding the cognitive processes on the molecular and systems level is therefore of utmost importance to help elucidate the pathophysiology and develop better medications.

During the last decade, despite the complexity and challenging nature of the problem, a number of studies provided compelling insights into the molecular mechanisms responsible for the acquisition and storage of new memories in the mammalian brain. This achievement can be attributed to the interdisciplinary and methodological convergence of biology research, and ranges from mouse genetics to diverse electrophysiological techniques and behavioral studies.

A number of molecular signaling cascades have since been established as important regulators of learning and memory processes. For example, N-methyl-aspartate receptor (NMDAR) signaling to gene expression has proved to be crucial for both learning and long-term potentiation (LTP) in rodent models of hippocampal functioning. Moreover, the formation of stable hippocampal long-term memory and LTP requires a prolonged activity of the protein kinases such as cyclic AMP-dependent protein kinase (PKA) (Abel et al., 1997; Kandel, 2001) or MAP kinase singaling (Atkins et al., 1998) protein kinase C (PKC) (Selcher et al., 2002) which triggers a molecular pathways that, for example, lead to the phosphorylation of the cAMP response element (CREB) (Alberini, 1999), a transcription factor regulating expression of genes implicated in long-term memory such as early growth response gene 1 (Egr-1) (Malkani and Rosen, 2000) and the immediate early gene cFOS (Radulovic et al., 1998). In addition to PKA, calcium/calmodulin-dependent protein kinase (CaMK) (Silva et al., 1992) and extracellularly regulated kinases (Erk-1/2) are also required for the formation of long-term memory and LTP. For a detailed summary of the molecular processes involved in learning and memory, we refer to more specific reviews (e.g., Kandel, 2001; Roberson and Sweatt, 1999; Levenson et al., 2006; Barco et al., 2006).

The most influential theories suggest that de novo gene expression is a commonality among all forms of long-term memory investigated thus far and that alterations in patterns of gene expression underlie the long-lasting changes in the strength of synaptic connections between neurons responsible for the encoding of memories in the nervous system (Kandel, 2001). Most data suggest that altered gene expression and new protein synthesis serve as a trigger for memory formation and is a prerequisite for synaptic tagging, a mechanism that is believed to underlie memory acquisition (Klann and Sweatt, 2007; Govindarajan et al., 2006; Frey and Morris, 1998).

It is important to note that different forms of memories can be distinguished and linked to certain brain regions. One area that has gained major attention is the hippocampus, which has been critically implicated in the encoding of memory in humans and rodents (Kim and Fanselow, 1992; Scoville and Milner, 1957). Notably, the hippocampus is a prime target of diverse forms of neuro-degeneration associated with cognitive impairments and memory loss, such as

AD and Huntington's disease (Scheff and Price, 2006; Spires and Hannan, 2007). Damage in the hippocampus and the temporal lobe affects what is now called *declarative* or explicit memory, that is memory for facts and events, whereas other forms of memory, such as learned motor skills, are spared with these lesions and are, therefore, referred to as *non-declarative* or implicit memories.

As for the pathogenesis of neuronal diseases, important advances in human genetics and clinical research in the last decade have led to the identification of genes responsible for various learning and memory disorders. In parallel, progress in mouse genetics and molecular biology has elucidated the signaling cascades that underlie the formation of memories in the brain. The convergence of clinical and basic research science is now providing new insight into brain function and, therefore, its dysfunction. The use of animal models, and in particular, the use of genetically modified mice to model neurological and neurodegenerative diseases have been critical as it clarified the molecular etiology of these conditions and opened up new therapeutic avenues for their treatments (Watase and Zoghbi, 2003; Fischer et al., 2007).

In summary, the orchestrated regulation and intracellular signaling events and its coupling of synaptic signals to gene expression is believed to be a prerequisite for synaptic plasticity and memory consolidation. Conversely, deregulation of such processes is the bases for neurodegenerative and mental diseases.

Cdk5 is a Novel Regulator of Synaptic Plasticity

Cyclin-dependent kinase 5 (Cdk5) is unique among the protein family of Cdks, which are important regulators of the cell cycle progression. Although Cdk5 was initially identified because of its homology to the cell cycle regulator Cdc2, so far no role of Cdk5 during cell cycle has been demonstrated. Moreover, unlike other Cdks, Cdk5 is not activated by association with cyclins, but by binding to its activator proteins p35 or p39. The activity of Cdk5 is mainly restricted to post-mitotic neurons, because p35 and p39 expression is highly enriched in CNS tissue (Dhavan and Tsai, 2001).

Cdk5 is best known for its role during neuronal development. Mice lacking the Cdk5 gene die at birth and display an inverted layering of the cortex, indicating defects in neuronal migration (Ohshima et al., 1996). Since then, multiple studies demonstrated the intimate involvement of Cdk5 activity in migrating neurons (Ayala et al., 2007; Xie et al., 2006). Interestingly, while p39-deficient mice show no overt phenotype (Ko et al., 2001), mice lacking the p35 gene are viable but resemble many of the phenotypes seen in Cdk5 knock-out mice (Chae et al., 1997). Notably, the phenotype of p35/p39 compound knockout mice is indistinguishable from Cdk5-deficient mice, demonstrating that p35 and p39 are sufficient and essential to activate Cdk5 (Ko et al., 2001).

Although it was known for years that postnatal Cdk5 activity remained high in certain brain regions, only recently has the role of Cdk5 in synaptic plasticity, learning, and memory began to emerge (Fischer et al., 2003). Early studies found that Cdk5 phosphorylates the major pre-synaptic proteins amphiphysin (Floyd et al., 2001) and synapsin I (Matsubara et al., 1996), indicating that Cdk5 may regulate neurotransmitter release. Moreover, a Cdk5-mediated phosphorylation of Munc-18 decreased the affinity of Munc-18 to the target soluble N-ethylmaleimide sensitive factor attachment protein receptor (t-SNARE) protein syntaxin 1A, suggesting that Cdk5 can regulate vesicle docking, thereby likely promote transmitter release (Shuang et al., 1998; Fletcher et al., 1999).

However, the first strong evidence that Cdk5 plays an important role in synaptic plasticity, learning, and memory came from the observations that Cdk5 function is associated with adaptive changes in the brain related to cocaine addiction (Bibb et al., 2001). Subsequent work from Li et al. (2001) and Fischer et al. (2002) demonstrated a clear role of Cdk5 activity in synaptic plasticity, learning, and memory. These studies showed that Cdk5 phosphorylates the NR2A subunit of the NMDA receptor and that pharmacological inhibition of Cdk5 impairs hippocampal LTP (Li et al., 2001). In addition, the levels of Cdk5 and p35 are dynamically regulated during associative learning in mice, and pharmacological inhibition of septal or hippocampal Cdk5 activity significantly impaired learning and memory (Fischer et al., 2002, 2003).

Since then, a plethora of studies began to unravel the molecular mechanisms by which Cdk5 activity regulates cognitive processes.

In agreement with its role during learning and memory, Cdk5 has been implicated in a number of pre- and postsynaptic processes. It was shown that inhibition of Cdk5/p35 by roscovitine enhanced neurotransmitter release. The effect on transmitter release was explained by Cdk5/p35-mediated phosphorylation of P/Q-type voltage-dependent calcium channels (VDCC) which inhibit the interactions between VDDC and the tSNARE (Tomizawa et al., 2002). It should be noted that roscovitin by itself acts as a potent activator of VDCC channels in the absence of Cdk5 activity (Yan et al., 2002), raising the question for the role of Cdk5 in neurotransmitter release. However, other work showed that Cdk5/p39 regulates insulin secretion in beta-cells via Munc-1-dependent mechanisms (Lilja et al., 2004). In addition, mice that lack the p35 or the p39 gene display impaired paired pulse facilitation, a form of pre-synaptic short-term plasticity (Beffert et al., 2004), while depletion of the reserve synaptic vesicle pool is facilitated in the hippocampus of transgenic mice that display a 2-fold increase in Cdk5 activity (Fischer et al., 2005). While in summary those studies point toward an important role of Cdk5 in pre-synaptic plasticity, further studies should take advantage of the available Cdk5 conditional mutant mice to address the role of Cdk5 in the exocytosis machinery in neurons.

A number of studies indicate that Cdk5 activity regulates synaptic vesicle endocytosis (Bibb et al. 2001) by phosphorylation of dephosphins (Tan et al., 2003; Tomizawa et al., 2003), thereby raising the possibility of Cdk5 being a major counteractor of calcineurin SVE (Lee et al., 2005). However, at first view, those studies come to somewhat opposite conclusions whether Cdk5 is necessary or inhibitory for synaptic vesicle endocytosis (Tan et al., 2003; Tomizawa et al., 2003). A closer look reveals that Cdk5 activity may inhibit the first round of SVE but could be necessary for the second round (Nguyen and Bibb, 2003). These data reiterate an important issue in that synaptic and neuronal processes are rather complex. Assigning the roles of key regulators of synaptic plasticity to certain biological processes as simply "necessary" or "inhibitory" is often a too simplified view.

In summary, the present data hint at an important role of Cdk5 in presynaptic signaling. It is however obvious that additional experiments are needed to delineate the precise role of Cdk5/p35 in the synaptic vesicle cycle.

A number of postsynaptic proteins have been identified as Cdk5 substrates (Cheung et al., 2006). To this end, Cdk5 was shown to phosphorylate the NR2A subunit of NMDA receptors on serine 1232 (Li et al., 2001; Wang et al., 2003) which resulted in an increased channel conductivity of the NMDA receptor (Wang et al., 2003). In line with this observation, pharmacological inhibition of Cdk5 impaired LTP in hippocampal slices (Li et al., 2001). Moreover, increased phosphorylation of $NR2A_{S1232}$ in the p25 transgenic mice that display elevated hippocampal Cdk5 activity correlates with facilitated NMDA receptor currents and LTP (Fischer et al., 2005). Additionally, Cdk5-dependent phosphorylation of the N-terminal domain of PSD-95 was suggested to regulate the clustering of PSD-95 and NMDA receptors at synapses, thus providing a possible mechanism for rapid changes in density and/or number of receptors at synapses (Morabito et al., 2004). A recent study reported that the conditional knockout of Cdk5 in the adult mouse brain improves learning in the fear-conditioning and water maze paradigm (Hawasli et al., 2007). This effect was accompanied by facilitated hippocampal plasticity and explained by reduced Cdk5-mediated degradation of the NR2B subunit of the NMDA receptor (Hawasli et al., 2007). More importantly, this effect was independent of the kinase activity of Cdk5. Taken together, the present data suggest that Cdk5 activity, depending on the experimental system, can either facilitate or impair synaptic function. In conclusion, the current data establish Cdk5 as an important regulator for pre- and postsynaptic signaling. It is likely that Cdk5 is a major kinase at the synapse, and future studies should further reveal key Cdk5 substrates that impact synaptic morphogenesis and plasticity. Equally important is that the current data on Cdk5 function and substrate phosphorylation will be investigated in a physiological *in vivo* situation, thereby clarifying under what circumstances Cdk5 activity acts to either promote or impair synaptic function, learning, and memory (Sananbenesi et al., 2007).

Currently, the general assumption is that the learning process involves structural changes within the neuronal network including strengthening or de novo formation of synapses. It is therefore of particular interest that at the structural level, Cdk5 activity has been implicated with the regulation of synapse and dendrite formation. To this end, it was shown that at the developing neuromuscular junction, Cdk5 regulates the expression of acetylcholine (Ach) receptors and ErbB activation in a neuregulin-dependent manner (Soriano et al., 2001). Another study found that Cdk5 regulates the Ach-induced dispersion of Ach-receptor clusters suggesting that Cdk5 negatively regulates the formation of neuromuscular synapses (Lin et al., 2005). In addition, Cdk5 activity seems to promote the formation of CNS synapses in the adult brain. The first indication that Cdk5 positively affects synaptogenesis in the adult brain stems from the observation that cocaine-induced increase in dendritic spine density of striatal neurons critically depends on Cdk5 activity (Norrholm et al., 2003). Furthermore, increased Cdk5 activity in the hippocampus significantly elevated the density of dendritic spines and synapse on CA1 pyramidal neurons (Fischer et al., 2005). The increase in synapse number was correlated with facilitated learning behavior and elevated hippocampal LTP (Fischer et al., 2005). Finally, mice deficient in the p35 protein had reduced baseline density of dendritic spines on hippocmpal CA1 neurons. They also exhibited severe learning impairments and altered synaptic plasticity (Fu et al., 2007; Fischer et al., 2005; Ohshima et al., 2005; Patel et al., 2004). Thus, Cdk5 plays a crucial role in maintaining the density of dendritic spines and synapse in the hippocampus. It is however interesting to note that a recent report found that Cdk5-mediated phosphorylation of WAVE1, a protein that regulates actin dynamics via the Arp2/3 complex, contributed to the retraction of dendritic spines in hippocampal neurons (Kim et al., 2006). Similarly, Cdk5 activity has been implicated in the retraction of dendritic spines in an ephexin-1-dependent manner (Fu et al., 2007), whereas Cdk5 regulates brain-derived growth factor (BDNF)-induced dendrite formation in hippocampal neurons (Cheung et al., 2006, 2007)

In summary, those data suggest that Cdk5 critically regulates the refining of the synaptic network of the adult brain. Whether Cdk5 activity promotes the formation or strengthening of new synapses or the retraction of existing dendritic spines seems to depend on the cellular context and is likely tightly regulated *in vivo*.

Obviously more work is needed to unravel the complex role of Cdk5 activity on synapse and dendrite formation, synaptic plasticity, and learning behavior.

It is however interesting to note that the formation, retraction, and modulation of synapses and dendrites strongly depends on the dynamics of the actin cytoskeleton (Matus, 2000). In fact, a dynamic actin cytoskeleton is a prerequisite for dendritic spine motility (Halpain, 2000), LTP (Krucker et al., 2000), and learning (Fischer et al., 2004). Interestingly, a number of Cdk5 substrates or binding partners are important regulators of actin dynamics (Nikolic et al., 1998; Dhavan et al., 2002a; Xie et al., 2006). Moreover, p39 can directly bind to actin, although the physiologic relevance of such interaction is not known (Humbert et al., 2000). We previously reported that p35 and p39 bind to the actin regulator protein alpha-actinin in a Ca^{2+}-dependent manner (Dhavan et al., 2002a) which may influence glutamate signaling and CamKII function (Dhavan et al., 2002a). An indirect way by which a Cdk5/p35 complex might influence actin dynamics is the phosphorylation of p21-activated kinase (Pak-1). Pak-1 is a major regulatory protein mediating actin dynamics through Lim Kinase-1 (Edwards et al., 1999) and cofilin (Arber et al., 1998). More importantly, Pak-1 activity is implicated in learning and memory and the formation of dendritic spines (Hayashi et al., 2004). Cdk5/p35 phosphorylates Pak-1, and p35 directly interacts with Pak-1 in a Rac-1-dependent manner (Nikolic et al., 1998; Rashid et al., 2001) Whereas the functional consequence of the phosphorylation event is not entirely clear, the data indicate that Cdk5/p35 action directly or indirectly inhibit Pak-1 activity (Rashid et al., 2001). It is therefore reasonable to speculate that Cdk5 mediates its effect on synapse formation, retraction, or morphology at least in part via Pak-1.

It is also interesting to note that Cdk5 was shown to phosphorylate and thereby inhibit myocyte enhancer factor 2 (MEF-2) (Gong et al., 2003). Although this process has been implicated in apoptotic cell death (Smith et al., 2006; Gong et al., 2003), inhibition of MEF-2-mediated transcription was shown to increase the number of excitatory synapses (Flavell et al., 2006) and regulate postsynaptic differentiation (Shalizi et al., 2006). Future studies will unravel the potential role of Cdk5-mediated MEF-2 phosphorylation during synaptic function.

Another mechanism by which Cdk5 might affect synaptic morphology and plasticity is the regulation of cell adhesion molecules. In the developing brain, synaptic contacts are mainly formed through the interaction of homo- and heterophilic adhesion molecules (Brose, 1999). More importantly, the formation of novel synaptic contacts during synaptogenesis is also believed to play a major role during learning processes in the adult brain. Numerous *in vitro* studies demonstrated that growth of dendritic spines, which is believed to be an important process for synaptogenesis in the adult brain, occurs in response to membrane depolarization (Halpain, 2000). Increased spine density was also described *in vivo* in response to trace eye blink conditioning (Leuner et al., 2003)

Cdk5 activity decreases N-cadherin-dependent cell adhesion by phosphorylation of β-catenin (Kwon et al., 2000). β-catenin mediates the interactions between cadherins and the actin cytoskeleton (Perez-Moreno et al., 2003). It was also shown to elicit intracellular signaling (Bamji et al., 2003). More importantly, recent *in vitro* studies demonstrated that membrane depolarization leads to the redistribution of β-catenin to dendritic spines where it interacts with the cadherins (Murase et al., 2002). Similarly, inhibition of Cdk5 activity by roscovitine both increased the shift of β-catenin into dendritic spines as well as increased the amount of β-catenin that co-immunoprecipitated with cadherin (Schuman and Murase, 2003). These interactions may be crucial for the formation of new synapses as well as the strengthening of old synaptic connections during learning and memory, as indicated by the findings that blockage of N-cadherin results in alterations of dendritic spine morphology (Togashi et al., 2002) and hippocampal LTP (Bonhoeffer and Yuste, 2002).

In conclusion, Cdk5 activity is intimately involved in the regulation of synaptogenesis, synaptic plasticity, and refinement of the synaptic network in

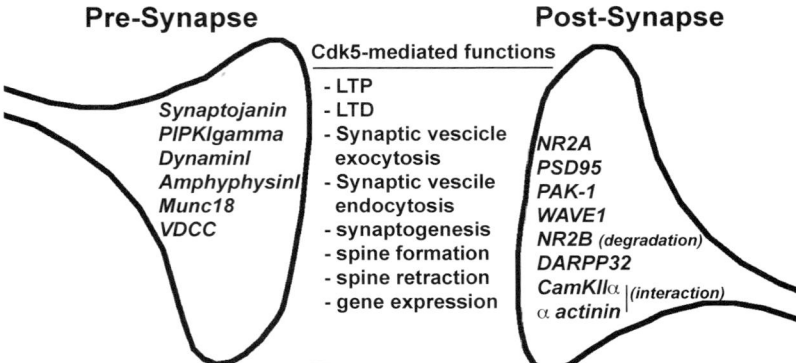

Fig. 1 Multiple functions of Cdk5 at the synapse. Cdk5 activity has been implicated with the regulation of various synaptic mechanisms (Li et al., 2001; Fischer et al., 2002, 2005; Norrholm et al., 2003; Ohshima et al., 2005). Cdk5 mediates its action mainly through phosphorylation of pre- and postsynaptic proteins (Lee et al., 2004, 2005; Tan et al., 2003; Tomizawa et al., 2003; Floyd et al., 2001; Shuang et al., 1998; Tomizawa et al., 2002; Nguyen et al., 2007; Li et al., 2001; Morabito et al., 2004; Nikolic et al., 1998; Sananbenesi et al., 2007; Kim et al., 2006; Fu et al., 2007). In addition, it has also been shown that the Cdk5/p35 complex can regulate synaptic mechanisms such as NR2B degradation in a phosphorylation-independent manner (Hawasli et al., 2007) and interact with synaptic proteins such as α-actinin or CamKIIα (Dhavan et al., 2002b) without phosphorylating them. Furthermore, Cdk5 activity affects gene expression such as the acetylcholine receptor at the neuromuscular junction (Fu et al., 2001)

the adult brain. It should be pointed out that the majority of data in regard to the role of Cdk5 are derived from excitatory neurons of the hippocampus, and it would be important to investigate other neuronal circuits as well. It is also obvious that for many of the cellular functions related to synaptic plasticity, Cdk5 is implicated in either promoting or inhibiting those processes. It is therefore of utmost importance to translate *in vitro* and *in vivo* findings into physiological experimental models in order to understand the diverse role of Cdk5 during learning (Fig. 1).

Deregulated Cdk5 Activity is Implicated in Neurodegenerative Diseases

The apparent Jekyll and Hyde character of Cdk5 is even more intriguing, considering its role in neurodegenerative diseases. Despite its role in synaptic plasticity and memory formation, Cdk5 activity is implicated in various neurodegenerative diseases. This apparent paradox is explained by the fact that under certain conditions, the Cdk5 activator p35 is cleaved to the N-terminal-truncated p25 protein. In contrast to p35, p25 lacks the myristoylation site important for the membrane localization of p35. In addition, p25 is more stable than p35. Thus, p25 causes aberrant activation of Cdk5 (Cruz and Tsai, 2004). More importantly, p25/Cdk5 induces hyperphosphorylation of tau (Cruz et al., 2003), while *in vitro* and *in vivo* data demonstrated that p35/Cdk5 is a poor tau kinase. For instance, triple transgenic mice expressing Cdk5, p35, and the human tau protein showed no hyperphosphorylation of tau and neurodegenerative phenotype (Van den Haute et al., 2001). In contrast, bi-transgenic mice expressing p25 under the control of the neuron-specific enolase (NSE) promoter and the human tau protein containing the P301L mutation found in front temporal dementia and parkinsonism that is linked to chromosome 17 (FTDP-17) developed neurofibrillary tangle pathology (Noble et al., 2003). In addition, p25 transgenic mice that inducibly express p25 under the control of the CamKII promoter display neurofibrillary pathology in the absence of deregulated human tau (Cruz et al., 2003).

Expression of p25 in various cell types including neurons is toxic and induces cell death (Tsai et al., 2004). More importantly, elevated p25 levels have been implicated in a number of neurodegenerative diseases and brain injuries such as AD, amyotrophic lateral sclerosis, Niemann–Pick Type C, Parkinson's disease, progressive supranuclear palsy, hippocampal sclerosis, and ischemia (Patrick et al., 1998; Bu et al., 2002; Borghi et al., 2002; Nguyen et al., 2001; Cruz and Tsai, 2004; Wang et al., 2003; Qu et al., 2007; Sen et al., 2006, 2007).

In most of those diseases, deregulated Cdk5 activity has been linked to abnormal phosphorylation of tau and deregulation of the cytoskeleton (Cruz et al., 2003; Nguyen et al., 2001; Bu et al., 2002; Borghi et al., 2002). However, it is clear that p25 mediates cell death via multiple actions. For example, the cell death of dopaminergic neurons in Parkinson's disease has been linked to p25/Cdk5-mediated phosphorylation of Parkin and Prx2, thereby most likely deregulating mitochondrial function and causing an increase in reactive oxygen species (ROS) (Smith et al., 2004; Avraham et al., 2007; Qu et al., 2007; Weishaupt et al., 2003). Similarly, hyperphosphorylation of the NMDA receptor subunit NR2A leads to deregulated calcium homeostasis and cell death of CA1 hippocampal neurons in ischemia (Wang et al., 2003). Especially since neurodegenerative diseases are often caused by multiple factors, it is likely that all of those mechanisms contribute to cell death in the various diseases and that additional p25/cdk5 dependent degenerative pathways will be discovered (see Table 1).

Most importantly, increased p25 levels and Cdk5 activity has been observed not only in *in vitro* experiments or animal models for neurodegenerative diseases but also in post-mortem or biopsy material from human patients (Sen et al., 2006, 2007; Patrick et al., 1999; Swatton, 2004; Tseng et al., 2002; Nakamura et al., 1997; Brion and Couck, 1995; Borghi et al., 2002).

For example p25 levels and Cdk5 activity were elevated in post-mortem brain tissues from AD patients (Patrick et al., 1999; Swatton, 2004; Tseng et al., 2002). It is noteworthy that other studies found p25 levels unchanged in post-mortem AD brains (Tandon et al., 2003; Yoo and Lubec, 2001). This discrepancy has been attributed to post-mortem processes that generate p25 (Taniguchi et al., 2001), suggesting that only results from short post-mortem

Table 1 The role of Cdk5/p25 in neurodegenerative diseases

Diseases	Cdk5 Deregulation/p25 Generation in Human Patients	Cdk5 Deregulation/p25 Generation in Animal Model	Proposed Mechanism for p25-Mediated Cell Death
Alzheimer's disease	Yes	Yes	Hyperphosphorylation of tau, inceased amyloid pathology, reentry to cell cycle (Cruz et al., 2006b; Cruz and Tsai, 2004; Fischer et al., 2005)
Amyotrophic lateral sclerosis	Nd	Yes	Hyperphosphorylation of tau and neurofillaments, cytoskeleton abnormalities (Nguyen et al., 2001; Kim et al., 2007)
Niemann–Pick Type C	Nd	Yes	Hyperphosphorylation of tau, MAP2 and neurofillaments, cytoskeleton abnormalities (Bu et al., 2002; Zhang et al., 2004; Hallows et al., 2006)
Parkinson	Yes	Yes	Phosphorylation of Parkin, and/or Prx2, mitochondrial dysfunction, decreased capacitiy to eliminate ROS (Nakamura et al., 1997; Brion and Couck, 1995; Smith et al., 2004; Qu et al., 2007; Avraham et al., 2007; Weishaupt et al., 2003)
Hippocampal sclerosis	Yes	Nd	Unclear (Sen et al., 2006, 2007)
Ischemia	Yes	Yes	Hyperphoshorylation of NR2A, mitochondrial dysfunction (Mitsios et al., 2007; Wang et al., 2003)
Progressive supranuclear palsy	Yes	Nd	Hyperphosphorylation of tau (Borghi et al., 2002)

Nd: not done.

delay samples are interpretable. Interestingly, one study that failed to detect increased p25 levels in the cortex of AD patients reported decreased Cdk5 levels in individuals with sporadic AD (Tandon et al., 2003). In summary, those studies indicate that the role of p25 in the pathogenesis of neurodegenerative diseases might be much more complex than initially expected. To this end, an interesting work by Sen et al. (2006) investigated hippocampal biopsy material from patients of hippocampal sclerosis. More importantly, the p25/p35 ratio was significantly increased in the hippocampus when compared to that of the adjacent temporal lobe, providing compelling evidence that increased p25 levels are associated with neurodegeneration in human patients. Even more striking was the observation that substantial amounts of p25 were detected in the hippocampi of patients displaying no neuronal loss. This suggests that p25 is produced at significant levels in young humans that do not suffer from neuro-degeneration (Sen et al., 2006).

What would be the role of p25 in healthy individuals? An interesting solution stems from the analysis of transgenic mice expressing p25 protein in an induci-ble manner (Fischer et al., 2005).

The first transgenic mice that modeled p25 pathology employed constitutive expression of p25 under the control of NSE or Thy1.2 promoter. Considering the pivotal role that Cdk5 plays in brain development, these results were some-what difficult to interpret (Ahlijanian et al., 2000; Bian et al., 2002). To circum-vent those problems, we generated bi-transgenic mice (CK-p25 mice) in which forebrain-restricted expression of p25 is under the control of the tet-off system (Cruz et al., 2003). This system allows for the selective induction of p25 through simple dietary alteration; removing bi-transgenic mice from a doxycycline-supplemented diet will derepress the transgene and allow for its expression (Fischer et al., 2005). Strikingly, CK-p25 mice develop severe neuronal loss in the cortex and hippocampus upon 6–8 weeks of induction (Cruz et al., 2003; Fischer et al., 2005). Notably, the progression of neurodegeneration is directly

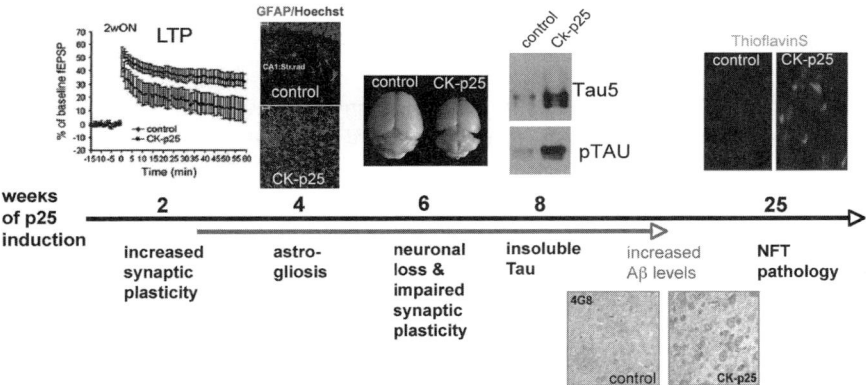

Fig. 2 The phenotypes of CK-p25 Tg mice. In bi-transgenic CK-p25 mice, inducible GFP-tagged human p25 expression is mediated by the CamK promoter (CK)-regulated tet-off system. As such p25 expression is repressed in the presence of doxycycline. We reported that these mice exhibit increased hippocampal-dependent learning and synaptic plasticity such as facilitated LTP and NMDA conductivity upon 2 weeks of p25 expression. The first pathological hallmark, hippocampal astrogliosis, is detected upon 4 weeks of induction followed by significant neuronal loss after 6 weeks of p25 expression. Neuronal loss is accompanied by severe learning and memory impairments and synaptic plasticity defects. Insoluble tau is first detected after 8 weeks of induction and neurofibrillary tangles (NFT) occur after 25 weeks. Interestingly, an increase in amyloid-beta peptides is detectable from 2 to 9 weeks of p25 induction (Cruz et al., 2003, 2006a; Fischer et al., 2005, 2007). (*See* Color Insert)

correlated with the duration of p25 expression, providing compelling evidence that p25 is capable of inducing neurodegeneration *in vivo*. This loss of neurons is accompanied by a reduction in synapses and dendrites in addition to severely impaired associative and spatial learning behavior (Fischer et al., 2005). The surprise came when mice that expressed p25 for only 2 weeks were analyzed. At this time point, no overt signs of neurodegeneration were observed. Instead, CK-p25 mice showed facilitated learning, elevated hippocampal LTP, and an increased number of synapses (Fischer et al., 2005) (Fig. 2). Similar findings were obtained in another line of p25 transgenic mice expressing low levels of p25 under the CamKII promoter in a non-inducible manner (Angelo et al., 2003; Ris et al., 2005). It is interesting to note that, depending on the strain background, the latter studies reported enhanced learning and hippocampal LTP in female p25 transgenic mice (Ris et al., 2005). In summary, these very unexpected data suggest that both low and high levels of transient p25 activity in the adult brain might initially be beneficial and even increase synaptic function and learning behavior. In contrast, prolonged p25 expression leads to Alzheimer's-like neurodegeneration.

p25 as a Plasticity Factor

The finding that low or even high levels of transiently elevated p25 may increase synaptic plasticity and facilitate learning raises questions in regard to the physiological role of the molecule. Since p25 is generated by calpain-mediated cleavage of the p35 protein, it is interesting to note that calpain is implicated in synaptic function and memory consolidation (Toth et al., 1996; Shimizu et al., 2007). Moreover, p35 is membrane localized and enriched at the synapse, suggesting that p25 generation could serve as a rapid mechanism to locally increase Cdk5 activity at the synapse when local Ca^{2+} concentration rises. It is therefore intriguing to speculate that p25 might be generated during learning processes. This could explain why substantial p25 levels are observed in biopsy material from young individuals that do not show neurodegeneration. We therefore propose that p25 is a plasticity factor that normally facilitates neuronal functions. However, p25 also contributes to neurodegeneration when chronically elevated.

What is the mechanism that underlies elevation of p25 to a detrimental level? A milestone study from Lu et al. (2004) investigated the gene expression profile in the aging human brain. They found that a certain group of genes involved in synaptic plasticity was downregulated in aged individuals likely due to DNA damage in the promoter region of the gene (Lu et al., 2004). One of the most drastically downregulated genes during human aging is p35 (Lu et al., 2004). This obervation suggests that Cdk5 activity decreases during aging. Since Cdk5 is a critical factor for synaptic plasticity, and the loss of p35 is associated with synaptic loss and learning impairment (Fischer et al., 2005; Ohshima et al.,

2005), it can be speculated that p25 is generated as a compensatory process to maintain physiological levels of Cdk5 activity with less p35 protein available.

This view is in line with the homeostatic plasticity model stating that similar to other regulation networks, the neuronal network has a great potential for compensation and is rather robust and flexible (Marder and Goaillard, 2006). For example, although the total number of neurons and synapses is reduced, the synaptic contact size is increased in brain samples of Alzheimer's patients (Scheff, 2003). Similarly, the expression of certain proteins related to synaptic plasticity such as GABA and glutamate receptors is elevated in AD brains (Armstrong et al., 2003). In addition "reactive synaptogenesis" takes place in response to brain injury (Kaplan, 1988; Scheff, 2003). These phenomena have been interpreted as compensatory responses that occur when neuroplasticity is impaired in order to retain the integrity of the neuronal network (Cotman and Anderson, 1988; Scheff, 2003). Thus, it can be speculated that p25-mediated synaptogenesis and increased plasticity may initially represent a compensatory response during aging and neurodegenerative diseases.

If we assume that p25 is produced to compensate for the loss of Cdk5 activity during aging, additional factors must eventually drive p25 to chronically high levels that contribute to the pathogenesis of AD. Interestingly, p25 expression is enhanced by several risk factors for AD, such as exitotoxicity, excessive Aβ-peptides, and ischemia (Cruz and Tsai, 2004; Lee et al., 2000; Wang et al., 2003). This effect can be enhanced in people carrying SNPs in the Cdk5 gene that is linked to sporadic AD (Rademakers et al., 2005) or are exposed to other AD risk factors, so that p25 eventually reaches critical levels and causes neuronal loss and cognitive decline. This could also help to explain the discrepancy among the studies investigating p25 levels in post-mortem human material. Notably, all studies analyzing p25 so far employed tissue from patients with later stages of AD and substantial brain atrophy. As our data indicate that p25 production might be a compensatory phenomenon in response to AD risk factors, p25 production may be a rather early event in AD pathogenesis. As such, it might be helpful to analyze p25 levels in distinct brain regions of patients with pre-clinical AD and mild cognitive impairment without substantial neuronal loss.

Notably, elevated p25 levels were also observed in animal models of neurodegenerative diseases (Nguyen et al., 2001; Saura et al., 2004; Oakley et al., 2006; Otth et al., 2002; Wu et al., 2007). For example, increased p25 levels were observed in mice lacking presenilin 1 and 2 (Saura et al., 2004; Wu et al., 2007) or expressing the APP and PS1 genes that harbor a combination of five different mutations associated with familial AD (Oakley et al., 2006). Although increased p25 levels in those models have been implicated with neuronal loss, it is important to note that p25 is detected in APPswe mice that show impaired learning and memory but do not develop neurodegeneration (Otth et al., 2002), indicating that elevated p25 levels do not always correlate with neuronal loss. It is intriguing to speculate the APPswe mice would perform even worse in learning and memory tests without elevated levels of p25. It is obvious that there seems to be a critical threshold for the p25 levels in the brain that determines cell death and detrimental effects.

More importantly, chronic expression of low p25 levels in the forebrain of transgenic mice does not cause neuronal loss (Angelo et al., 2003). Taken together, these data suggest that p25 production *in vivo* is not detrimental per se but can lead to neuronal cell death when p25 levels are chronically high. This view is also supported by the fact that p25 expression under the NSE promoter can compensate for the phenotype of p35-deficient mice (Patzke et al., 2003).

It is important to note that p25 generation at high levels has been implicated not only with neuronal and synaptic loss but also with Tau pathology (Cruz and Tsai, 2004), the generation of Aβ-peptides (Cruz et al., 2006b; Lee et al., 2003), and learning impairment (Fischer et al., 2005). To this end, p25 is unique in that its generation at high levels can contribute to all major hallmarks of AD. This is particularly exciting since a unifying theory for the pathogenesis of AD is not yet available.

In summary, it has been postulated that a key event during the pathogenesis of AD, which manifests over several years, is that compensatory mechanisms initially enhancing neuroplasticity may eventually become maladaptive when chronically activated (Arendt, 2004; Mesulam, 1999). Such a scenario can indeed be envisioned for p25. (Fig. 3).

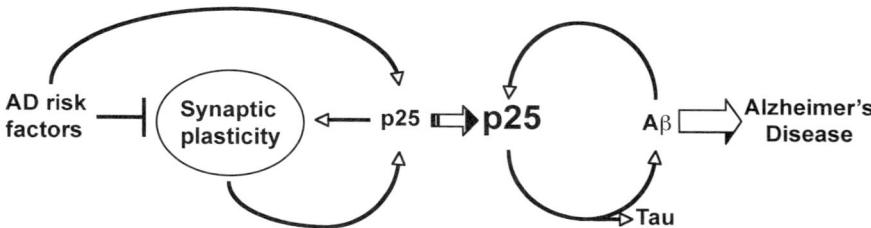

Fig. 3 The Jekyl and Hyde role of p25 in the pathogenesis of sporadic Alzheimer's disease. A number of risk factors for AD, such as aging, challenge neuroplasticity, and impair learning and memory. In response, the brain is capable to initiate a compensatory response when neuroplasticity is challenged. We propose that increased p25 levels are part of this compensatory response for the following reasons: 1. AD risk factors such as aging decrease the expression of the p35 gene but increase p25 levels (Lu et al., 2004; Patzke et al., 2003). 2. Animal studies suggest that low levels of p25 or only transient increase in p25 can facilitate synaptic plasticity. However, when p25 levels become chronically elevated, this will cause the major phenotypes observed during the pathogenesis of AD, such as amyloid and tau pathology. In summary, we propose a complex feedback mechanism with Cdk5/p25 as a key factor in the onset of sporadic AD that links amyloid, tau pathology, neurodegeneration, and learning impairment. Therefore, p25 would be a suitable drug target in AD. However, because Cdk5 activity critically regulates learning and memory, targeting p25/Cdk5 is difficult. Since p25 is a plasticity factor and may have physiological functions, we speculate that mechanisms for p25 degradation exist and that identifying such mechanisms will be of utmost importance and may provide a powerful drug target.

Acknowledgments We would like to thank Dr. Benjamin Samuals, Dr.Farahnaz Sananbe-nesi, Matthew Dobbin, and Jessica Wittnam for reading the manuscript and critical discussion. This work was supported by a EURYI award to AF. The ENI-Goettingen is jointly funded by the Max Planck Society and the Medical School, Georg-August University, Göttingen, Germany. L-HT is an investigator of the Howard Hughes Medical Institute, RIKEN-MIT Neuroscience Research Center, director of Neurobiology Program at Stanley Center for Psychiatric Research, Cambridge, MA, USA.

References

Abel, T., Nguyen, P. V., Barad, M., Deuel, T. A., Kandel, E. R., and Bourtchouladze, R. (1997). Genetic demonstration of a role for PKA in the late phase of LTP and in hippocampus-based long-term memory. Cell *88*, 615–626.

Ahlijanian, M. K., Barrezueta, N. X., Williams, R. D., Jakowski, A., Kowsz, K. P., McCarthy, S., Coskran, T., Carlo, A., Seymour, P. A., Burkhardt, J. E., et al. (2000). Hyperphosphorylated tau and neurofilament and cytoskeletal disruptions in mice overexpressing human p25, an activator of cdk5. Proc Natl Acad Sci U S A *97*, 2910–2915.

Alberini, C. M. (1999). Genes to remember. J Exp Biol *202*, 2887–2891.

Angelo, M., Plattner, F., Irvine, E. E., and Giese, K. P. (2003). Improved reversal learning and altered fear conditioning in transgenic mice with regionally restricted p25 expression. Eur J Neurosci *18*, 423–431.

Arber, S., Barbayannis, F. A., H., H., Schneider, C., Stanyon, C. A., Bernard, O., and Caroni, P. (1998). Regulation of actin dynamics through phosphorylation of cofilin by LIM-kinase. Nature *393*, 739–740.

Arendt, T. (2004). Neurodegeneration and plasticity. Int J Dev Neurosci *22*, 507–514.

Armstrong, D. M., Sheffield, R., Mishizen-Eberz, A. J., Carter, T. L., Rissman, R. A., Mizukami, K., and Ikonomovic, M. D. (2003). Plasticity of glutamate and GABAA receptors in the hippocampus of patients with Alzheimer's disease. Cell Mol Neurobiol *23*, 491–505.

Atkins, C. M., Selcher, J. C., Petraitis, J. J., Trzaskos, J. M., and Sweatt, J. D. (1998). The MAPK cascade is required for mammalian associative learning. Nat Neurosci *1*, 602–609.

Avraham, E., Rott, R., Liani, E., Szargel, R., and Engelender, S. (2007). Phosphorylation of Parkin by the cyclin-dependent kinase 5 at the linker region modulates its ubiquitin-ligase activity and aggregation. J Biol Chem *282*, 12842–12850.

Ayala, R., Shu, T., and Tsai, L. H. (2007). Trekking across the brain: the journey of neuronal migration. Cell *128*, 29–43.

Bamji, S. X., Shimazu, K., Kimes, N., Huelsken, J., Birchmeier, W., Lu, B., and Reichardt, L. F. (2003). Role of beta-catenin in synaptic vesicle localization and presynaptic assembly. Neuron *40*, 719–731.

Barco, A., Bailey, C. H., and Kandel, E. R. (2006). Common molecular mechanisms in explicit and implicit memory. J Neurochem *97*, 1520–1533.

Beffert, U., Weeber, E. J., Morfini, G., Ko, J., Brady, S. T., Tsai, L. H., Sweatt, J. D., and Herz, J. (2004). Reelin and cyclin-dependent kinase 5-dependent signals cooperate in regulating neuronal migration and synaptic transmission. J Neurosci *24*, 1897–1906.

Bian, F., Nath, R., Sobocinski, G., Booher, R. N., Lipinski, W. J., Callahan, M. J., Pack, A., Wang, K. K., and Walker, L. C. (2002). Axonopathy, tau abnormalities, and dyskinesia, but no neurofibrillary tangles in p25-transgenic mice. J Comp Neurol *446*, 257–266.

Bibb, J. A., Chen, J., Taylor, J. R., Svenningsson, P., Nishi, A., Snyder, G. L., Yan, Z., Sagawa, Z. K., Ouimet, C. C., and Nairn, A. C., et al. (2001). Effects of chronic exposure to cocaine are regulated by the neuronal protein Cdk5. Nature *410*, 376–380.

Bonhoeffer, T., and Yuste, R. (2002). Spine motility. Phenomenology, mechanisms, and function. Neuron *35*, 1019–1027.

Borghi, R., Giliberto, L., Assini, A., Delacourte, A., Perry, G., Smith, M. A., Strocchi, P., Zaccheo, D., and Tabaton, M. (2002). Increase of cdk5 is related to neurofibrillary pathology in progressive supranuclear palsy. Neurology *58*, 589–592.

Brion, J. P., and Couck, A. M. (1995). Cortical and brainstem-type Lewy bodies are immunoreactive for the cyclin-dependent kinase 5. Am J Pathol *147*, 1465–1476.

Brose, N. (1999). Synaptic cell adhesion proteins and synaptogenesis in the mammalian central nervous system. Naturwissenschaften *86*, 516–524.

Bu, B., Li, J., Davies, P., and Vincent, I. (2002). Deregulation of cdk5, hyperphosphorylation, and cytoskeletal pathology in the Niemann-Pick type C murine model. J Neurosci *22*, 6515–6525.

Chae, T., Kwon, Y. T., Bronson, R., Dikkes, P., Li, E., and Tsai, L. H. (1997). Mice lacking p35, a neuronal specific activator of Cdk5, display cortical lamination defects, seizures, and adult lethality. Neuron *18*, 29–42.

Cheung, Z. H., Chin, W. H., Chen, Y., Ng, Y. P., and Ip, N. Y. (2007). Cdk5 Is Involved in BDNF-stimulated dendritic growth in hippocampal neurons. PLos Biol *5*, e63.

Cheung, Z. H., Fu, A. K., and Ip, N. Y. (2006). Synaptic roles of Cdk5: implications in higher cognitive functions and neurodegenerative diseases. Neuron *50*, 13–18.

Cotman, C. W., and Anderson, K. J. (1988). Synaptic plasticity and functional stabilization in the hippocampal formation: possible role in Alzheimer's disease. Adv Neurol *47*, 12–35.

Cruz, J. C., Kim, D., Moy, L. Y., Dobbin, M. M., Sun, X., Bronson, R. T., and Tsai, L. H. (2006a). Free Full Text p25/cyclin-dependent kinase 5 induces production and intraneuronal accumulation of amyloid beta in vivo. J Neurosci *26*, 10536–10541.

Cruz, J. C., Kim, D., Moy, L. Y., Dobbin, M. M., Sun, X., Bronson, R. T., and Tsai, L. H. (2006b). p25/cyclin-dependent kinase 5 induces production and intraneuronal accumulation of amyloid beta in vivo. J Neurosci *26*, 10536–10541.

Cruz, J. C., and Tsai, L. H. (2004). Jekyll and Hyde kinase: roles for Cdk5 in brain development and disease. Curr Opin Neurobiol *14*, 390–394.

Cruz, J. C., Tseng, H. C., Goldman, J. A., Shih, H., and Tsai, L. H. (2003). Aberrant Cdk5 activation by p25 triggers pathological events leading to neurodegeneration and neurofibrillary tangles. Neuron *40*, 471–483.

Dhavan, R., Greer, P. L., Morabito, M. A., Orlando, L. R., and Tsai, L. H. (2002a). The cyclin-dependent kinase 5 activators p35 and p39 interact with the alpha-subunit of Ca2+/calmodulin-dependent protein kinase II and alpha-actinin-1 in a calcium-dependent manner. J Neurosci *22*, 7879–7891.

Dhavan, R., Greer, P. L., Morabito, M. A., Orlando, L. R., and Tsai, L. H. (2002b). The cyclin-dependent kinase 5 activators p35 and p39 interact with the alpha-subunit of Ca2+/calmodulin-dependent protein kinase II and alpha- actinin-1 in a calcium-dependent manner. J Neurosci *22*, 7879–7891.

Dhavan, R., and Tsai, L. H. (2001). A decade of CDK5. Nat Rev Mol Cell Biol *2*, 749–759.

Edwards, D. C., Sanders, L. C., Bokoch, G. M., and Gill, G. N. (1999). Activation of LIM-kinase by Pak1 couples Rac/Cdc42 GTPase signalling to actin cytoskeletal dynamics. Nat Cell Biol *1*, 253–259.

Fischer, A., Sananbenesi, F., Pang, P. T., Lu, B., and Tsai, L. H. (2005). Opposing roles of transient and prolonged expression of p25 in synaptic plasticity and hippocampus-dependent memory. Neuron *48*, 825–838.

Fischer, A., Sananbenesi, F., Schrick, C., Spiess, J., and Radulovic, J. (2002). Cyclin-dependent kinase 5 is required for associative learning. J Neurosci *22*, 3700–3707.

Fischer, A., Sananbenesi, F., Schrick, C., Spiess, J., and Radulovic, J. (2004). Distinct roles of hippocampal de novo protein synthesis and actin rearrangement in extinction of contextual fear. J Neurosci *24*, 1962–1966.

Fischer, A., Sananbenesi, F., Spiess, J., and Radulovic, J. (2003). Cdk5 in the adult non-demented brain. Curr Drug Targets CNS Neurol Disord 6, 375–381.

Fischer, A., Sananbenesi, F., Wang, X., Dobbin, M., and Tsai, L. H. (2007). Recovery of learning and memory after neuronal loss is associated with chromatin remodeling. Nature 447, 178–182

Flavell, S. W., Cowan, C. W., Kim, T. K., Greer, P. L., Lin, Y., Paradis, S., Griffith, E. C., Hu, L. S., Chen, C., and Greenberg, M. E. (2006). Activity-dependent regulation of MEF2 transcription factors suppresses excitatory synapse number. Science 311, 1008–1012.

Fletcher, A. I., Shuang, R., Giovannucci, D. R., Zhang, L., Bittner, M. A., and Stuenkel, E. L. (1999). Regulation of exocytosis by cyclin-dependent kinase 5 via phosphorylation of Munc18. J Biol Chem 274, 4027–4035.

Floyd, S. R., Porro, E. B., Slepnev, V. I., Ochoa, G. C., Tsai, L. H., and De Camilli, P. (2001). Amphiphysin 1 binds the cyclin-dependent kinase (cdk) 5 regulatory subunit p35 and is phosphorylated by cdk5 and cdc2. J Biol Chem 276, 8104–8110.

Frey, U., and Morris, R. G. (1998). Synaptic tagging: implications for late maintenance of hippocampal long-term potentiation. Trends Neurosci 21, 181–188.

Fu, A. K., Fu, W. Y., Cheung, J., Tsim, K. W., Fanny, F. C. Ip., Wang, J. H., and Ip, N. Y. (2001). Cdk5 is involved in neuregulin-induced AChR expression at the neuromuscular junction. Nat Neurosci 4, 374–381.

Fu, W. Y., Chen, Y., Sahin, M., Zhao, X. S., Shi, L., Bikoff, J. B., Lai, K. O., Yung, W. H., Fu, A. K., Greenberg, M. E., and Ip, N. Y.. (2007). Cdk5 regulates EphA4-mediated dendritic spine retraction through an ephexin1-dependent mechanism. Nat Neurosci 10, 67–76.

Gong, X., Tang, X., Wiedmann, M., Wang, X., Peng, J., Zheng, D., Blair, L. A., Marshall, J., and Mao, Z. (2003). Cdk5-mediated inhibition of the protective effects of transcription factor MEF2 in neurotoxicity-induced apoptosis. Neuron 38, 33–46.

Govindarajan, A., Kelleher, R. J., and Tonegawa, S. (2006). A clustered plasticity model of long-term memory engrams. Nat Rev Neurosci 7, 575–583.

Hallows, J. L., Iosif, R. E., Biasell, R. D., and Vincent, I. (2006). p35/p25 is not essential for tau and cytoskeletal pathology or neuronal loss in Niemann-Pick type C disease. J Neurosci 26, 2738–2744.

Halpain, S. (2000). Actin and the agile spine: how and why do dendritic spines dance? Trends Neurosci 23, 141–146.

Hawasli, A. H., Benavides, D. R., Nguyen, C., Kansy, J. W., Hayashi, K., Chambon, P., Greengard, P., Powell, C. M., Cooper, D. C., and Bibb, J. A. (2007). Cyclin-dependent kinase 5 governs learning and synaptic plasticity via control of NMDAR degradation. Nat Neurosci 10,:880–806.

Hayashi, M. L., Choi, S. Y., Rao, B. S., Jung, H. Y., Lee, H. K., Zhang, D., Chattarji, S., Kirkwood, A., and Tonegawa, S. (2004). Altered cortical synaptic morphology and impaired memory consolidation in forebrain- specific dominant-negative PAK transgenic mice. Neuron 42, 773–787.

Humbert, S., Dhavan, R., and Tsai, L. (2000). p39 activates cdk5 in neurons, and is associated with the actin cytoskeleton. J Cell Sci 113, 975–983.

Kandel, E. R. (2001). The molecular biology of memory storage: a dialogue between genes and synapses. Science 294, 1030–1038.

Kaplan, M. S. (1988). Plasticity after brain lesions: contemporary concepts. Arch Phys Med Rehabil 69, 984–991.

Kim, D., Nguyen, M. D., Dobbin, M. M., Fischer, A., Sananbenesi, F., Rodgers, J. T., Delalle, I., Baur, J. A., Sui, G., Armour, S. M., et al. (2007). SIRT1 deacetylase protects against neurodegeneration in models for Alzheimer's disease and amyotrophic lateral sclerosis. EMBO J 26, 3169–3179.

Kim, J. J., and Fanselow, M. S. (1992). Modality-specific retrograde amnesia of fear. Science 256, 675–677.

Kim, Y., Sung, J. Y., Ceglia, I., Lee, K. W., Ahn, J. H., Halford, J. M., Kim, A. M., Kwak, S. P., Park, J. B., Ho Ryu, S., et al. (2006). Phosphorylation of WAVE1 regulates actin polymerization and dendritic spine morphology. Nature *442*, 814–817.

Klann, E., and Sweatt, J. D. (2007). Altered protein synthesis is a trigger for long-term memory formation. Neurobiol Learn Mem, *89,*:247–259.

Ko, J., Humbert, S., Bronson, R. T., Takahashi, S., Kulkarni, A. B., Li, E., and Tsai, L. H. (2001). p35 and p39 are essential for cyclin-dependent kinase 5 function during neurodevelopment. J Neurosci *21*, 6758–6771.

Krucker, T., Siggins, G. R., and Halpain, S. (2000). Dynamic actin filaments are required for stable long-term potentiation (LTP) in area CA1 of the hippocampus. Proc Natl Acad Sci U S A *97*, 6856–6861.

Kwon, Y. T., Gupta, A., Zhou, Y., Nikolic, M., and Tsai, L. H. (2000). Regulation of N-cadherin-mediated adhesion by the p35-Cdk5 kinase. Curr Biol *10*, 363–372.

Lee, M. S., Kao, S. C., Lemere, C. A., Xia, W., Tseng, H. C., Zhou, Y., Neve, R., Ahlijanian, M. K., and Tsai, L. H. (2003). APP processing is regulated by cytoplasmic phosphorylation. J Cell Biol *163*, 83–95.

Lee, M. S., Kwon, Y. T., Li, M., Peng, J., Friedlander, R. M., and Tsai, L. H. (2000). Neurotoxicity induces cleavage of p35 to p25 by calpain. Nature *405*, 360–364.

Lee, S. Y., Voronov, S., Letinic, K., Nairn, A. C., Di Paolo, G., and De Camilli, P. (2005). Regulation of the interaction between PIPKI gamma and talin by proline-directed protein kinases. J Cell Biol *168*, 789–799.

Lee, S. Y., Wenk, M. R., Kim, Y., Nairn, A. C., and De Camilli, P. (2004). Regulation of synpatojanin 1 by cyclin-dependent kinase 5 at synapses. Proc Natl Acad Sci U S A *101*, 546–551.

Leuner, B., Falduto, J., and Shors, T. J. (2003). Associative memory formation increases the observation of dendritic spines in the hippocampus. J Neurosci *23*, 659–665.

Levenson, J. M., Roth, T. L., Lubin, F. D., Miller, C. A., Huang, I. C., Desai, P., Malone, L. M., and Sweatt, J. D. (2006). Evidence that DNA (cytosine-5) methyltransferase regulates synaptic plasticity in the hippocampus. J Biol Chem *281*, 15763–15773.

Li, B. S., Sun, M. K., Zhang, L., Takahashi, S., Ma, W., Vinade, L., Kulkarni, A. B., Brady, R. O., and Pant, H. C. (2001). Regulation of NMDA receptors by cyclin-dependent kinase-5. Proc Natl Acad Sci U S A *98*, 12742–12747.

Lilja, L., Johansson, J. U., Gromada, J., Mandic, S. A., Fried, G., Berggren, P. O., and Bark, C. (2004). Cyclin-dependent kinase 5 associated with p39 promotes Munc18-1 phosphorylation and Ca(2+)-dependent exocytosis. J Biol Chem *279*, 29534–29541.

Lin, W., Dominguez, B., Yang, J., Aryal, P., Brandon, E. P., Gage, F. H., and Lee, K. F. (2005). Neurotransmitter acetylcholine negatively regulates neuromuscular synapse formation by a Cdk5-dependent mechanism. Neuron *46*, 141–150.

Lu, T., Pan, Y., Kao, S. Y., Li, C., Kohane, I., Chan, J., and Yankner, B. A. (2004). Gene regulation and DNA damage in the ageing human brain. Nature *429*, 883–891.

Malkani, S., and Rosen, J. B. (2000). Specific induction of early growth response gene 1 in the lateral nucleus of the amygdala following contextual fear conditioning in rats. Neuroscience *102*, 853–861.

Marder, E., and Goaillard, J. M. (2006). Variability, compensation and homeostasis in neuron and network function. Nat Rev Neurosci *7*, 563–574.

Matsubara, M., Kusubata, M., Ishiguro, K., Uchida, T., Titani, K., and Taniguchi, H. (1996). Site-specific phosphorylation of synapsin I by mitogen-activated protein kinase and Cdk5 and its effects on physiological functions. J Biol Chem *271*, 21108–21113.

Matus, A. (2000). Actin-based plasticity in dendritic spines. Science *290*, 754–758.

Mesulam, M. M. (1999). Neuroplasticity failure in Alzheimer's disease: bridging the gap between plaques and tangles. Neuron *24*, 521–529.

Mitsios, N., Pennucci, R., Krupinski, J., Sanfeliu, C., Gaffney, J., Kumar, P., Kumar, S., Juan-Babot, O., and Slevin, M. (2007). Expression of cyclin-dependent kinase 5 mRNA and protein in the human brain following acute ischemic stroke. Brain Pathol *17*, 11–23.

Morabito, M. A., Sheng, M., and Tsai, L. H. (2004). Cyclin-dependent kinase 5 phosphorylates the N-terminal domain of the postsynaptic density protein PSD-95 in neurons. J Neurosci *24*, 865–876.

Murase, S., Mosser, E., and Schuman, E. M. (2002). Depolarization drives beta-Catenin into neuronal spines promoting changes in synaptic structure and function. Neuron *35*, 91–105.

Nakamura, S., Kawamoto, Y., Nakano, S., Akiguchi, I., and Kimura, J. (1997). p35nck5a and cyclin-dependent kinase 5 colocalize in Lewy bodies of brains with Parkinson's disease. Acta Neuropathol (Berl) *94*, 153–157.

Nguyen, C., and Bibb, J. A. (2003). Cdk5 and the mystery of synaptic vesicle endocytosis. J Cell Biol *163*, 697–699.

Nguyen, C., Hosokawa, T., Kuroiwa, M., Ip, N. Y., Nishi, A., Hisanaga, S., and Bibb, J. A. (2007). Differential regulation of the Cdk5-dependent phosphorylation sites of inhibitor-1 and DARPP-32 by depolarization. J Neurochem *103*, 1582–1593.

Nguyen, M. D., Lariviere, R. C., and Julien, J. P. (2001). Deregulation of Cdk5 in a mouse model of ALS: toxicity alleviated by perikaryal neurofilament inclusions. Neuron *30*, 135–147.

Nikolic, M., Chou, M. M., Lu, W., Mayer, B. J., and Tsai, L. H. (1998). The p35/Cdk5 kinase is a neuron-specific Rac effector that inhibits Pak1 activity. Nature *395*, 194–198.

Noble, W., Olm, V., Takata, K., Casey, E., Mary, O., Meyerson, J., Gaynor, K., LaFrancois, J., Wang, L., Kondo, T., et al. (2003). Cdk5 is a key factor in tau aggregation and tangle formation in vivo. Neuron *38*, 555–565.

Norrholm, S. D., Bibb, J. A., Nestler, E. J., Ouimet, C. C., Taylor, J. R., and Greengard, P. (2003). Cocaine-induced proliferation of dendritic spines in nucleus accumbens is dependent on the activity of cyclin-dependent kinase-5. Neuroscience *116*, 19–22.

Oakley, H., Cole, S. L., Logan, S., Maus, E., Shao, P., Craft, J., Guillozet-Bongaarts, A., Ohno, M., Disterhoft, J., Van Eldik, L., et al. (2006). ntraneuronal beta-amyloid aggregates, neurodegeneration, and neuron loss in transgenic mice with five familial Alzheimer's disease mutations: potential factors in amyloid plaque formation. J Neurosci *26*, 10129–10140.

Ohshima, T., Ogura, H., Tomizawa, K., Hayashi, K., Suzuki, H., Saito, T., kamei, H., Nishi, A., Bibb, J. A., Hisanaga, S., et al. (2005). Impairment of hippocampal long-term depression and defective spatial learning and memory in p35–/– mice. Journal of Neurochemistry *10*, 4159–4168.

Ohshima, T., Ward, J. M., Huh, C. G., Longenecker, G., Veeranna, Pant, H. C., Brady, R. O., Martin, L. J., and Kulkarni, A. B. (1996). Targeted disruption of the cyclin-dependent kinase 5 gene results in abnormal corticogenesis, neuronal pathology and perinatal death. Proc Natl Acad Sci U S A *93*, 11173–11178.

Otth, C., Concha, II, Arendt, T., Stieler, J., Schliebs, R., Gonzalez-Billault, C., and Maccioni, R. B. (2002). AbetaPP induces cdk5-dependent tau hyperphosphorylation in transgenic mice Tg2576. J Alzheimers Dis *4*, 417–430.

Patel, L. S., Wenzel, H. J., and Schwartzkroin, P. A. (2004). Physiological and morphological characterization of dentate granule cells in the p35 knock-out mouse hippocampus: evidence for an epileptic circuit. J Neurosci *24*, 9005–9014.

Patrick, G. N., Zhou, P., Kwon, Y. T., Howley, P. M., and Tsai, L. H. (1998). p35, the neuronal-specific activator of cyclin-dependent kinase 5 (Cdk5) is degraded by the ubiquitin-proteasome pathway. J Biol Chem *273*, 24057–24064.

Patrick, G. N., Zukerberg, L., Nikolic, M., de la Monte, S., Dikkes, P., and Tsai, L. H. (1999). Conversion of p35 to p25 deregulates Cdk5 activity and promotes neurodegeneration. Nature *402*, 615–622.

Patzke, H., Maddineni, U., Ayala, R., Morabito, M., Volker, J., Dikkes, P., Ahlijanian, M. K., and Tsai, L. H. (2003). Partial rescue of the p35-/- brain phenotype by low expression of a neuronal-specific enolase p25 transgene. J Neurosci *23*, 2769–2778.

Perez-Moreno, M., Jamora, C., and Fuchs, E. (2003). Sticky business: orchestrating cellular signals at adherens junctions. Cell *112*, 535–548.

Qu, D., Rashidian, J., Mount, M. P., Aleyasin, H., Parsanejad, M., Lira, A., Haque, E., Zhang, Y., Callaghan, S., Daigle, M., et al. (2007). Role of Cdk5-mediated phosphorylation of Prx2 in MPTP toxicity and Parkinson's disease. Neuron *55*, 37–52.

Rademakers, R., Sleegers, K., Theuns, J., Van den Broeck, M., Bel Kacem, S., Nilsson, L. G., Adolfsson, R., van Duijn, C. M., Van Broeckhoven, C., and Cruts, M. (2005). Association of cyclin-dependent kinase 5 and neuronal activators p35 and p39 complex in early-onset Alzheimer's disease. Neurobiol Aging *8*, 1145–1151.

Radulovic, J., Kammermeier, J., and Spiess, J. (1998). Relationship between fos production and classical fear conditioning: effects of novelty, latent inhibition, and unconditioned stimulus preexposure. J Neurosci *18*, 7452–7461.

Rashid, T., Banerjee, M., and Nikolic, M. (2001). Phosphorylation of Pak1 by the p35/Cdk5 kinase affects neuronal morphology. J Biol Chem *276*, 49043–49052.

Ris, L., Angelo, M., Plattner, F., Capron, B., Errington, M. L., Bliss, T. V., Godaux, E., and Giese, K. P. (2005). Sexual dimorphisms in the effect of low-level p25 expression on synaptic plasticity and memory. Eur J Neurosci *21*, 3023–3033.

Roberson, E. D., and Sweatt, J. D. (1999). A biochemical blueprint for long-term memory. Learn Mem *6*, 399–416.

Sananbenesi, F., Fischer, A., Wang, X., Schrick, C., Neve, R., Radulovic, J., and Tsai, L. H. (2007). A hippocampal Cdk5 pathway regulates extinction of contextual fear. Nat Neurosci *10*, 1012–1019.

Saura, C. A., Choi, S. Y., Beglopoulos, V., Malkani, S., Zhang, D., Shankaranarayana Rao, B. S., Chattarji, S., Kelleher, R. J., 3rd, Kandel, E. R., Duff, K., et al. (2004). Loss of presenilin function causes impairments of memory and synaptic plasticity followed by age-dependent neurodegeneration. Neuron *42*, 23–36.

Scheff, S. (2003). Reactive synaptogenesis in aging and Alzheimer's disease: lessons learned in the Cotman laboratory. Neurochem Res *11*, 1625–1630.

Scheff, S. W., and Price, D. A. (2006). Alzheimer's disease-related alterations in synaptic density: neocortex and hippocampus. J Alzheimers Dis *9*, 101–115.

Schuman, E. M., and Murase, S. (2003). Adherins and synaptic plasticity: activity-dependent cyclin-dependent kinase 5 regulation of synaptic beta-catenin-cadherin interactions. Philos Trans R Soc Lond B Biol Sci *358*, 749–756.

Scoville, W. B., and Milner, B. (1957). Loss of recent memory after bilateral hippocampal lesions. Neuropsychiatry Clin Neurosci 2000 (classical article) *1*, 103–113.

Selcher, J. C., Weeber, E. J., Varga, A. W., Sweatt, J. D., and Swank, M. (2002). Protein kinase signal transduction cascades in mammalian associative conditioning. Neuroscientist *8*, 122–131.

Sen, A., Thom, M., Martinian, L., Jacobs, T., Nikolic, M., and Sisodiya, S. M. (2006). Deregulation of cdk5 in Hippocampal sclerosis. J Neuropathol Exp Neurol *65*, 55–66.

Sen, A., Thom, M., Martinian, L., Yogarajah, M., Nikolic, M., and Sisodiya, S. M. (2007). Increased immunoreactivity of cdk5 activators in hippocampal sclerosis. Neuroreport *18*, 511–516.

Shalizi, A., Gaudilliere, B., Yuan, Z., Stegmuller, J., Shirogane, T., Ge, Q., Tan, Y., Schulman, B., Harper, J. W., and Bonni, A. (2006). A calcium-regulated MEF2 sumoylation switch controls postsynaptic differentiation. Science *311*, 1012–1017.

Shimizu, K., Phan, T., Mansuy, I. M., and Storm, D. R. (2007). Proteolytic degradation of SCOP in the hippocampus contributes to activation of MAP kinase and memory. Cell *128*, 1219–1229.

Shuang, R., Zhang, L., Fletcher, A., Groblewski, G. E., Pevsner, J., and Stuenkel, E. L. (1998). Regulation of Munc-18/syntaxin 1A interaction by cyclin-dependent kinase 5 in nerve endings. J Biol Chem 273, 4957–4966.

Silva, A. J., Paylor, R., Wehner, J. M., and Tonegawa, S. (1992). Impaired spatial learning in alpha-calcium-calmodulin kinase II mutant mice. Science 257, 206–211.

Smith, P. D., Crocker, S. J., Jackson-Lewis, V., Jordan-Sciutto, K. L., Hayley, S., Mount, M. P., O'Hare, M. J., Callaghan, S., Slack, R. S., Przedborski, S., et al. (2004). Cyclin-dependent kinase 5 is a mediator of dopaminergic neuron loss in a mouse model of Parkinson's disease. Proc Natl Acad Sci U S A 100, 13650–13655.

Smith, P. D., Mount, M. P., Shree, R., Callaghan, S., Slack, R. S., Anisman, H., Vincent, I., Wang, X., Mao, Z., and Park, D. S. (2006). Calpain-regulated p35/cdk5 plays a central role in dopaminergic neuron death through modulation of the transcription factor myocyte enhancer factor 2. J Neurosci 26, 440–447.

Soriano, S., Kang, D. E., Fu, M., Pestell, R., Chevallier, N., Zheng, H., and Koo, E. H. (2001). Presenilin 1 negatively regulates beta-catenin/T cell factor/lymphoid enhancer factor-1 signaling independently of beta-amyloid precursor protein and notch processing. J Cell Biol 152, 785–794.

Spires, T. L., and Hannan, A. J. (2007). Molecular mechanisms mediating pathological plasticity in Huntington's disease and Alzheimer's disease. J Neurochem 100, 874–882.

Swatton, J. E., Sellers, L. A., Faull, R. L., Holland, A., Iritani, S., and Bahn, S. (2004). Increased MAP kinase activity in Alzheimer's and Down syndrome but not in schizophrenia human brain. Eur J Neurosci 19, 2711–2719.

Tan, T. C., Valova, V. A., Malladi, C. S., Graham, M. E., Berven, L. A., Jupp, O. J., Hansra, G., McClure, S. J., Sarcevic, B., Boadle, R. A., et al. (2003). Cdk5 is essential for synaptic vesicle endocytosis. Nat Cell Biol 5, 701–710.

Tandon, A., Yu, H., Wang, L., Rogaeva, E., Sato, C., Chishti, M. A., Kawarai, T., Hasegawa, H., Chen, F., Davies, P., et al. (2003). Brain levels of CDK5 activator p25 are not increased in Alzheimer's or other neurodegenerative diseases with neurofibrillary tangles. J Neurochem 86, 572–583.

Taniguchi, S., Fujita, Y., Hayashi, S., Kakita, A., Takahashi, H., Murayama, S., Saido, T. C., Hisanaga, S., Iwatsubo, T., and Hasegawa, M. (2001). Calpain-mediated degradation of p35 to p25 in postmortem human and rat brains. FEBS Lett 489, 46–50.

Togashi, H., Abe, K., Mizoguchi, A., Takaoka, K., Chisaka, O., and Takeichi, M. (2002). Cadherin regulates dendritic spine morphogenesis. Neuron 35, 77–89.

Tomizawa, K., Ohta, J., Matsushita, M., Moriwaki, A., Li, S. T., Takei, K., and Matsui, H. (2002). Cdk5/p35 regulates neurotransmitter release through phosphorylation and down-regulation of P/Q-type voltage-dependent calcium channel activiy. J Neurosci 22, 2590–2597.

Tomizawa, K., Sunada, S., Lu, Y. F., Oda, Y., Kinuta, M., Ohshima, T., Saito, T., Wei, F. Y., Matsushita, M., Li, S. T., et al. (2003). Cophosphorylation of amphiphysin I and dynamin I by Cdk5 regulates clathrin-mediated endocytosis of synaptic vesicles. J Cell Biol 163, 813–824.

Toth, E., Bruin, J. P., Heinsbroek, R. P., and Joosten, R. N. (1996). Spatial learning and memory in calpastatin-deficient rats. Neurobiol Learn Mem 66, 230–235.

Tsai, L. H., Lee, M. S., and Cruz, J. (2004). Cdk5, a therapeutic target for Alzheimer's disease? Biochim Biophys Acta 1697, 137–142.

Tseng, H. C., Zhou, Y., Shen, Y., and Tsai, L. H. (2002). A survey of Cdk5 activator p35 and p25 levels in Alzheimer's disease brains. FEBS Lett 523, 58–62.

Van den Haute, C., Spittaels, K., Van Dorpe, J., Lasrado, R., Vandezande, K., Laenen, I., Geerts, H., and Van Leuven, F. (2001). Coexpression of human cdk5 and its activator p35 with human protein tau in neurons in brain of triple transgenic mice. Neurobiol Dis 8, 32–44.

Wang, J., Liu, S., Fu, Y., Wang, J. H., and Lu, Y. (2003). Cdk5 activation induces hippo-campal CA1 cell death by directly phosphorylating NMDA receptors. Nat Neurosci *6*, 1039–1047.

Watase., K., and Zoghbi, H. Y. (2003). Modelling brain diseases in mice: the challenges of design and analysis. Nat Rev Genet *4*, 296–307.

Weishaupt, J. H., Kussmaul, L., Grotsch, P., Heckel, A., Rohde, G., Romig, H., Bahr, M., and Gillardon, F. (2003). Inhibition of CDK5 is protective in necrotic and apoptotic paradigms of neuronal cell death and prevents mitochondrial dysfunction. Mol Cell Neurosci *2*, 489–502.

Wu, P., Shen, Q., Dong, S., Xu, Z., Tsien, J. Z., and Hu, Y. (2007). Calorie restriction ameliorates neurodegenerative phenotypes in forebrain-specific presenilin-1 and preseni-lin-2 double knockout mice. Neurobiol Aging *Epub ahead of print*.

Xie, Z., Samuels, B. A., and Tsai, L. H. (2006). Cyclin-dependent kinase 5 permits efficient cytoskeletal remodeling–a hypothesis on neuronal migration. Cereb Cortex *16*, 64–68.

Yan, Z., Chi, P., Bibb, J. A., Ryan, T. A., and Greengard, P. (2002). Roscovitine: a novel regulator of P/Q-type calcium channels and transmitter release in central neurons. J Physiol *540*, 761–770.

Yoo, B. C., and Lubec, G. (2001). p25 protein in neurodegeneration. Nature *411*, 763–764; discussion 764–765.

Zhang, M., Li, J., Chakrabarty, P., Bu, B., and Vincent, I. (2004). Cyclin-dependent kinase inhibitors attenuate protein hyperphosphorylation, cytoskeletal lesion formation, and motor defects in Niemann-Pick Type C mice. Am J Pathol *165*, 843–853.

Cdk5 as a Drug Target for Alzheimer's Disease

Lit-Fui Lau and Carol D. Hicks

Abstract Neurofibrillary tangles (NFTs) are a pathological hallmark in the Alzheimer's disease (AD) brain and are well correlated with progression of the disease. Since NFTs consist mainly of hyperphosphorylated tau, tau kinases have been suggested as drug targets to slow the progression of AD. This notion has further been supported by recent studies showing the importance of tau and its phosphorylation in neurodegeneration and cognitive deficits in animal models. Among the different putative tau kinases, cyclin-dependent kinase 5 (Cdk5) remains an appealing drug target for AD. Activation of Cdk5 in animal models recapitulates many AD features including tau hyperphosphorylation, NFTs, neurodegeneration, cognitive impairments, and increase in Aβ levels. In addition, inhibition of Cdk5 activities is neuroprotective and appears to enhance long-term potentiation and improve learning and memory—all are potential beneficial features for the treatment of AD.

Introduction

Alzheimer's disease (AD) is the most common form of dementia accounting for 50–70% of all dementia cases. It is the fifth leading cause of death among elderly, aged 65 and older. The statistics released from the Alzheimer's Association in 2007 (http://www.alz.org/national/documents/Report_2007FactsAnd Figures.pdf) is sobering: It is estimated that in every 72 s, someone in the United States develops AD. The number of AD patients has recently risen to 5.1 million and will climb to 16 million by 2050. Combination of direct and indirect cost for the treatment and care of AD patients is a staggering $148 billion a year in the United States and $315.4 billion worldwide. The need to find a treatment for this disease is ever more urgent.

L.-F. Lau
Neurodegeneration Research, GlaxoSmithKline R&D China, No. 3 Building, 898 Halei Road, Zhangjiang Hi-Tech Park, Pudong, Shanghai 201203, China
e-mail: lit-fui.1.lau@gsk.com

N.Y. Ip, L.-H. Tsai (eds.), *Cyclin Dependent Kinase 5 (Cdk5)*,
DOI: 10.1007/978-0-387-78887-6_18, © Springer Science+Business Media, LLC 2008

AD is a chronic progressive neurodegenerative disease with an insidious onset. AD patients usually begin with an inability to remember recent events followed by aphasia, disorientation, impaired judgment, and personality changes. In advanced stages, AD patients lose their motor functions (e.g., feeding, dressing) and thus ability for self-care. Nearly all AD patients eventually need to move into long-term care facilities despite the desire of family members to keep them at home.

The exact cause of AD is unknown. The prevailing view is the amyloid cascade hypothesis (Hardy and Selkoe, 2002; Hardy, 2006). It postulates that the β-amyloid (Aβ) peptides are the initial culprit in triggering a cascade of harmful events ultimately leading to AD. Aβ is generated from the amyloid precursor protein (APP) after the proteolytic action of two aspartyl proteases: BACE1 and γ-secretase. Elevation of Aβ levels as a result of mutations of APP or presenilins (components of the γ-secretase complex) is believed to cause some cases of familial AD. Accumulation of Aβ peptides in senile plaques in brains is a diagnostic feature of AD. Deposition of Aβ and cognitive deficits can be recapitulated in transgenic mice overexpressing APP. However, other key AD pathologies—neurofibrillary tangles (NFTs) and neuronal cell loss—are conspicuously missing in APP transgenic mice. This incomplete representation of AD pathologies questions the amyloid cascade hypothesis. Nevertheless, recent evidence shows that Aβ can augment expression of NFTs under specific experimental conditions. (Gotz et al., 2001; Lewis et al., 2001; Tomidokoro et al., 2001; Oddo et al., 2003).

Although tau is believed to be downstream of Aβ in AD pathogenesis, its significance in neurodegeneration and cognitive deficits in AD should not be undermined. Progression of AD is well correlated with progression of tau pathologies (Grober et al., 1999; Tiraboschi et al., 2004). Elevated total tau and/or phosphotau in the cerebrospinal fluid correlates with incidence (Sunderland et al., 2003; Hampel et al., 2004; Fagan et al., 2007) and progression (Hampel et al., 2005) of AD. Familial mutations in tau have been found to be sufficient to induce neurodegeneration in frontotemporal dementia and Parkinsonism linked to chromosome 17 (FTDP-17) (Hutton et al., 1998; Poorkaj et al., 1998; Spillantini and Goedert, 1998). Most recently, a genetic polymorphism of the GAB2 gene has been shown to increase the risk of late-onset AD by 4-fold (Reiman et al., 2007). Interference with GAB2 expression increases tau phosphorylation (Reiman et al., 2007). Transgenic animal models of tau (Lewis et al., 2000; Wittmann et al., 2001; Andorfer et al., 2005; Santacruz et al., 2005; Yoshiyama et al., 2007), tau kinase, or tau kinase activator (Ahlijanian et al., 2000; Bian et al., 2002; Jackson et al., 2002; Cruz et al., 2003; Engel et al., 2006b) exhibit clear neurodegeneration and/or cognitive deficits. Interestingly, knocking out tau expression rescues cultured hippocampal neurons from Aβ-induced toxicities (Rapoport et al., 2002) and the life span and cognitive deficits of APP transgenic mice (Roberson et al., 2007), underscoring the significance of tau as a mediator of Aβ-induced deficits.

The mechanisms by which abnormal tau in AD causes neuronal toxicities are not completely understood. Since NFTs are mainly composed of hyperphosphorylated tau, tau hyperphosphorylation has been suggested to be a source of toxicities. Tau is normally a microtubule-binding and stabilizing protein. Hyperphosphorylated tau loses these functions (Sengupta et al., 1998; Evans et al., 2000), subsequently leading to disruption of axonal transport and dying back of neurons. In addition, hyperphosphorylated tau tends to oligomerize (Ruben et al., 1997; Alonso et al., 2001) and may cause gain of toxic functions (Lee et al., 2001). The exact identity of toxic tau species is unknown and has been suggested to be tau multimers (Berger et al., 2007). The significance of tau phosphorylation in neurodegeneration has been demonstrated in a recent study of transgenic Drosophila overexpressing tau (Steinhilb et al., 2007). Fruit flies overexpressing tau develop typical tau pathologies and neurodegeneration reminiscent of AD pathologies. Interestingly, when all the proline-directed phosphorylation sites on tau were mutated and removed, these flies no longer developed the AD-like pathologies (Steinhilb et al., 2007). Administration of kinase inhibitors has also been shown to mitigate pathologies and/or behavioral deficits in tau transgenic mice (Noble et al., 2005; Engel et al., 2006a; Le Corre et al., 2006). Taken together, these results suggest that tau phosphorylation plays an important role in AD pathogenesis.

CDK5 in Tau Hyperphosphorylation

Proline-directed serine/threonine protein kinase prefers the presence of a proline residue on the C-terminal side of its substrate phosphorylation site. Proline-directed protein kinases that have been shown to phosphorylate tau include Cdk5, glycogen synthase kinase 3β (GSK3β), and mitogen-activated protein kinases. Although it is unclear which of the above kinases is/are primarily responsible for the abnormal tau hyperphosphorylation in AD patients, there is a multitude of evidence suggesting that Cdk5 is a prime candidate for this role.

Like other cyclin-dependent protein kinases, the kinase activity of Cdk5 depends on the presence of an activator, in this case, p35 (Lew et al., 1994; Tsai et al., 1994) or p39 (Tang et al., 1995; Humbert et al., 2000). p35 and p39 are myristoylated proteins anchored to cellular membranes. It has been shown that calcium-activated calpain cleaves p35 and p39 to p25 and p29, respectively (Lee et al., 2000; Nath et al., 2000; Patzke and Tsai, 2002). The longer half-lives of p25 and p29 than those of their parent molecules may lead to an accumulation of these Cdk5 activators and increase in Cdk5 kinase activity (Patrick et al., 1999). p25 is intrinsically more effective than p35 in activating Cdk5 to phosphorylate tau (Hashiguchi et al., 2002). Further activation of Cdk5 can be achieved through phosphorylation of its tyrosine 15 residue (Zukerberg et al., 2000). Conflicting results have been reported on Cdk5 activity by phosphorylation of its serine 159 residue (Sharma et al., 1999; Tarricone et al., 2001).

Since identification of p25/Cdk5 as a tau protein kinase in the 1990 s (Ishiguro et al., 1994), its role as a tau kinase has been confirmed and extended in multiple experimental systems by multiple laboratories. These experimental systems range from *in vitro* enzyme assays to cell-based assays to animal models. Purified Cdk5 can phosphorylate tau directly *in vitro* on multiple epitopes as determined by different methods including mass spectrometry, 2D gel electrophoresis plus radioisotope labeling, and tau phosphorylation– dependent antibodies (Paudel et al., 1993; Lund et al., 2001; Liu et al., 2002; Sakaue et al., 2005). When Cdk5 and its activators are overexpressed in cell lines and neurons, they are able to induce tau phosphorylation on different epitopes (Patrick et al., 1999; Shelton et al., 2004; Jamsa et al., 2006). In addition, exposure of cultured cells to toxic stimuli, for example, Aβ or glutamate, can lead to elevation of p25, activation of Cdk5, and tau hyperphosphorylation (Town et al., 2002; Han et al., 2005). This increase in tau phosphorylation can be prevented by anti-sense oligonucleotides against the Cdk5 activator, p35 (Town et al., 2002), or by the Cdk5 inhibitor, roscovitine (Han et al., 2005).

In recent years these *in vitro* findings have been observed *in vivo* as well. Overexpression of the Cdk5 activator, p25, in mice can lead to tau hyperphosphorylation, formation of tau filament and NFTs, neurodegeneration, and/or impairment of cognitive functions (Ahlijanian et al., 2000; Bian et al., 2002; Cruz et al., 2003; Fischer et al., 2005). Overexpression of p25 in the P301L mutant tau line accelerated the tau pathologies (Noble et al., 2003). However, not all p25 transgenic mice display tau pathologies (Takashima et al., 2001; Angelo et al., 2003). Reasons for these disparities are not immediately apparent. Other genetically engineered mice that display increased levels of p25 and Cdk5 activity display a concomitant increase in tau phosphorylation. These include the amyotrophic lateral sclerosis model, SOD-1^{G37R} mice (Nguyen et al., 2001), and Niemann–Pick disease model, npc-1 mice (Bu et al., 2002). Tau hyperphosphorylation in the npc-1 mice is susceptible to inhibition by intracerebroventricular injection of Cdk5 inhibitors, roscovitine and olomoucine (Zhang et al., 2004), suggesting a causal role of p25–Cdk5 in hyperphosphorylation of tau. Interestingly, a subsequent study shows tau phosphorylation is not diminished when the npc-1 mice is crossed with the p35 knockout mice in which Cdk5 kinase activity is reduced by 78% (Hallows et al., 2006). Based on these data, the authors argue that tau hyperphosphorylation in the npc-1 mice is independent of Cdk5 kinase activity. However, it is difficult to conclusively eliminate any role of Cdk5 in tau hyperphosphorylation here in light of the fact that 22% residual Cdk5 kinase activity in this animal model remains and that another p35 knockout shows compensatory elevation of p39 expression and mislocalization of Cdk5 (Takahashi et al., 2003).

Experimental conditions leading to elevation of endogenous p25 can lead to tau hyperphosphorylation. Chronic treatment of lipopolysaccharide (LPS) in the triple transgenic mice (PS1 × APP × P301L tau) induces elevation of interleukin-1β and p25 and activation of Cdk5 (Kitazawa et al., 2005). Such treatment has no effect on APP processing but enhances tau phosphorylation on the AT8 and AT-180 epitopes. Intracerebroventricular injection of roscovitine prevents the increase in tau phosphorylation, suggesting that p25–Cdk5 is responsible for the induction of tau phosphorylation. LPS-induced p25 increase and tau hyperphosphorylation are not dependent on the FTDP-17 mutant tau nor on the overexpression of APP since these effects can be replicated in wild-type mice. In addition to inflammatory stimuli, ischemic conditions have also been shown to result in Cdk5 activation and subsequent tau hyperphosphorylation in wild-type animals. Transient cerebral ischemia induces the production of p25, an increase in immunoprecipitated Cdk5 kinase activity, and tau hyperphosphorylation (Wen et al., 2004, 2007). The elevated tau phosphorylation can be significantly reduced by intracerebroventricular injection of roscovitine, suggesting that the enhanced p25/Cdk5 activity mediates the induced tau phosphorylation (Wen et al., 2004). Similar findings have been observed by Morioka et al. (Morioka et al., 2006). They show that transient forebrain ischemia increases tau phosphorylation that is inhibited by intracerebroventricular injection of another Cdk5 inhibitor, olomoucine. Neurons injured by ischemia leaking glutamate into the extracellular milieu may give rise to excitotoxicity of surrounding neurons. This can be modeled by intraperitoneal injection of a glutamate agonist, kainate. Mice treated with kainate not only experience the neuronal damage expected of excitotoxicity but also display increased p25 levels and tau hyperphosphorylation (Crespo-Biel et al., 2007) consistent with those induced by ischemic treatment.

Support for the involvement of Cdk5 in AD pathogenesis partly comes from studies in postmortem brains of AD patients. Cdk5 kinase activity is increased in AD brains (Lee et al., 1999; Patrick et al., 1999). Cdk5 protein is elevated in brains of patients with mild cognitive impairment (Sultana and Butterfield, 2007), some of which have been suggested to be an early form of AD. Cdk5 immunoreactivity is detected in neurons bearing pre-NFTs and/or NFTs (Yamaguchi et al., 1996; Pei et al., 1998; Takahashi et al., 2000; Augustinack et al., 2002). In addition, there are reports showing elevation of p25 in AD brains (Patrick et al., 1999; Tseng et al., 2002; Swatton et al., 2004) although such elevation is not observed by others (Takashima et al., 2001; Taniguchi et al., 2001; Yoo and Lubec, 2001; Tandon et al., 2003). Stimuli that induce p25 production and Cdk5 activation are present in AD brains. These include increased Aβ levels, inflammation, and calpain activation (Saito et al., 1993; Grynspan et al., 1997; Taniguchi et al., 2001). Finally, genetic polymorphisms in the Cdk5 gene have been shown to be associated with an elevated risk of developing early-onset AD (Rademakers et al., 2005). These data suggest that aberrant activation of Cdk5 occurs in AD and may contribute to tau pathology and neurodegeneration.

Beyond Tau Hyperphosphorylation

In addition to its role in AD pathogenesis as a tau kinase, hyperactivation of Cdk5 may regulate APP processing and compromise neuronal survival through pro-apoptotic phosphorylation of transcription factors. A recent report (Cruz et al., 2006) suggests that Cdk5 may contribute to the other pathological hallmark of AD, namely, accumulation of Aβ peptides. It shows that intracellular Aβ levels are elevated in the p25-inducible mice when compared to uninduced controls. Similarly, intracellular Aβ levels are increased in the double transgenic mice crossed between the p25 inducible and APP transgenic mice (PDAPP) when compared to the APP transgenic mice alone. These investigators propose that aberrant axonal transport, increased BACE1 levels, and increased phosphorylation of APP on Thr668 by p25–Cdk5 may account for the accumulation of Aβ peptides inside neurons. The augmentation of Aβ levels by Cdk5 have independently been observed by Karen Duff's laboratory at Columbia University (Wen et al., 2008) and Sul-Hee Chung and colleagues at Inje University (Son et al., 2005). Phosphorylation of APP on Thr668 by Cdk5 has been suggested to increase APP processing and Aβ production by *in vitro* studies (Lee et al., 2003; Liu et al., 2003). However, phosphorylation of Thr668 by Cdk5 appears to be an unlikely mechanism for regulating physiological levels of Aβ40 and 42 in mouse brains since neither Aβ40 nor Aβ42 is altered in knockin mice in which the Thr668 residue on APP has been mutated to the nonphosphorylatable alanine residue (Sano et al., 2006). Other Cdk5 substrates that are directly or indirectly involved in APP processing (e.g., presenilin-1 (Lau et al., 2002)) could be responsible.

Cdk5 has been implicated in contributing to neuronal cell death and deficits in a number of experimental models. Aβ has been shown to induce p25 levels (Otth et al., 2002, 2003; Oakley et al., 2006). Aβ-induced cell death can be reduced by inhibition of Aβ-induced p25 elevation (Li et al., 2003) or by direct inhibition of Cdk5 using Cdk5 inhibitors (Alvarez et al., 1999; Wei et al., 2002; Zheng et al., 2005) or antisense oligonucleotide against Cdk5 (Alvarez et al., 1999). In cultured hippocampal slices, Aβ-impaired LTP can be prevented by Cdk5 inhibitors, roscovitine, or butyrolactone (Wang et al., 2004). In a rat model of transient forebrain ischemia, p25 production is increased with a concomitant elevation in Cdk5 kinase activity (Wang et al., 2003). Activated Cdk5 phosphorylates and activates the NMDA receptor, resulting in excitotoxicity. Inhibition of Cdk5 by viral infection of a dominant-negative form of Cdk5 reduces neuronal cell death in the CA1 region of the hippocampus (Wang et al., 2003). Similarly, p25 levels are increased in the MPTP model of Parkinson's disease (Smith et al., 2003, 2006), SOD1^{G37R} model of amyotrophic lateral sclerosis (Nguyen et al., 2001), and npc-1 model of Niemann–Pick's disease (Bu et al., 2002). In the MPTP model, it is suggested that p25/Cdk5 phosphorylates and inactivates the prosurvival factor, myocyte enhancer factor 2 (MEF2), and eventually induces death of the dopaminergic neurons in the striatum. Significantly, overexpression of a dominant-negative Cdk5 (Smith et al., 2003) or MEF2-lacking Cdk5 phosphorylation sites (Smith et al., 2006) attenuates MPTP-induced neuronal cell loss. Effects of Cdk5 on MEF2 phosphorylation and inactivation have been demonstrated *in vitro* as well (Gong et al., 2003). Finally, Cdk5 has been shown to phosphorylate and increase the expression level of another transcription factor, the tumor suppressor protein p53, and may lead to p53-induced apoptosis (Zhang et al., 2002). Therefore, inhibition of Cdk5 activity may halt neurodegeneration through multiple mechanisms.

Safety of Cdk5 Inhibitors

A key issue of developing a Cdk5 inhibitor for the treatment of AD is a poor understanding of the safety risks associated with Cdk5 inhibition in humans. While non-selective Cdk inhibitors with activities against Cdk5 have been profiled in preclinical toxicology and clinical studies, selective inhibitors of Cdk5 have not been tested in clinical trials yet. Furthermore, it is unclear how much selectivity over other Cdk family members is needed to safely administer a Cdk5 inhibitor. However, data collected with non-selective Cdk inhibitors can be used to help understand potential safety issues that may be associated with inhibition of Cdk5 or inhibition of other Cdk family members (Fischer and Gianella-Borradori, 2005). Data to date suggest that the pancreas and eye may be of some concern for possible mechanism-based side effects, whereas the lymphoid organs and gastrointestinal tract may be affected by inhibitors with low selectivity over other Cdks. In addition, a thorough understanding of the role of Cdk5 in regulating normal cognition will be important to evaluate the effects of Cdk5 inhibitors on cognition.

In pancreatic islet cells, use of non-selective Cdk inhibitors or modulation of Cdk5 complex expression has provided conflicting results on insulin secretion (Lilja et al., 2001, 2004; Wei et al., 2005). Cdk5, p35, and p39 are expressed in pancreatic β-cells and β-derived cell lines. p35/Cdk5 expression and activity are enhanced by glucose (Lilja et al., 2004; Ubeda et al., 2004). Overexpression of p39/Cdk5 can elevate insulin expression and exocytosis (Lilja et al., 2004; Ubeda et al., 2004). Suppression of p35/Cdk5, not p39, using antisense oligonucleotides or overexpression of a dominant-negative Cdk5 protein, results in an inhibition of insulin exocytosis in primary pancreatic β-cells (Lilja et al., 2001, 2004). Similarly, a Cdk5 inhibitor, roscovitine, inhibits basal and evoked insulin secretion of primary ob/ob pancreatic β cells in the presence of high glucose (Lilja et al., 2001). These data suggest that p39/Cdk5 promotes insulin secretion. On the other hand, inhibition of Cdk5 by olomoucine elevates glucose-stimulated insulin release in MIN6 and primary rodent pancreatic β-cells (Wei et al., 2005). Similar increase in insulin release is observed using β-cell isolated from the p35 knockout mice (Wei et al., 2005). Which of these *in vitro* results is predictive of the *in vivo* effects of a Cdk5 inhibitor is unknown.

In clinical studies, different Cdk inhibitors have also produced inconsistent effects on pancreatic function. Treatment with UCN-01 or flavopiridol results in hyperglycemia, consistent with the notion that Cdk5 inhibition may downregulate pancreatic function. Experiments with Cdk4 knockout mice suggest that inhibition of Cdk4 can lead to degeneration of the endocrine pancreas and hyperglycemia (Rane et al., 1999; Tsutsui et al., 1999). However, hyperglycemic effects have not been generally reported for other pan-Cdk inhibitors that have entered clinical development (Roscovitine, Indisulam, BMS-387032) (Senderowicz and Sausville, 2000; Zhai et al., 2002; Fischer and Gianella-Borradori, 2005), suggesting these effects by flavopiridol may be off-target effects specific to this compound. Interestingly, another pan-Cdk inhibitor AG-012986 has been reported to cause apoptosis in the exocrine, but not the endocrine pancreas where insulin-secreting β-cells reside and where Cdk4 and Cdk5 are thought to play a role in β-cell maintenance and function (Rane et al., 1999; Tsutsui et al., 1999; Ramiro-Ibanez et al., 2005). As AG-012986 shows a prolonged half-life in the pancreas, elevated and/or prolonged drug exposure may have caused exocrine degeneration through an off-target activity, although a role in maintenance of exocrine cell survival by Cdks cannot be ruled out.

The other possible target organ for undesirable mechanism-based effects of Cdk inhibition is the eye. Retinal degeneration has been described in the cyclin D1 knockout and in the inducible Rb knockout mice, suggesting deregulation of Cdks can cause retinal degeneration (Ma et al., 1998; MacPherson et al., 2004). Recently, high doses of the pan-Cdk inhibitor, AG-012986, are found to exacerbate naturally occurring retinal degeneration in the CD1/ICR mice (Illanes et al., 2006). In contrast to the above studies, retinal toxicity has not been generally reported with other non-selective Cdk inhibitors in both pre-clinical and clinical studies (Fischer and Gianella-Borradori, 2005). Therefore, it is unclear if the reported retinal toxicity of AG-012986 is due to its Cdk inhibition profile, off-target effects, or a unique prolonged drug distribution property in the eye similar to what was described in the pancreas and stomach (Ramiro-Ibanez et al., 2005). On the contrary, intra-ocular injection of olomoucine, roscovitine, or butyrolactone has been shown to protect retinal ganglion cells from axotomy-induced degeneration (Lefevre et al., 2002). In addition, roscovitine and olomoucine show neuroprotective activity in cultured rat retinal ganglion cells (Maas et al., 1998). Cdk5 has been shown *in vitro* and in frog retinal ganglion cells to inhibit PDE6 activity through phosphorylation of the γ-regulatory subunit (Hayashi et al., 2000; Matsuura et al., 2000). These data suggest that a Cdk5 inhibitor would result in activation of PDE6 and depletion of retinal cGMP. It is unknown if activation of PDE6 is neuroprotective although it is known that inhibition of retinal cGMP production and activity may prevent retinal degeneration in retinitis pigmintosa (Vallazza-Deschamps et al., 2005). Conversely, loss of function of PDE6 has been shown to result in retinal degeneration (Lem et al., 1992; Tsang et al., 1996; Hayashi et al., 2000; Matsuura et al., 2000; Chang et al., 2007; Gargini et al., 2007). As both the pro-apoptotic and neuroprotective activities proposed for Cdk5 in the retinal cell populations has been demonstrated with non-selective Cdk inhibitors, it is difficult to determine the precise role and consequences of Cdk5 inhibition in the retina. Follow-up experiments with RNAi, inducible knockouts, or selective Cdk5 inhibitors (Zheng et al., 2005) are needed to determine if Cdk5 inhibition presents a safety liability or a possible therapeutic approach for retinal degeneration.

In addition to concerns regarding peripheral mechanism-based toxicities, there has been some concern regarding adverse reactions from Cdk5 inhibitors on cognitive functions based on implications from recent studies of Cdk5's role in synaptic plasticity (Cheung et al., 2006). Studies utilizing various neuronal and non-neuronal cell lines and tissue preparations have shown that Cdk5 can phosphorylate a variety of presynaptic and postsynaptic proteins. They have produced conflicting data regarding the role of Cdk5 in vesicle endocytosis, exocytosis, neurotransmitter receptor function, and synaptic plasticity (Angelo et al., 2006). To address the functional consequences of Cdk5 modulation at the synapse, behavioral studies have been performed using p25 transgenic mice, direct injection of non-selective Cdk inhibitors, or analysis of p35 knockout mice. Studies with p25 transgenic mice have suggested that the effects of Cdk5 activation depend on the duration and level of p25 expression. Mice with a low level of forebrain-specific expression of p25 behave normally in the Morris water maze and contextual fear conditioning, but show an enhancement in reversal learning and cued fear conditioning (Angelo et al., 2003). Enhanced performance in the contextual fear-conditioning and Morris water maze is observed with transient elevation of Cdk5 activity in the inducible p25 transgenic mouse (Fischer et al., 2005). In contrast, long-lasting elevation of p25 in the inducible p25 mouse has resulted in impairment in synaptic plasticity and learning and memory using the contextual fear-conditioning and Morris water maze assays (Fischer et al., 2005). Collectively, these data suggest that transient or low levels of Cdk5 hyperactivity may improve cognitive performance, but that prolonged activation of Cdk5 is neurotoxic. While these studies address the functional consequences of elevating Cdk5 activity, they do not address the role that endogenous Cdk5 may play in synaptic function and cognition. Pharmacological studies utilizing stereotactic injection of the non-selective Cdk inhibitors butyrolactone I and roscovitine into the brain have shown that Cdk inhibition can disrupt contextual, but not cued, fear conditioning (Fischer et al., 2002, 2003). These results are consistent with a requirement for Cdk5 in normal cognitive function. Interestingly, selective loss of Cdk5, through inducible knockout, leads to an enhancement in performance in the contextual fear-conditioning assay (Hawasli et al., 2007), suggesting that inhibition of activities other than that of Cdk5 may be responsible for the cognitive impairment observed in the pharmacological studies. Analysis of p35 knockout mice have also shown impairments in synaptic plasticity and cognitive performance in the Morris water maze (Ohshima et al., 2005). However, p35 knockout mice also show defects in hippocampal architecture and elevated p39 expression, suggesting that due to the developmental role of Cdk5 and the compensatory changes in p39 expression, the p35 knockout mouse may be an inappropriate model for predicting effects anticipated by selective Cdk5 inhibition (Takahashi et al., 2003; Ohshima et al., 2005). Further studies are warranted to better understand the role of Cdk5 in normal neurotransmission, and whether Cdk5 inhibition can result in either undesirable CNS effects, or if Cdk5 inhibition may provide a symptomatic benefit for AD patients in addition to preventing p25-induced neurodegeneration.

References

Ahlijanian MK, Barrezueta NX, Williams RD, Jakowski A, Kowsz KP, McCarthy S, Coskran T, Carlo A, Seymour PA, Burkhardt JE, Nelson RB, McNeish JD (2000) Hyperphosphorylated tau and neurofilament and cytoskeletal disruptions in mice overexpressing human p25, an activator of cdk5. Proc Natl Acad Sci USA 97:2910–2915.

Alonso A, Zaidi T, Novak M, Grundke-Iqbal I, Iqbal K (2001) Hyperphosphorylation induces self-assembly of tau into tangles of paired helical filaments/straight filaments. Proc Natl Acad Sci USA 98:6923–6928.

Alvarez A, Toro R, Caceres A, Maccioni RB (1999) Inhibition of tau phosphorylating protein kinase cdk5 prevents beta-amyloid-induced neuronal death. FEBS Lett 459: 421–426.

Andorfer C, Acker CM, Kress Y, Hof PR, Duff K, Davies P (2005) Cell-cycle reentry and cell death in transgenic mice expressing nonmutant human tau isoforms. J Neurosci 25:5446–5454.

Angelo M, Plattner F, Giese KP (2006) Cyclin-dependent kinase 5 in synaptic plasticity, learning and memory. J Neurochem 99:353–370.

Angelo M, Plattner F, Irvine EE, Giese KP (2003) Improved reversal learning and altered fear conditioning in transgenic mice with regionally restricted p25 expression. Eur J Neurosci 18:423–431.

Augustinack JC, Sanders JL, Tsai LH, Hyman BT (2002) Colocalization and fluorescence resonance energy transfer between cdk5 and AT8 suggests a close association in pre-neurofibrillary tangles and neurofibrillary tangles. J Neuropathol Exp Neurol 61:557–564.

Berger Z, Roder H, Hanna A, Carlson A, Rangachari V, Yue M, Wszolek Z, Ashe K, Knight J, Dickson D, Andorfer C, Rosenberry TL, Lewis J, Hutton M, Janus C (2007) Accumulation of pathological tau species and memory loss in a conditional model of tauopathy. J Neurosci 27:3650–3662.

Bian F, Nath R, Sobocinski G, Booher RN, Lipinski WJ, Callahan MJ, Pack A, Wang KK, Walker LC (2002) Axonopathy, tau abnormalities, and dyskinesia, but no neurofibrillary tangles in p25-transgenic mice. J Comp Neurol 446:257–266.

Bu B, Li J, Davies P, Vincent I (2002) Deregulation of cdk5, hyperphosphorylation, and cytoskeletal pathology in the Niemann-Pick type C murine model. J Neurosci 22:6515–6525.

Chang B, Hawes NL, Pardue MT, German AM, Hurd RE, Davisson MT, Nusinowitz S, Rengarajan K, Boyd AP, Sidney SS, Phillips MJ, Stewart RE, Chaudhury R, Nickerson JM, Heckenlively JR, Boatright JH (2007) Two mouse retinal degenerations caused by missense mutations in the beta-subunit of rod cGMP phosphodiesterase gene. Vision Res 47:624–633.

Cheung ZH, Fu AK, Ip NY (2006) Synaptic roles of Cdk5: implications in higher cognitive functions and neurodegenerative diseases. Neuron 50:13–18.

Crespo-Biel N, Canudas AM, Camins A, Pallas M (2007) Kainate induces AKT, ERK and cdk5/GSK3beta pathway deregulation, phosphorylates tau protein in mouse hippocampus. Neurochem Int 50:435–442.

Cruz JC, Kim D, Moy LY, Dobbin MM, Sun X, Bronson RT, Tsai LH (2006) p25/cyclin-dependent kinase 5 induces production and intraneuronal accumulation of amyloid beta in vivo. J Neurosci 26:10536–10541.

Cruz JC, Tseng HC, Goldman JA, Shih H, Tsai LH (2003) Aberrant Cdk5 activation by p25 triggers pathological events leading to neurodegeneration and neurofibrillary tangles. Neuron 40:471–483.

Engel T, Goni-Oliver P, Lucas JJ, Avila J, Hernandez F (2006a) Chronic lithium administration to FTDP-17 tau and GSK-3beta overexpressing mice prevents tau hyperphosphorylation and neurofibrillary tangle formation, but pre-formed neurofibrillary tangles do not revert. J Neurochem 99:1445–1455.

Engel T, Lucas JJ, Gomez-Ramos P, Moran MA, Avila J, Hernandez F (2006b) Cooexpression of FTDP-17 tau and GSK-3beta in transgenic mice induce tau polymerization and neurodegeneration. Neurobiol Aging 27:1258–1268.

Evans DB, Rank KB, Bhattacharya K, Thomsen DR, Gurney ME, Sharma SK (2000) Tau phosphorylation at serine 396 and serine 404 by human recombinant tau protein kinase II inhibits tau's ability to promote microtubule assembly. J Biol Chem 275:24977–24983.

Fagan AM, Roe CM, Xiong C, Mintun MA, Morris JC, Holtzman DM (2007) Cerebrospinal fluid tau/beta-amyloid(42) ratio as a prediction of cognitive decline in nondemented older adults. Arch Neurol 64:343–349.

Fischer A, Sananbenesi F, Pang PT, Lu B, Tsai LH (2005) Opposing roles of transient and prolonged expression of p25 in synaptic plasticity and hippocampus-dependent memory. Neuron 48:825–838.

Fischer A, Sananbenesi F, Schrick C, Spiess J, Radulovic J (2002) Cyclin-dependent kinase 5 is required for associative learning. J Neurosci 22:3700–3707.

Fischer A, Sananbenesi F, Schrick C, Spiess J, Radulovic J (2003) Regulation of contextual fear conditioning by baseline and inducible septo-hippocampal cyclin-dependent kinase 5. Neuropharmacology 44:1089–1099.

Fischer PM, Gianella-Borradori A (2005) Recent progress in the discovery and development of cyclin-dependent kinase inhibitors. Expert Opin Investig Drugs 14:457–477.

Gargini C, Terzibasi E, Mazzoni F, Strettoi E (2007) Retinal organization in the retinal degeneration 10 (rd10) mutant mouse: a morphological and ERG study. J Comp Neurol 500:222–238.

Gong X, Tang X, Wiedmann M, Wang X, Peng J, Zheng D, Blair LA, Marshall J, Mao Z (2003) Cdk5-mediated inhibition of the protective effects of transcription factor MEF2 in neurotoxicity-induced apoptosis. Neuron 38:33–46.

Gotz J, Chen F, van Dorpe J, Nitsch RM (2001) Formation of neurofibrillary tangles in P301l tau transgenic mice induced by Abeta 42 fibrils. Science 293:1491–1495.

Grober E, Dickson D, Sliwinski MJ, Buschke H, Katz M, Crystal H, Lipton RB (1999) Memory and mental status correlates of modified Braak staging. Neurobiol Aging 20:573–579.

Grynspan F, Griffin WR, Cataldo A, Katayama S, Nixon RA (1997) Active site-directed antibodies identify calpain ii as an early-appearing and pervasive component of neurofibrillary pathology in Alzheimers-disease. Brain Res 763:145–158.

Hallows JL, Iosif RE, Biasell RD, Vincent I (2006) p35/p25 is not essential for tau and cytoskeletal pathology or neuronal loss in Niemann-Pick type C disease. J Neurosci 26:2738–2744.

Hampel H, Burger K, Pruessner JC, Zinkowski R, DeBernardis J, Kerkman D, Leinsinger G, Evans AC, Davies P, Moller HJ, Teipel SJ (2005) Correlation of cerebrospinal fluid levels of tau protein phosphorylated at threonine 231 with rates of hippocampal atrophy in Alzheimer disease. Arch Neurol 62:770–773.

Hampel H, Buerger K, Zinkowski R, Teipel SJ, Goernitz A, Andreasen N, Sjoegren M, DeBernardis J, Kerkman D, Ishiguro K, Ohno H, Vanmechelen E, Vanderstichele H, McCulloch C, Moller HJ, Davies P, Blennow K (2004) Measurement of phosphorylated tau epitopes in the differential diagnosis of Alzheimer disease: a comparative cerebrospinal fluid study. Arch Gen Psychiatry 61:95–102.

Han P, Dou F, Li F, Zhang X, Zhang YW, Zheng H, Lipton SA, Xu H, Liao FF (2005) Suppression of cyclin-dependent kinase 5 activation by amyloid precursor protein: a novel excitoprotective mechanism involving modulation of tau phosphorylation. J Neurosci 25:11542–11552.

Hardy J (2006) Alzheimer's disease: the amyloid cascade hypothesis: an update and reappraisal. J Alzheimers Dis 9:151–153.

Hardy J, Selkoe DJ (2002) The amyloid hypothesis of Alzheimer's disease: progress and problems on the road to therapeutics. Science 297:353–356.

Hashiguchi M, Saito T, Hisanaga S, Hashiguchi T (2002) Truncation of CDK5 activator p35 induces intensive phosphorylation of Ser202/Thr205 of human tau. J Biol Chem 277:44525–44530.

Hawasli AH, Benavides DR, Nguyen C, Kansy JW, Hayashi K, Chambon P, Greengard P, Powell CM, Cooper DC, Bibb JA (2007) Cyclin-dependent kinase 5 governs learning and synaptic plasticity via control of NMDAR degradation. Nat Neurosci 10:880–886.

Hayashi F, Matsuura I, Kachi S, Maeda T, Yamamoto M, Fujii Y, Liu H, Yamazaki M, Usukura J, Yamazaki A (2000) Phosphorylation by cyclin-dependent protein kinase 5 of the regulatory subunit of retinal cGMP phosphodiesterase. II. Its role in the turnoff of phosphodiesterase in vivo. J Biol Chem 275:32958–32965.

Humbert S, Dhavan R, Tsai L (2000) p39 activates cdk5 in neurons, and is associated with the actin cytoskeleton. J Cell Sci 113(Pt 6):975–983.

Hutton M, Lendon CL, Rizzu P, Baker M, Froelich S, Houlden H, Pickering-Brown S, Chakraverty S, Isaacs A, Grover A, Hackett J, Adamson J, Lincoln S, Dickson D, Davies P, Petersen RC, Stevens M, de Graaff E, Wauters E, van Baren J, Hillebrand M, Joosse M, Kwon JM, Nowotny P, Che LK, Norton J, Morris JC, Reed LA, Trojanowski J, Basun H, Lannfelt L, Neystat M, Fahn S, Dark F, Tannenberg T, Dodd PR, Hayward N, Kwok JB, Schofield PR, Andreadis A, Snowden J, Craufurd D, Neary D, Owen F, Oostra BA, Hardy J, Goate A, van Swieten J, Mann D, Lynch T, Heutink P (1998) Association of missense and 5'-splice-site mutations in tau with the inherited dementia FTDP-17. Nature 393:702–705.

Illanes O, Anderson S, Niesman M, Zwick L, Jessen BA (2006) Retinal and peripheral nerve toxicity induced by the administration of a pan-cyclin dependent kinase (cdk) inhibitor in mice. Toxicol Pathol 34:243–248.

Ishiguro K, Kobayashi S, Omori A, Takamatsu M, Yonekura S, Anzai K, Imahori K, Uchida T (1994) Identification of the 23 kDa subunit of tau protein kinase II as a putative activator of cdk5 in bovine brain. FEBS Lett 342:203–208.

Jackson GR, Wiedau-Pazos M, Sang TK, Wagle N, Brown CA, Massachi S, Geschwind DH (2002) Human wild-type tau interacts with wingless pathway components and produces neurofibrillary pathology in Drosophila. Neuron 34:509–519.

Jamsa A, Backstrom A, Gustafsson E, Dehvari N, Hiller G, Cowburn RF, Vasange M (2006) Glutamate treatment and p25 transfection increase Cdk5 mediated tau phosphorylation in SH-SY5Y cells. Biochem Biophys Res Commun 345:324–331.

Kitazawa M, Oddo S, Yamasaki TR, Green KN, LaFerla FM (2005) Lipopolysaccharide-induced inflammation exacerbates tau pathology by a cyclin-dependent kinase 5-mediated pathway in a transgenic model of Alzheimer's disease. J Neurosci 25:8843–8853.

Lau KF, Howlett DR, Kesavapany S, Standen CL, Dingwall C, McLoughlin DM, Miller CC (2002) Cyclin-dependent kinase-5/p35 phosphorylates Presenilin 1 to regulate carboxy-terminal fragment stability. Mol Cell Neurosci 20:13–20.

Le Corre S, Klafki HW, Plesnila N, Hubinger G, Obermeier A, Sahagun H, Monse B, Seneci P, Lewis J, Eriksen J, Zehr C, Yue M, McGowan E, Dickson DW, Hutton M, Roder HM (2006) An inhibitor of tau hyperphosphorylation prevents severe motor impairments in tau transgenic mice. Proc Natl Acad Sci USA 103:9673–9678.

Lee KY, Clark AW, Rosales JL, Chapman K, Fung T, Johnston RN (1999) Elevated neuronal Cdc2-like kinase activity in the Alzheimer disease brain. Neurosci Res 34:21–29.

Lee MS, Kao SC, Lemere CA, Xia W, Tseng HC, Zhou Y, Neve R, Ahlijanian MK, Tsai LH (2003) APP processing is regulated by cytoplasmic phosphorylation. J Cell Biol 163:83–95.
Lee MS, Kwon YT, Li M, Peng J, Friedlander RM, Tsai LH (2000) Neurotoxicity induces cleavage of p35 to p25 by calpain. Nature 405:360–364.
Lee VM, Goedert M, Trojanowski JQ (2001) Neurodegenerative tauopathies. Annu Rev Neurosci 24:1121–1159.
Lefevre K, Clarke PG, Danthe EE, Castagne V (2002) Involvement of cyclin-dependent kinases in axotomy-induced retinal ganglion cell death. J Comp Neurol 447:72–81.
Lem J, Flannery JG, Li T, Applebury ML, Farber DB, Simon MI (1992) Retinal degeneration is rescued in transgenic rd mice by expression of the cGMP phosphodiesterase beta subunit. Proc Natl Acad Sci USA 89:4422–4426.
Lew J, Huang QQ, Qi Z, Winkfein RJ, Aebersold R, Hunt T, Wang JH (1994) A brain-specific activator of cyclin-dependent kinase 5. Nature 371:423–426.
Lewis J, Dickson DW, Lin WL, Chisholm L, Corral A, Jones G, Yen SH, Sahara N, Skipper L, Yager D, Eckman C, Hardy J, Hutton M, McGowan E (2001) Enhanced neurofibrillary degeneration in transgenic mice expressing mutant tau and APP. Science 293:1487–1491.
Lewis J, McGowan E, Rockwood J, Melrose H, Nacharaju P, Van Slegtenhorst M, Gwinn-Hardy K, Paul Murphy M, Baker M, Yu X, Duff K, Hardy J, Corral A, Lin WL, Yen SH, Dickson DW, Davies P, Hutton M (2000) Neurofibrillary tangles, amyotrophy and progressive motor disturbance in mice expressing mutant (P301L) tau protein. Nat Genet 25:402–405.
Li G, Faibushevich A, Turunen BJ, Yoon SO, Georg G, Michaelis ML, Dobrowsky RT (2003) Stabilization of the cyclin-dependent kinase 5 activator, p35, by paclitaxel decreases beta-amyloid toxicity in cortical neurons. J Neurochem 84:347–362.
Lilja L, Johansson JU, Gromada J, Mandic SA, Fried G, Berggren PO, Bark C (2004) Cyclin-dependent kinase 5 associated with p39 promotes Munc18-1 phosphorylation and Ca(2+)-dependent exocytosis. J Biol Chem 279:29534–29541.
Lilja L, Yang SN, Webb DL, Juntti-Berggren L, Berggren PO, Bark C (2001) Cyclin-dependent kinase 5 promotes insulin exocytosis. J Biol Chem 276:34199–34205.
Liu F, Iqbal K, Grundke-Iqbal I, Gong CX (2002) Involvement of aberrant glycosylation in phosphorylation of tau by cdk5 and GSK-3beta. FEBS Lett 530:209–214.
Liu F, Su Y, Li B, Zhou Y, Ryder J, Gonzalez-DeWhitt P, May PC, Ni B (2003) Regulation of amyloid precursor protein (APP) phosphorylation and processing by p35/Cdk5 and p25/Cdk5. FEBS Lett 547:193–196.
Lund ET, McKenna R, Evans DB, Sharma SK, Mathews WR (2001) Characterization of the in vitro phosphorylation of human tau by tau protein kinase II (cdk5/p20) using mass spectrometry. J Neurochem 76:1221–1232.
Ma C, Papermaster D, Cepko CL (1998) A unique pattern of photoreceptor degeneration in cyclin D1 mutant mice. Proc Natl Acad Sci USA 95:9938–9943.
Maas Jr. JW, Horstmann S, Borasio GD, Anneser JM, Shooter EM, Kahle PJ (1998) Apoptosis of central and peripheral neurons can be prevented with cyclin-dependent kinase/mitogen-activated protein kinase inhibitors. J Neurochem 70:1401–1410.
MacPherson D, Sage J, Kim T, Ho D, McLaughlin ME, Jacks T (2004) Cell type-specific effects of Rb deletion in the murine retina. Genes Dev 18:1681–1694.
Matsuura I, Bondarenko VA, Maeda T, Kachi S, Yamazaki M, Usukura J, Hayashi F, Yamazaki A (2000) Phosphorylation by cyclin-dependent protein kinase 5 of the regulatory subunit of retinal cGMP phosphodiesterase. I. Identification of the kinase and its role in the turnoff of phosphodiesterase in vitro. J Biol Chem 275: 32950–32957.
Morioka M, Kawano T, Yano S, Kai Y, Tsuiki H, Yoshinaga Y, Matsumoto J, Maeda T, Hamada J, Yamamoto H, Fukunaga K, Kuratsu J (2006) Hyperphosphorylation at serine 199/202 of tau factor in the gerbil hippocampus after transient forebrain ischemia. Biochem Biophys Res Commun 347:273–278.

Nath R, Davis M, Probert AW, Kupina NC, Ren X, Schielke GP, Wang KK (2000) Processing of cdk5 activator p35 to its truncated form (p25) by calpain in acutely injured neuronal cells. Biochem Biophys Res Commun 274:16–21.

Nguyen MD, Lariviere RC, Julien JP (2001) Deregulation of Cdk5 in a mouse model of ALS: toxicity alleviated by perikaryal neurofilament inclusions. Neuron 30:135–147.

Noble W, Olm V, Takata K, Casey E, Mary O, Meyerson J, Gaynor K, LaFrancois J, Wang L, Kondo T, Davies P, Burns M, Veeranna, Nixon R, Dickson D, Matsuoka Y, Ahlijanian M, Lau LF, Duff K (2003) Cdk5 is a key factor in tau aggregation and tangle formation in vivo. Neuron 38:555–565.

Noble W, Planel E, Zehr C, Olm V, Meyerson J, Suleman F, Gaynor K, Wang L, LaFrancois J, Feinstein B, Burns M, Krishnamurthy P, Wen Y, Bhat R, Lewis J, Dickson D, Duff K (2005) Inhibition of glycogen synthase kinase-3 by lithium correlates with reduced tauopathy and degeneration in vivo. Proc Natl Acad Sci USA 102:6990–6995.

Oakley H, Cole SL, Logan S, Maus E, Shao P, Craft J, Guillozet-Bongaarts A, Ohno M, Disterhoft J, Van Eldik L, Berry R, Vassar R (2006) Intraneuronal beta-amyloid aggregates, neurodegeneration, and neuron loss in transgenic mice with five familial Alzheimer's disease mutations: potential factors in amyloid plaque formation. J Neurosci 26:10129–10140.

Oddo S, Caccamo A, Kitazawa M, Tseng BP, LaFerla FM (2003) Amyloid deposition precedes tangle formation in a triple transgenic model of Alzheimer's disease. Neurobiol Aging 24:1063–1070.

Ohshima T, Ogura H, Tomizawa K, Hayashi K, Suzuki H, Saito T, Kamei H, Nishi A, Bibb JA, Hisanaga S, Matsui H, Mikoshiba K (2005) Impairment of hippocampal long-term depression and defective spatial learning and memory in p35 mice. J Neurochem 94: 917–925.

Otth C, Concha, II, Arendt T, Stieler J, Schliebs R, Gonzalez-Billault C, Maccioni RB (2002) AbetaPP induces cdk5-dependent tau hyperphosphorylation in transgenic mice Tg2576. J Alzheimers Dis 4:417–430.

Otth C, Mendoza-Naranjo A, Mujica L, Zambrano A, Concha, II, Maccioni RB (2003) Modulation of the JNK and p38 pathways by cdk5 protein kinase in a transgenic mouse model of Alzheimer's disease. Neuroreport 14:2403–2409.

Patrick GN, Zukerberg L, Nikolic M, de la Monte S, Dikkes P, Tsai LH (1999) Conversion of p35 to p25 deregulates Cdk5 activity and promotes neurodegeneration. Nature 402: 615–622.

Patzke H, Tsai LH (2002) Calpain-mediated cleavage of the cyclin-dependent kinase-5 activator p39 to p29. J Biol Chem 277:8054–8060.

Paudel HK, Lew J, Ali Z, Wang JH (1993) Brain proline-directed protein kinase phosphorylates tau on sites that are abnormally phosphorylated in tau associated with Alzheimer's paired helical filaments. J Biol Chem 268:23512–23518.

Pei JJ, Grundke-Iqbal I, Iqbal K, Bogdanovic N, Winblad B, Cowburn RF (1998) Accumulation of cyclin-dependent kinase 5 (cdk5) in neurons with early stages of Alzheimer's disease neurofibrillary degeneration. Brain Res 797:267–277.

Poorkaj P, Bird TD, Wijsman E, Nemens E, Garruto RM, Anderson L, Andreadis A, Wiederholt WC, Raskind M, Schellenberg GD (1998) Tau is a candidate gene for chromosome 17 frontotemporal dementia. Ann Neurol 43:815–825.

Rademakers R, Sleegers K, Theuns J, Van den Broeck M, Bel Kacem S, Nilsson LG, Adolfsson R, van Duijn CM, Van Broeckhoven C, Cruts M (2005) Association of cyclin-dependent kinase 5 and neuronal activators p35 and p39 complex in early-onset Alzheimer's disease. Neurobiol Aging 26:1145–1151.

Ramiro-Ibanez F, Trajkovic D, Jessen B (2005) Gastric and pancreatic lesions in rats treated with a pan-CDK inhibitor. Toxicol Pathol 33:784–791.

Rane SG, Dubus P, Mettus RV, Galbreath EJ, Boden G, Reddy EP, Barbacid M (1999) Loss of Cdk4 expression causes insulin-deficient diabetes and Cdk4 activation results in beta-islet cell hyperplasia. Nat Genet 22:44–52.

Rapoport M, Dawson HN, Binder LI, Vitek MP, Ferreira A (2002) Tau is essential to beta -amyloid-induced neurotoxicity. Proc Natl Acad Sci USA 99:6364–6369.

Reiman EM, Webster JA, Myers AJ, Hardy J, Dunckley T, Zismann VL, Joshipura KD, Pearson JV, Hu-Lince D, Huentelman MJ, Craig DW, Coon KD, Liang WS, Herbert RH, Beach T, Rohrer KC, Zhao AS, Leung D, Bryden L, Marlowe L, Kaleem M, Mastroeni D, Grover A, Heward CB, Ravid R, Rogers J, Hutton ML, Melquist S, Petersen RC, Alexander GE, Caselli RJ, Kukull W, Papassotiropoulos A, Stephan DA (2007) GAB2 Alleles Modify Alzheimer's Risk in APOE varepsilon4 Carriers. Neuron 54:713–720.

Roberson ED, Scearce-Levie K, Palop JJ, Yan F, Cheng IH, Wu T, Gerstein H, Yu GQ, Mucke L (2007) Reducing endogenous tau ameliorates amyloid beta-induced deficits in an Alzheimer's disease mouse model. Science 316:750–754.

Ruben GC, Ciardelli TL, Grundke-Iqbal I, Iqbal K (1997) Alzheimer disease hyperphosphorylated tau aggregates hydrophobically. Synapse 27:208–229.

Saito K, Elce JS, Hamos JE, Nixon RA (1993) Widespread activation of calcium-activated neutral proteinase (calpain) in the brain in Alzheimer disease: a potential molecular basis for neuronal degeneration. Proc Natl Acad Sci USA 90:2628–2632.

Sakaue F, Saito T, Sato Y, Asada A, Ishiguro K, Hasegawa M, Hisanaga S (2005) Phosphorylation of FTDP-17 mutant tau by cyclin-dependent kinase 5 complexed with p35, p25, or p39. J Biol Chem 280:31522–31529.

Sano Y, Nakaya T, Pedrini S, Takeda S, Iijima-Ando K, Iijima K, Mathews PM, Itohara S, Gandy S, Suzuki T (2006) Physiological mouse brain Abeta levels are not related to the phosphorylation state of threonine-668 of Alzheimer's APP. PLoS ONE 1:e51.

Santacruz K, Lewis J, Spires T, Paulson J, Kotilinek L, Ingelsson M, Guimaraes A, DeTure M, Ramsden M, McGowan E, Forster C, Yue M, Orne J, Janus C, Mariash A, Kuskowski M, Hyman B, Hutton M, Ashe KH (2005) Tau suppression in a neurodegenerative mouse model improves memory function. Science 309:476–481.

Senderowicz AM, Sausville EA (2000) Preclinical and clinical development of cyclin-dependent kinase modulators. J Natl Cancer Inst 92:376–387.

Sengupta A, Kabat J, Novak M, Wu Q, Grundke-Iqbal I, Iqbal K (1998) Phosphorylation of tau at both Thr 231 and Ser 262 is required for maximal inhibition of its binding to microtubules. Arch Biochem Biophys 357:299–309.

Sharma P, Sharma M, Amin ND, Albers RW, Pant HC (1999) Regulation of cyclin-dependent kinase 5 catalytic activity by phosphorylation. Proc Natl Acad Sci USA 96:11156–11160.

Shelton SB, Krishnamurthy P, Johnson GV (2004) Effects of cyclin-dependent kinase-5 activity on apoptosis and tau phosphorylation in immortalized mouse brain cortical cells. J Neurosci Res 76:110–120.

Smith PD, Crocker SJ, Jackson-Lewis V, Jordan-Sciutto KL, Hayley S, Mount MP, O'Hare MJ, Callaghan S, Slack RS, Przedborski S, Anisman H, Park DS (2003) Cyclin-dependent kinase 5 is a mediator of dopaminergic neuron loss in a mouse model of Parkinson's disease. Proc Natl Acad Sci USA 100:13650–13655.

Smith PD, Mount MP, Shree R, Callaghan S, Slack RS, Anisman H, Vincent I, Wang X, Mao Z, Park DS (2006) Calpain-regulated p35/cdk5 plays a central role in dopaminergic neuron death through modulation of the transcription factor myocyte enhancer factor 2. J Neurosci 26:440–447.

Son M, Lee H, Kim M, Ha I, Chung S (2005) Enhancement of BACE1 activity by p25/cdk5-mediated phosphorylation in Alzheimer's disease. Soc Neurosci Abstr Program # 661.4.

Spillantini MG, Goedert M (1998) Tau protein pathology in neurodegenerative diseases. Trends Neurosci 21:428–433.

Steinhilb ML, Dias-Santagata D, Mulkearns EE, Shulman JM, Biernat J, Mandelkow EM, Feany MB (2007) S/P and T/P phosphorylation is critical for tau neurotoxicity in Drosophila. J Neurosci Res 85:1271–1278.

Sultana R, Butterfield DA (2007) Regional expression of key cell cycle proteins in brain from subjects with amnestic mild cognitive impairment. Neurochem Res 32:655–662.

Sunderland T, Linker G, Mirza N, Putnam KT, Friedman DL, Kimmel LH, Bergeson J, Manetti GJ, Zimmermann M, Tang B, Bartko JJ, Cohen RM (2003) Decreased beta-amyloid1-42 and increased tau levels in cerebrospinal fluid of patients with Alzheimer disease. JAMA 289:2094–2103.

Swatton JE, Sellers LA, Faull RL, Holland A, Iritani S, Bahn S (2004) Increased MAP kinase activity in Alzheimer's and Down syndrome but not in schizophrenia human brain. Eur J Neurosci 19:2711–2719.

Takahashi M, Iseki E, Kosaka K (2000) Cdk5 and munc-18/p67 co-localization in early stage neurofibrillary tangles-bearing neurons in Alzheimer type dementia brains. J Neurol Sci 172:63–69.

Takahashi S, Saito T, Hisanaga S, Pant HC, Kulkarni AB (2003) Tau phosphorylation by cyclin-dependent kinase 5/p39 during brain development reduces its affinity for microtubules. J Biol Chem 278:10506–10515.

Takashima A, Murayama M, Yasutake K, Takahashi H, Yokoyama M, Ishiguro K (2001) Involvement of cyclin dependent kinase5 activator p25 on tau phosphorylation in mouse brain. Neurosci Lett 306:37–40.

Tandon A, Yu H, Wang L, Rogaeva E, Sato C, Chishti MA, Kawarai T, Hasegawa H, Chen F, Davies P, Fraser PE, Westaway D, St George-Hyslop PH (2003) Brain levels of CDK5 activator p25 are not increased in Alzheimer's or other neurodegenerative diseases with neurofibrillary tangles. J Neurochem 86:572–581.

Tang D, Yeung J, Lee KY, Matsushita M, Matsui H, Tomizawa K, Hatase O, Wang JH (1995) An isoform of the neuronal cyclin-dependent kinase 5 (Cdk5) activator. J Biol Chem 270:26897–26903.

Taniguchi S, Fujita Y, Hayashi S, Kakita A, Takahashi H, Murayama S, Saido TC, Hisanaga S, Iwatsubo T, Hasegawa M (2001) Calpain-mediated degradation of p35 to p25 in postmortem human and rat brains. FEBS Lett 489:46–50.

Tarricone C, Dhavan R, Peng J, Areces LB, Tsai LH, Musacchio A (2001) Structure and regulation of the CDK5-p25(nck5a) complex. Mol Cell 8:657–669.

Tiraboschi P, Hansen LA, Thal LJ, Corey-Bloom J (2004) The importance of neuritic plaques and tangles to the development and evolution of AD. Neurology 62: 1984–1989.

Tomidokoro Y, Ishiguro K, Harigaya Y, Matsubara E, Ikeda M, Park JM, Yasutake K, Kawarabayashi T, Okamoto K, Shoji M (2001) Abeta amyloidosis induces the initial stage of tau accumulation in APP(Sw) mice. Neurosci Lett 299:169–172.

Town T, Zolton J, Shaffner R, Schnell B, Crescentini R, Wu Y, Zeng J, DelleDonne A, Obregon D, Tan J, Mullan M (2002) p35/Cdk5 pathway mediates soluble amyloid-beta peptide-induced tau phosphorylation in vitro. J Neurosci Res 69:362–372.

Tsai LH, Delalle I, Caviness Jr. VS, Chae T, Harlow E (1994) p35 is a neural-specific regulatory subunit of cyclin-dependent kinase 5. Nature 371:419–423.

Tsang SH, Gouras P, Yamashita CK, Kjeldbye H, Fisher J, Farber DB, Goff SP (1996) Retinal degeneration in mice lacking the gamma subunit of the rod cGMP phosphodiesterase. Science 272:1026–1029.

Tseng HC, Zhou Y, Shen Y, Tsai LH (2002) A survey of Cdk5 activator p35 and p25 levels in Alzheimer's disease brains. FEBS Lett 523:58–62.

Tsutsui T, Hesabi B, Moons DS, Pandolfi PP, Hansel KS, Koff A, Kiyokawa H (1999) Targeted disruption of CDK4 delays cell cycle entry with enhanced p27(Kip1) activity. Mol Cell Biol 19:7011–7019.

Ubeda M, Kemp DM, Habener JF (2004) Glucose-induced expression of the cyclin-dependent protein kinase 5 activator p35 involved in Alzheimer's disease regulates insulin gene transcription in pancreatic beta-cells. Endocrinology 145:3023–3031.

Vallazza-Deschamps G, Cia D, Gong J, Jellali A, Duboc A, Forster V, Sahel JA, Tessier LH, Picaud S (2005) Excessive activation of cyclic nucleotide-gated channels contributes to neuronal degeneration of photoreceptors. Eur J Neurosci 22:1013–1022.

Wang J, Liu S, Fu Y, Wang JH, Lu Y (2003) Cdk5 activation induces hippocampal CA1 cell death by directly phosphorylating NMDA receptors. Nat Neurosci 6:1039–1047.

Wang Q, Walsh DM, Rowan MJ, Selkoe DJ, Anwyl R (2004) Block of long-term potentiation by naturally secreted and synthetic amyloid beta-peptide in hippocampal slices is mediated via activation of the kinases c-Jun N-terminal kinase, cyclin-dependent kinase 5, and p38 mitogen-activated protein kinase as well as metabotropic glutamate receptor type 5. J Neurosci 24:3370–3378.

Wei FY, Nagashima K, Ohshima T, Saheki Y, Lu YF, Matsushita M, Yamada Y, Mikoshiba K, Seino Y, Matsui H, Tomizawa K (2005) Cdk5-dependent regulation of glucose-stimulated insulin secretion. Nat Med 11:1104–1108.

Wei W, Wang X, Kusiak JW (2002) Signaling events in amyloid beta-peptide-induced neuronal death and insulin-like growth factor I protection. J Biol Chem 277:17649–17656.

Wen Y, Planel E, Herman M, Figueroa HY, Wang L, Liu L, Lau LF, Yu WH, Duff K (2008) Interplay between cyclin-dependent kinase 5 and glycogen synthase kinase 3 beta mediated by neuregulin signaling leads to differential effects on tau phosphorylation and amyloid precursor protein processing. J Neurosci 28:2624–2632.

Wen Y, Yang S, Liu R, Brun-Zinkernagel AM, Koulen P, Simpkins JW (2004) Transient cerebral ischemia induces aberrant neuronal cell cycle re-entry and Alzheimer's disease-like tauopathy in female rats. J Biol Chem 279:22684–22692.

Wen Y, Yang SH, Liu R, Perez EJ, Brun-Zinkernagel AM, Koulen P, Simpkins JW (2007) Cdk5 is involved in NFT-like tauopathy induced by transient cerebral ischemia in female rats. Biochim Biophys Acta 1772:473–483.

Wittmann CW, Wszolek MF, Shulman JM, Salvaterra PM, Lewis J, Hutton M, Feany MB (2001) Tauopathy in Drosophila: neurodegeneration without neurofibrillary tangles. Science 293:711–714.

Yamaguchi H, Ishiguro K, Uchida T, Takashima A, Lemere CA, Imahori K (1996) Preferential labeling of Alzheimer neurofibrillary tangles with antisera for tau protein kinase (TPK) I/glycogen synthase kinase-3 beta and cyclin-dependent kinase 5, a component of TPK II. Acta Neuropathol (Berl) 92:232–241.

Yoo BC, Lubec G (2001) p25 protein in neurodegeneration. Nature 411:763–764; discussion 764–765.

Yoshiyama Y, Higuchi M, Zhang B, Huang SM, Iwata N, Saido TC, Maeda J, Suhara T, Trojanowski JQ, Lee VM (2007) Synapse loss and microglial activation precede tangles in a P301S tauopathy mouse model. Neuron 53:337–351.

Zhai S, Senderowicz AM, Sausville EA, Figg WD (2002) Flavopiridol, a novel cyclin-dependent kinase inhibitor, in clinical development. Ann Pharmacother 36:905–911.

Zhang J, Krishnamurthy PK, Johnson GV (2002) Cdk5 phosphorylates p53 and regulates its activity. J Neurochem 81:307–313.

Zhang M, Li J, Chakrabarty P, Bu B, Vincent I (2004) Cyclin-dependent kinase inhibitors attenuate protein hyperphosphorylation, cytoskeletal lesion formation, and motor defects in Niemann-Pick Type C mice. Am J Pathol 165:843–853.

Zheng YL, Kesavapany S, Gravell M, Hamilton RS, Schubert M, Amin N, Albers W, Grant P, Pant HC (2005) A Cdk5 inhibitory peptide reduces tau hyperphosphorylation and apoptosis in neurons. Embo J 24:209–220.

Zukerberg LR, Patrick GN, Nikolic M, Humbert S, Wu CL, Lanier LM, Gertler FB, Vidal M, Van Etten RA, Tsai LH (2000) Cables links Cdk5 and c-Abl and facilitates Cdk5 tyrosine phosphorylation, kinase upregulation, and neurite outgrowth. Neuron 26:633–646.

Index

Printed in the United States of America